HANDBOOK OF LARVAL AMPHIBIANS OF THE UNITED STATES AND CANADA

HANDBOOK OF LARVAL AMPHIBIANS OF THE UNITED STATES AND CANADA

Ronald Altig

Roy W. McDiarmid

Foreword by Aaron M. Bauer

COMSTOCK PUBLISHING ASSOCIATES

a division of

CORNELL UNIVERSITY PRESS

ITHACA AND LONDON

First published 2015 by Cornell University Press

Printed in the United States of America

Library of Congress Cataloging-in-Publication Data

Altig, Ronald, author.
    Handbook of larval amphibians of the United States and Canada / Ronald Altig and Roy W. McDiarmid.
        page cm
    Includes bibliographical references and index.
    ISBN 978-0-8014-3943-8 (cloth : alk. paper)
    1. Amphibians—Larvae—United States. 2. Amphibians—Larvae—Canada. I. McDiarmid, Roy W., author. II. Title.
    QL641.A45 2015
    597.813'92—dc23   2014030532

Cornell University Press strives to use environmentally responsible suppliers and materials to the fullest extent possible in the publishing of its books. Such materials include vegetable-based, low-VOC inks and acid-free papers that are recycled, totally chlorine-free, or partly composed of nonwood fibers. For further information, visit our website at www.cornellpress.cornell.edu.

Cloth printing   10 9 8 7 6 5 4 3 2 1

CONTENTS

Order Anura: Frogs and Toads   *143*

Plates follow page 222.

FOREWORD

The transition from larva to adult in amphibians is often dramatic. One can truly say that larvae, especially tadpoles, are entirely different animals from their corresponding adults. Certainly their appearance, lifestyles, and ecological roles are radically different. For biologists, larvae are moving targets — changing their appearance rapidly, sometimes on a scale of days or even hours, and with these changes transitioning from one set of ecological interactions to another. The techniques employed to study conspicuous, loudly calling frogs, for example, do not apply to tadpoles occupying the murky world beneath the pond's surface. Indeed, it is extremely difficult to observe the actions of free-living larvae without disturbing them and much of what we know of their biology is based on studies staged in artificial containers, large or small, or on anecdotal observations from the wild.

For some amphibian biologists, larvae may be viewed simply as stepping stones on the way to the "definitive" adult, or as characters of the organisms they will become. Although appreciation for the larval life has increased among herpetologists and naturalists in general, few biologists have truly devoted themselves to the study of this life stage. Ronald Altig and Roy W. McDiarmid have each spent a lifetime learning to know and appreciate larval amphibians for their own sake, contributing, individually and collectively, to the growing literature in the field and developing some of the most useful and often cited keys to North American larval amphibians. In 1999 they edited a groundbreaking volume, *Tadpoles: The Biology of Anuran Larvae*, that synthesized much of the diverse research on tadpoles globally, and later they turned their attentions to the much neglected topic of amphibian eggs (Altig and McDiarmid 2007).

For more than eighty years the Comstock Classic Handbook series has served as a critical source for the scattered information about the North American herpetofauna. In the two original amphibian volumes—*Frogs and Toads* by Anna Allen Wright and Albert Hazen Wright (1933) and *Salamanders* by Sherman C. Bishop (1943)—larval forms were treated in admirable detail, but these accounts are now woefully out of date. In addition, the application of genetic approaches to systematics has revised our view of species boundaries, resulting in the resurrection or description of many species and necessitating a reassessment of the taxa to which some older larval observations should be attributed.

Although the intervening decades have seen the publication of many books treating North American amphibians, no one has synthesized the premetamorphic stages of these animals in a single source. In this volume Altig and McDiarmid have compiled a richly illustrated, comprehensive overview of the eggs, embryos, and larvae of the salamanders, frogs, and toads of the United States and Canada, complete with keys, ranges, identifying features, and natural history data. That this work is dedicated exclusively to larval amphibians not only reflects the sheer volume of material on the subject but also acknowledges the fact that larvae are biologically distinct from adult amphibians, not mere footnotes to the lives of the salamanders and frogs with which they share genetic identity.

AARON M. BAUER

Villanova, Pennsylvania
24 May 2014

PREFACE

We grew up at about the same time but almost 2000 miles apart. Each of us was fascinated by what we found in mud puddles and roadside ditches in Illinois (RA) or in the water hazards of California golf courses (RWM). We both spent many days looking for amphibians and reptiles. New discoveries were the most memorable, whether found under a log or piece of tin or squirming among leaves in the bottom of a homemade dip net. We learned the names of common species by matching the adults to pictures in the few books we had, but batches of eggs and larvae usually led to more questions and additional field time. We learned that the size and depth of the knowledge gaps in the basic natural history of many species of amphibians and reptiles were substantial.

Beginning in the 1930s, the Handbook Series published by Comstock Publishing of Cornell University Press summarized what was known about the natural history of North American species of amphibians (frogs and toads, A. A. Wright and Wright 1933, A. H. Wright and Wright 1949; salamanders, S. C. Bishop 1947) and reptiles (lizards, H. M. Smith 1946; turtles, Carr 1952; snakes, A. H. Wright and Wright 1957). These landmark publications provided a starting point, but many questions were still unanswered and details were too often lacking. It was this realization that has served as a guiding principle for most of our careers. While progress has been made, especially in understanding of the biology of adults, we still do not know a lot about the early stages in the life histories of most amphibians.

In the early 1990s, reports on declining amphibians (e.g., Barinaga 1990; Blaustein and Wake 1990; Borchelt 1990) renewed interest in the need for reliable field data and development of standardized methods to facilitate monitoring programs. Publications designed to meet these needs (e.g., Heyer et al. 1994) relied primarily on data for adults. Because eggs and larvae are often available in field sites for longer periods than most adults, the need to incorporate these early life history stages into monitoring projects was great. Reliable keys for their identification as well as data on their natural history and developmental ecology were lacking. As a result, the need for a review and synthesis of available information on the larval ecology of amphibians was acute.

In the mid-1990s we began two major projects to rectify this problem. The reference volume on all aspects of tadpole biology (McDiarmid and Altig 1999) has been well received. This second effort focuses on the early life history stages of the amphibian fauna of the United States and Canada. Because the morphology and natural history of eggs, spermatophores, hatchlings, and larvae are much less accurately documented than those of adults, the study of these forms is more difficult, and the preparation of a summary treatise was simultaneously exciting and frustrating. Some stages of most species remain poorly known, and immense amounts of geographic and ontogenetic variation are not understood. The scattered presence of cryptic taxa throughout the fauna adds another level of difficulty. Accordingly, we strongly encourage researchers to preserve pertinent specimens, deposit them in accessible collections, take appropriate pictures, and publish their data. Only a concerted effort within the amphibian research community will allow significant progress to be made. Those persons who willfully tackle the nasty trios within each group (i.e., *Ambystoma, Eurycea,* and *Desmognathus* in salamanders; *Anaxyrus, Pseudacris,* and *Lithobates* among frogs) will definitely need extraordinary abilities and perseverance.

ACKNOWLEDGMENTS

We are grateful for numerous kinds of help and time-consuming favors received from many people during preparation of this book, and we hope we have not forgotten anyone during the long tenure of this project. The following persons assisted in obtaining specimens to photograph: E. C. Akers, S. C. Anderson, A. Asquith, C. K. Beachy, S. Bennett, D. F. Bradford, R. A. Brandon, E. Britzke, R. C. Bruce, T. Bryan, D. C. Cannatella, S. Corn, B. I. Crother, G. H. Dayton, S. M. Deban, P. Delis, L. V. Diller, R. C. Drewes, L. A. Fitzgerald, G. W. Folkerts, L. S. Ford, R. Franz, M. Geraud, E. W. A. Gergus, S. W. Gotte, A. Graybeal, D. G. Hokit, T. H. Holder, R. D. Jennings, J. B. Jensen, J. C. Jones, S. A. Johnson, J. M. Kiesecker, S. Kuchta, C. Luke, D. Lynn, P. Marino, A. McCready, W. E. Meshaka Jr., J. S. Miller, P. E. Moler, H. Mueller, R. A. Newman, D. Paddock, P. Pister, A. H. Price, L. Powell, J. K. Reaser, M. K. Redmer, D. Richardson, S. C. Richter, M. A. Robertson, F. L. Rose, E. Routman, R. L. Saunders, A. H. Savitsky, C. K. Sherman, A. Sih, E. Simandle, D. Spicer, S. Sweet, R. B. Thomas, S. G. Tilley, C. R. Tracy, B. Turner, D. J. Wear, H. H. Welsh, E. Wildy, R. F. Wilkinson Jr., D. Wilson, T. Wood, and J. W. Wright.

In addition to the authors (Altig, RA; McDiarmid, RWM), the following people, noted by their initials where their pictures appear, graciously submitted photographs for our consideration: A. Asquith (AA), S. L. Barten (SLB), J. P. Bogart (JPB), J. Bond (JB), R. A. Brandon (RAB), E. D. Brodie, Jr. (EDB), R. C. Bruce (RCB), J. F. Bunnell (JFB), D. Chamberlain (DAC), J. P. Collins (JPC), C. C. Corkran (CCC), S. M. Deban (SMD), D. M. Dennis (DMD), L. V. Diller (LVD), D. L. Drake (DLD), E. Ervin (EE), J. Forman (JF), J. Fries (JNF), M. García-París (MGP), J. S. Godley, G. Grall (GG), R. L. Grasso (RLG), R. W. Hansen (RWH), D. M. Hillis (DMH), R. H. Humbert (RHH), R. D. Jennings (RDJ), J. B. Jensen (JBJ), E. L. Jockusch (ELJ), G. N. Johnson (GNJ), S. A. Johnson (SAJ), T. R. Johnson (TRJ), G. F. Johnston (GFJ), T. R. Kahn (TRK), J. R. Lee (JRL), P. Licht (PL), B. Mansell and P. E. Moler (BM), D. L. Martin (DLM), J. Martinez (JM), K. R. McAllister, D. B. Means (DBM), J. C. Murphy (JCM), W. Meinzer (WM), B. T. Miller (BTM), H. Mueller (HM) R. Norton, L. O'Donnell (LO), K. Ovaska (KO), D. Paddock (DP), K. R Pawlik (KRP), M. Penuel-Matthews (MPM), C. R. Peterson (CRP), D. W. Pfennig (DWP), D. J. Printiss, R. A. Pyles (RAP), M. Redmer (MR), A. M. Richmond (AMR), J. M. Romansic (JMR), J. C. Rorabaugh (JCR), N. J. Scott Jr. (NJS),

R. D. Semlitsch (RDS), G. Sievert (GS), M. Sredl (MS), R. M. Storm (RMS), J. E. Tkach (JET), S. E. Trauth (SET), L. O. Tubbs (LOT), M. D. Venesky (MDV), K. D. Wells (KDW), L. West (LW), D. W. Zaff (DWZ), and especially, because of the large number of slides they submitted, W. P. Leonard (WPL), D. J. Stevenson (DJS), and R. W. Van Devender (RWV).

J. D'Ambrosio (JD), D. Karges (DK), K. Spencer (KS), and P. C. Ustach (PCU) produced various line drawings. Preserved specimens from the California Academy of Sciences (CAS, R. C. Drewes) and the National Museum of Natural History, Smithsonian Institution (USNM), were photographed; L. S. Ford, then at the American Museum of Natural History, and H. W. Greene, then at the University of California-Berkeley, also loaned specimens.

The following people also deserve our thanks for their assistance in various ways: H. L. Bart, J. Bernardo, R. A. Brandon, A. L. Braswell, E. D. Brodie Jr., A. Channing, P. T. Chippindale, D. L. Drake, G. W. Folkerts, D. C. Forester, M. S. Foster, R. Franz, J. W. Gibbons, E. W. A. Gergus, S. Hale, R. W. Hansen, M. P. Hayes, W. R. Heyer, R. Jones, S. Kuchta, G. Longley, W. E. Meshaka Jr., J. C. Mitchell, T. K. Pauley, J. W. Petranka, S. R. Reilly, F. L. Rose, R. D. Semlitsch, M. Sisson, M. J. Sredl, R. A. Thomas, D. S. Townsend, T. D. Tuberville, D. B. Wake, S. C. Walls, R. F. Wilkinson, and J. Wooten. At Mississippi State, S. V. Diehl and G. Thibaudeau helped in many ways, and in Washington, D.C., S. W. Gotte, J. A. Poindexter, and M. S. Foster were valuable assistants. M. J. Lannoo allowed us to peruse a book manuscript, and J. Bernardo, R. C. Bruce, C. D. Camp, D. B. Means, and S. G. Tilley aided with the accounts of the bewildering genus *Desmognathus*. J. A. Hall, L. Hallock, and W. P. Leonard clarified points on egg morphology of *Spea*. W. P. Leonard and R. B. Thomas kindly reviewed the manuscript at early stages. The National Biological Survey provided grant support to McDiarmid for two years (1996–1997) for a proposal titled "A Review and Synthesis of the Biology of North American Amphibian Larvae." Funds were used to cover costs of field travel and surveys, shipping of eggs and tadpoles from field sites to Mississippi, laboratory supplies, photography, illustrations, and other expenses associated with preparation of the manuscript.

HANDBOOK OF
LARVAL AMPHIBIANS
OF THE UNITED STATES
AND CANADA

INTRODUCTION

Amphibians of North America include salamanders (i.e., salamanders, newts, and waterdogs) and frogs (i.e., frogs, toads, treefrogs, and spadefoots), and many researchers seem to forget or ignore eggs and larvae. We list various keys as an introduction to the state of our present knowledge (see Table 1). Coverage of paedomorphic salamanders and a few amphibian larvae occurs in some field guides, but many larval amphibians of North American species have not been adequately described. Essentially none of the eggs has been described in a way useful to field biologists, as noted by Livezey and Wright (1947) more than 60 years ago. Spermatophores and hatchlings are also poorly studied and inadequately described. Within common taxa, the morphological diversity of larvae is relatively low but variable. Larvae of closely related species are often frustratingly similar, but cases of similar adults with distinctive larvae exist (e.g., *Taricha granulosa* versus *T. torosa*; *Hyla avivoca* versus *H. "versicolor"*).

Table 1. Literature summary of keys and other sources pertinent to the identification of amphibian eggs and larvae of the United States and Canada

Citation	Geographic area
	Eggs
Corkran and Thoms 1996	Pacific Northwest
Desroches and Rodrigue 2004	Québec and Maritimes
N. B. Green 1952	Frogs, West Virginia
Hunter et al. 1999	Maine
Livezey and Wright 1947	Frogs, United States
Maxell et al. 2003	Montana
Parmalee et al. 2002	Minnesota, Wisconsin, Iowa
Pickwell 1972	Pacific states
Preston 1982	Manitoba
Russell and Bauer1993	Alberta
Watermolen 1995	Wisconsin
A. H. Wright 1910	Frogs, Ithaca, New York
A. H. Wright 1932	Frogs, Okefenokee area, Georgia
A. H. Wright and Wright 1924	Frogs east of Mississippi River
A. H. Wright and Wright 1995	Frogs, United States
	Larvae
Larval salamanders	
Altig and Ireland 1984	United States
J. K. Baker 1961	Neotenic *Eurycea*, Texas

(*Continued*)

Table 1 (Continued)

Larval salamanders

Ballinger and Lynch 1983	Families of salamanders, United States
Bragg et al. 1950	Oklahoma
Brandon 1961	Five *Ambystoma*
Brandon 1964	Ohio
Corkran and Thoms 1996	Pacific Northwest
Desroches and Rodrigue 2004	Québec and Maritimes
Dundee and Rossman 1989	Louisiana
Gibbons and Semlitsch 1991	SREL site, South Carolina
Ireland 1981	Virginia
Ireland and Altig 1983	Ozark region
Maxell et al. 2003	Montana
Nussbaum et al. 1983	Pacific Northwest
Parmalee et al. 2002	Minnesota, Wisconsin, Iowa
Pfingsten and Downs 1989	Ohio
Pickwell 1972	Pacific states
Preston 1982	Manitoba
H. M. Smith 1934, 1956	Kansas
Valentine and Dennis 1964	Families of salamanders with gills or gill slits
Vogt 1981	Wisconsin
C. F. Walker 1946	Ohio
Watermolen and Gilbertson 1996	Wisconsin

Tadpoles

Altig 1970	United States
Altig et al. 1998	United States and Canada
Ballinger and Lynch 1983	Families of frogs and toads, United States
Bragg 1941b	*Scaphiopus, Spea*
Corkran and Thoms 1996	Pacific Northwest
Desroches and Rodrigue 2004	Québec and Maritimes
Dundee and Rossman 1989	Louisiana
Fanning 1966	Florida
Gibbons and Semlitsch 1991	SREL site, South Carolina
N. B. Green and Pauley 1987	West Virginia
Maxell et al. 2003	Montana
Nussbaum et al. 1983	Pacific Northwest
Orton 1939	New Hampshire
Orton 1952	United States genera
Parmalee et al. 2002	Minnesota, Wisconsin, and Iowa
Pickwell 1972	Pacific states
Preston 1982	Manitoba
Russell and Bauer 1993	Alberta
N. J. Scott and Jennings 1985	*L. pipiens* group, New Mexico
Siekmann 1949	Louisiana
H. M. Smith 1934, 1956	Kansas
Stebbins 1951	Western United States
Stevenson 1976	Florida
Storer 1925	California
Travis 1981a	North Carolina
Vogt 1981	Wisconsin
Watermolen and Gilbertson 1996	Wisconsin
A. H. Wright 1914	Upstate New York
A. H. Wright 1929	North America
A. H. Wright 1932	Okefenokee area, Georgia
A. H. Wright and Wright 1923	Okefenokee Swamp, Georgia
A. H. Wright and Wright 1995	United States

Amphibian larvae are small, fragile, and difficult to observe, and their characteristics are unfamiliar to most researchers. The sources and patterns of large ontogenetic and geographical variations are seldom understood. Shape and coloration are quite variable and change under different ecological conditions and in preservative. Extensive morphological differences modified by the interactions with coinhabitants (e.g., Relyea and Auld 2005) is a relatively recent discovery. All these factors contributed to our decision to prepare this book. We attempted to provide an improved means of identifying amphibian eggs and larvae and to compile the literature that documents their morphology and ecology. While we include pedotypic and paedomorphic salamanders, all endotrophs and metamorphosed, terrestrial adults were excluded. The available data force us to emphasize larval forms near the middle of their ontogeny. Material presented by McDiarmid and Altig (1999) on the ancillary subjects of collecting, photography, preservation, and rearing apply equally well to salamander larvae and tadpoles.

BACKGROUND AND SCOPE

In the taxonomic accounts we summarize the morphology and basic natural history and provide an introduction to published information for each species. New or revised keys are included for eggs, embryos, salamander larvae, and tadpoles. Color photographs of larvae of many species are presented, but one should not rely exclusively on picture matching for identification because ontogenetic and ecological variations are large. In the plate legends, some references to figures refer to closely related species. Tadpole mouthparts include major characters used in identifications, and we have included illustrations of a number of species. A glossary of terms as used in this book is appended.

Three additional factors must be noted. First, the value of well-preserved voucher specimens accompanied by detailed notes on coloration and ecological conditions cannot be overemphasized; without such verification, statements of occurrence and other related data are rendered nearly useless. Published data unsupported with vouchers will often be ignored, simply because verification of identification is unlikely. The preservation of additional material of all stages is important, but properly prepared specimens of eggs, spermatophores, and hatchlings are imperative. Second, systematic consensus has not yet been achieved for all taxa, and the common (i.e., Crother 2012) and scientific names (i.e., Frost 2011) we use in these pages should not be cited as a taxonomic authority. The reader should always compare information on related species in particularly confusing groups (e.g., *Ambystoma barbouri* and *A. texanum*, *Ambystoma mavortium* and *A. tigrinum*, the *Eurycea bislineata* group; *Anaxyrus americanus* group, *Pseudacris triseriata* group, and the *Lithobates pipiens* complex). Last, a working knowledge of geography of the United States and Canada will facilitate the use of the keys (see Fig. 1).

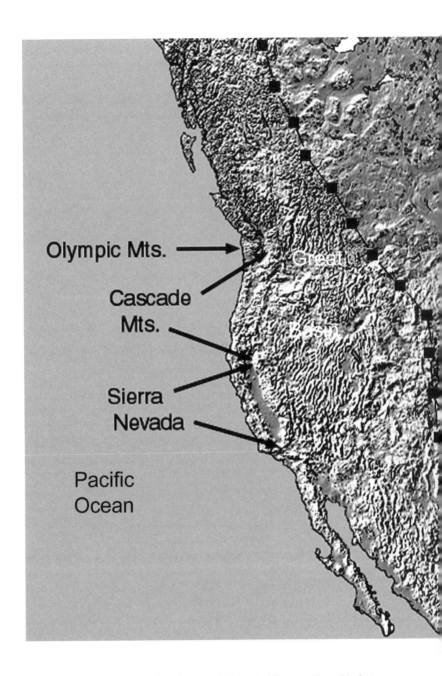

Figure 1. **Landform map of the United States and Canada**. Major physiographic features mentioned in the text; EP = Edwards Plateau, OU = Ouachita Mountains, OZ = Ozark Mountains, black dots = track of Mississippi River, black squares = track of crest of Rocky Mountains, white dots = boundary of the Coastal Plain (= Fall Line) and Mississippi Embayment (modified from a digital map prepared by Ray Sterner, Applied Physics Laboratory, Johns Hopkins University).

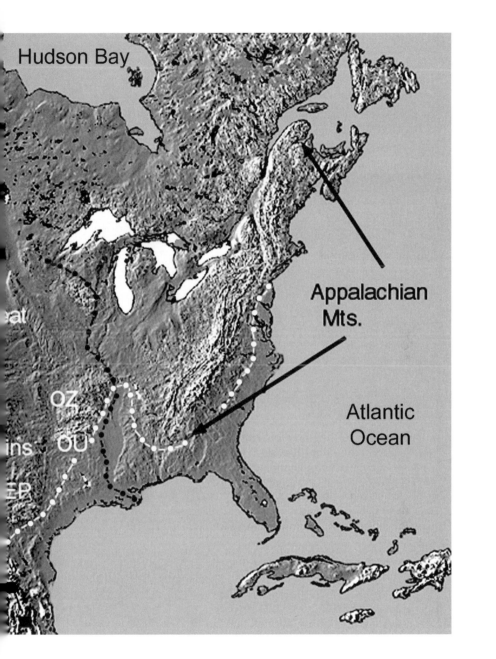

Hudson Bay

Appalachian
Mts.

Atlantic
Ocean

OZ

OU

ins

EP

eat

THE AMPHIBIAN FAUNA

SALAMANDERS, NEWTS, AND WATERDOGS

The salamanders of the United States and Canada that have free-swimming, feeding larvae (i.e., exotrophic) include 97 species in 17 genera and 8 families. There are no exotic salamanders, but larval forms of *Ambystoma mavortium* and *Desmognathus* spp. have been transported to new areas for use as fish bait. Because of the diversity among ambystomatids (*Dicamptodon*), cryptobranchids, most taxa of plethodontids, proteids, rhyacotritonids, and some western salamandrids, total larval diversity in flowing water is higher than in nonflowing water (e.g., amphiumids, most ambystomatids, eastern salamandrids, and sirenids). Ambystomatids plus plethodontids comprise almost 70% of the fauna, and within each family, many of the larvae are annoyingly similar. Hybridization and changes in ploidy in ambystomatids have produced interesting biological situations, and the occurrence and variability of different developmental morphotypes (e.g., cannibalistic/carnivorous) in *Ambystoma* surely has piqued research curiosities. The presence of an eft stage in the life history of many populations of the salamandrid genus *Notophthalmus* is unique.

Altig and Ireland (1984) provided the first key to all species of larval salamanders in the United States and Canada, and Petranka (1998) was the first author to treat larval salamanders in considerable detail since S. C. Bishop's classic work (1943; reprint 1994).

FROGS AND TOADS

The frogs of the United States and Canada that have free-living, feeding tadpoles include 103 species in 20 genera and 10 families. *Xenopus laevis* in southern California and Arizona; *Dendrobates auratus, Glandirana rugosa, Lithobates catesbeianus*, and *Rhinella marina* in Hawaii; and *Osteopilus septentrionalis* in most counties of the Florida Peninsula are established exotics. Populations of *Lithobates berlandieri, L. catesbeianus*, and *L. clamitans* thrive in many areas outside their native ranges on the mainland.

Tadpoles occur in all sorts of nonflowing water, and a few occur in flowing water. Those of *Ascaphus* are suctorial in fast-flowing water in the Pacific Northwest, and several ranids occur typically (e.g., *Rana boylii, R. muscosa*, and

Lithobates tarahumarae) or occasionally (e.g., some in the *Lithobates catesbeianus* and *L. pipiens* groups) in various kinds of stream habitats. The facultative generation of a profoundly different carnivorous morphotype in scaphiopodid tadpoles in the genus *Spea* is the only major deviation from a typical life cycle. These carnivorous tadpoles have modified mouthparts and jaw musculature relative to their omnivorous siblings.

The study of the biology of tadpoles in North America started with reports by Hinckley (1880, 1881, 1882a), and the study of morphological variation the likes of Nichols (1937) has never been repeated. The classic works by A. H. and A. A. Wright from 1914 to 1949 provide information on identification and natural history like no other. Based on information available from numerous preceding studies and considerable field experience, the first key that covered the tadpoles of all North American species (Altig 1970) was followed nearly 30 years later by a digital edition (Altig et al. 1998).

THE AMPHIBIAN LIFE CYCLE

DEVELOPMENTAL CATEGORIES

Larva(e) and *larval* are useful generic terms that can apply to a tadpole or sala-mander larva. *Adult* and *mature* denote sexual maturity and thus are inappropri-ate in most cases. Likewise, notions of "large/small" and "old/young" imply age or size and usually are not very accurate because of observer biases, different size maxima among species, and variations in growth rate caused by ecological fac-tors.

Staging allows comparisons among larvae of the same or different species without regard to age, size, or developmental period. The attainments of specific morphological features are the signals of interest. The hind legs of salamanders become fully developed soon after hatching, and development after hatching to metamorphosis involves mostly changes in size. This long period of relative mor-phological stasis makes staging impossible compared to that of a tadpole, and thus stage designations are not used; for example, the stages proposed for *Ambys-toma macrodactylum* (Watson and Russell 2000) cover the entire ontogeny, but most of larval life is in one stage. Most larvae are observed in the period after limb development and prior to metamorphosis, and thus total length is the only useful comparative descriptor. A summary of staging tables for both salamanders and tadpoles appears in Duellman and Trueb (1986:128).

Some salamanders become sexually mature while maintaining a larval mor-phology. A pedotype (Reilly et al. 1997) is an individual with a truncated ontog-eny relative to the normal ontogeny of that species. These individuals retain a larval morphology but become mature as long as amiable biotic and abiotic con-ditions for an aquatic existence persist. A paedomorph is an individual with a truncated ontogeny relative to the ancestral condition, and partial metamorphosis may be involved. Such individuals are usually refractory to natural and artificial metamorphic stimuli.

We recommend the following terms to describe the ontogenetic periods of sal-amanders: *embryo* (from fertilization to hatching); *hatchling* (from hatching to loss of balancers or a comparable stage in those that lack balancers and comple-tion of limb development; Figs. 15B, 56B); *larva* (from all limbs fully differenti-ated to the start of metamorphosis), *metamorph* (fins and gills starting to atrophy, eyelids forming, gular fold starting to close); *juvenile* (metamorphosis complete but sexually immature); and *adult* (sexually mature and usually metamorphosed).

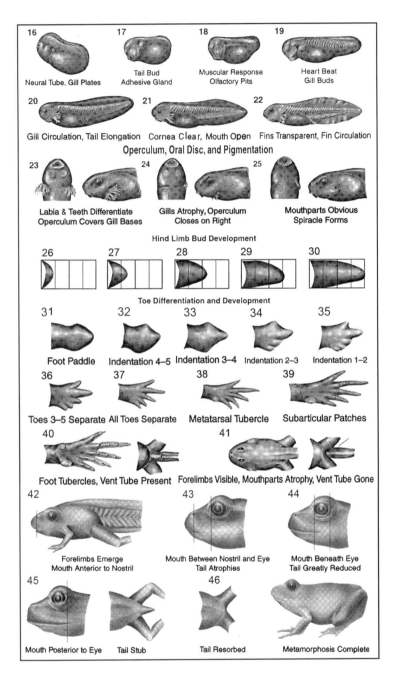

Figure 2. **Tadpole developmental stages.** Most specimens examined in the context of this book will be in stages 27–36 (modified from McDiarmid and Altig 1999).

In contrast to the early development of the limbs of salamanders, those of tadpoles develop throughout most of larval ontogeny, and gradual development of the exposed hind limbs provides the primary characters used in staging. Gosner's stages (1960; Fig. 2; also in Duellman and Trueb 1970:134 and Altig and McDiarmid 1999:10) are used most frequently. Qualitative terms denote longer developmental periods: embryo (stages 1 to about 20); hatchling (about 21–24); tadpole (25–41); metamorph (42–45); juvenile (= froglet, stage 46 until sexual maturity); and adult (sexually mature). The stage at hatching varies among species and environmental conditions, and it is hatching and not the morphological stage that is the absolute landmark. Recognition of this short hatchling period within the concept of embryos and tadpoles emphasizes the unique ecomorphology of hatchlings (= nonfeeding, hatched embryos that are motile by epidermal cilia). When metamorphic size is recorded, it is assumed to refer to snout-vent length at stages 42–46. A robust relationship between tadpole body length at stage 36 or greater and metamorphic snout-vent length allows one to estimate metamorphic size at these stages. Small tadpoles in stage 25 or greater are well past hatching and should not be referred to as hatchlings.

EGGS

The lack of descriptions of amphibian eggs in a format useful to field biologists makes field identifications particularly difficult. Clutches of eggs should be observed prior to removal from the deposition site to avoid the destruction of important aspects of their form. Detailed studies of newly laid clutches derived from known parents that include staining and special lighting are most useful. Staining egg jellies with 1% Toluidine Blue colors the outer surfaces, and soaking eggs in food coloring for a few minutes improves contrast by coloring the fluids within the various jelly layers. Transferring these eggs to clean water provides an improved understanding of basic clutch structure. The notion that a given species lays eggs in one mode or clutch structure is usually correct, but variations are known. Some ovipositional modes can blend from one to another, and observer acuity is a must in all cases. The mechanisms by which films float and whether the outer jellies are coherent or adherent differ among taxa. The diameters and other characteristics of egg jellies change through time (Thurow 1997b, Volpe et al. 1961) and are influenced by both the embryo and the environment. Eggs laid in containers in the laboratory should be viewed with caution because some of the conditions normally associated with oviposition are often absent. Finally, be sure of what you have. Over the years we have been asked to identify presumed amphibian eggs that in fact were eggs of insects, snails or fishes, colonial bryozoans as large as basketballs, and several kinds of algae.

The generic term *egg* (e.g., Altig and McDiarmid 2007) describes a gamete and vitelline membrane of ovarian origin and the jelly layers of oviductal origin; *egg diameter* (ED) is the distance across the outer jelly layer of a given egg. *Ovum* (ova) refers only to the gamete or early, spherical embryo (= *ovum diameter*, OD). A *jelly layer* is a visible jelly coat (i.e., that radius between two successive visible changes in optical density that may appear like discrete membranes or thicker amorphous layers), although proper staining will commonly show that more layers of jelly may be present than are detectable visually (Steinke and Benson 1970). *Clutch* refers to the total number of eggs deposited (i.e., number of ovulated ova is sometimes greater than number of ova oviposited; White and Pyke 2002) per ovulation event regardless of the ovipositional mode or number of ovipositional bouts; eggs oviposited during multiple bouts do not equal a clutch. If a female ovulates and oviposits more than once a year or in multiple years, she has produced multiple clutches. *Group* is a generic term we sometimes use to refer to a number of eggs without committing to taxon or ovipositional mode.

The capsular chamber is a subtle but distinctive difference (Salthe 1963) that distinguishes the eggs of salamanders from those of frogs, as covered in this book. Salamanders have a capsular chamber and frogs do not. The disintegration of the first jelly layer external to the vitelline membrane soon after fertilization forms this chamber in salamanders, and this dissolution releases the ovum from confinement. Because the yolk-laden vegetal pole of an egg is heavier than the animal pole, these ova turn upright immediately if the group is turned over, and the center of the ova is positioned slightly below the center of the inner jelly layers (e.g., Figs. 3D, 12A). Frog ova and early embryos remain centered in the egg jellies and take several minutes to turn over because they remain more confined by the lack of a capsular chamber (i.e., an intact first jelly layer). Unfertilized eggs remain in whatever position they are placed and soon degrade into gray, amorphous balls.

We use ovipositional mode as the major character in the egg key, but site and certain other features are incorporated into the definition when available. By using these definitions, one can usually identify eggs to genus within a local fauna. Suspected differences in ovipositional modes within a given species may be real, based on eggs of different ages, or reflect observer bias, and for these reasons some taxa occur in multiple places in the key. The lack of definitive data is almost universal.

Ovipositional modes are treated in five categories: independent eggs, linear arrangements, three-dimensional arrangements, films, and foam nests (Altig and McDiarmid 2007).

Independent eggs. Independent eggs are attached (Figs. 3G, 19D) to plants or free (Fig. 75A) on the bottom as singles or haphazard groups. The usual occurrence of outliers reduces the chance of confusing grouped singles with clumps or masses even when singles are placed adjacent or on top of each other. The small area where a variable number of jelly layers are involved in securing an egg to a substrate is termed the *pedicel.*

An *array* is a group of suspended eggs (Fig. 3C), each of which has an independent attachment point. Eggs in arrays may be pendent (Fig. 3A) or not (Fig. 3B), and they most often occur on the lower surfaces of aquatic substrates. The eggs are

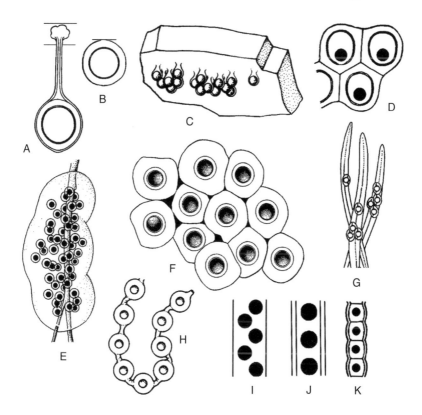

Figure 3. **Ovipositional modes and clutch structures.** (A) suspended, pendent egg, (B) suspended but not pendent egg, (C) array (see also Fig. 27B), (D) melded clump, (E) mass (see also Fig. 12A), (F) part of a clump with interstices blackened for emphasis, (G) single eggs attached to vegetation (see also Figs. 19D, 89A), (H) rosary (see also Figs. 21A, 22B), (I–K) strings of eggs that are (I) staggered in a unilayered tube, (J) uniserial in a bilayered tube, (K) uniserial in a bilayered tube with partitions and scalloped margins (A, B, and D modified from Stebbins 1985; C, E, G, and H modified from Pfingsten and Downs 1989; F modified from Liu 1950; I–K modified from A. H. Wright and Wright 1949).

usually placed in an area with a diameter that is as large or larger than the total length of the adult. Arrays attached to the lower surface of a single stone are usually orderly (Figs. 3C, 27B, 48A), but irregular arrangements on several stones or vegetation may be interpreted as singles. Also, a small number of eggs in an array may be interpreted as single eggs.

Linear arrangements. Rosaries, wrapped rosaries, and strings are modes that involve linear series of ova. A rosary (Figs. 3H, 21A, 22B) has the jelly between successive, usually widely-separated, large ova notably constricted and often twisted; all components of the tube appear to be continuous, but each ovum may have associated jelly layers. A wrapped rosary must be observed carefully because it may look like a clump or mass.

Strings occur only in the genus *Anaxyrus* (Figs. 3I–K, 62A, E, 63F–G) and have a uni- or biserial series of ova aligned linearly inside either a uni- or bilayered jelly tube. The entire clutch is usually laid as continuous, lengthy strings. The outer surface of the tube is not very sticky, and uncommonly it is so diaphanous as to be nearly invisible. For example, A. H. Wright and Wright (1949; reprint 1995) reported that some *Anaxyrus* eggs occur as short groups stuck together linearly without an encompassing tube (= bar; *Anaxyrus quercicus*). Volpe and Dobie (1959) verified that strings of *A. quercicus* eggs are in fact typical of *Anaxyrus* but the strings are short and the outer diaphanous tube can be easily missed; eggs of members of the *A. debilis* group are similar. *A. punctatus* eggs are oviposited as singles, although they may adhere in small groups temporarily.

An unknown tensile quality of the jelly often causes the strings to lie initially in a loose spiral, and slight constrictions between successive ova may cause the outer surface of the tube to appear scalloped (Fig. 3K). Partitions formed from juxtaposed inner jelly layers around successive ova may or may not occur. Movements of the ovipositing adults usually cause the strings to be wrapped around vegetation or irregularly about the bottom.

Literature references to the eggs of scaphiopodids range from confusing to inaccurate. Our recent observations of two species of *Scaphiopus* show that these frogs oviposit a wrapped rosary; the constrictions are less notable than in a rosary, the inter-egg portion is not twisted, and the whole assembly is wrapped around some support (Fig. 124A). As understood at the moment, the eggs of *Spea* are oviposited in small, flimsy clumps (*S. intermontana*; Fig. 127B), haphazardly placed singles (*S. bombifrons*; Fig. 125A), or as singles with elongate outer jelly layers placed in an array on vegetation (*S. hammondii* and *S. multiplicata*; Figs. 126A, 128B).

Three-dimensional arrangements. A *clump* is a group of irregularly arranged terrestrial or aquatic eggs without a common, surrounding surface, the jellies may be adherent or not, and interstices between eggs may occur (open clump, analogous to a stack of marbles) or not (melded clump with interstices obliterated by slumping jelly layers; Fig. 3D). Large, terrestrial eggs in a clump may slide into a monolayer. Because a similar situation can result from single eggs being placed close to or on top of each other, one must view the general construct and look for outliers which seldom exist if eggs were deposited as a clump.

Large, aquatic clumps as oviposited by *Rana* and most *Lithobates* (Figs. 3F, 113A–B, 114A, 115B) are formed by the adherence of individual, spherical jellies at their contact points. Interstices between eggs are present initially, and an entire clutch is commonly oviposited as one clump. The jellies eventually swell or sag enough to fill in the interstices, but a definite, lobular surface remains. A clump is typically rounded in top view, horizontally oblong, and commonly attached to vegetation off the bottom; clumps of stream forms are smaller than those that breed in ponds and often are at or near the bottom. Clumps with embryos near hatching often break loose from their supports and float to the surface. The multilayered construction of these flattened clumps and typical entrapment of bubbles in the jelly (Fig. 119A) helps distinguish these clumps from films.

A *cluster* is a small group of eggs that may be suspended and pendent or not and the attachments of all eggs are at the same or adjacent points. Each egg is oviposited singly. Within the scope of this book, clusters occur only in the genus *Desmognathus*. These clusters are not as obviously constructed as in some terrestrial-breeding *Plethodon*, and they commonly fall from their original suspension points. Because the support strands are short, the cluster can look like a clump if viewed superficially. One must push eggs aside or detach the clutch from the substrate for the best views.

A *mass* (Fig. 3E) is a group with the outer jelly layer of all eggs melded together to form what appears as a matrix (e.g., like marbles, the ova usually with one jelly layer, embedded in a volume of jelly, the matrix). Colored water poured over a mass flows around the periphery and not through interstices as would happen with an open clump. The matrix may be relatively sparse or voluminous relative to the total ovum volume, and thus the distribution of ova within the mass appears to differ among taxa. Because the outer jelly layers in masses of *Pseudacris* (Figs. 92C–D, 94A) are relatively thin, the ova appear to be distributed close to the surface of the mass and thus uniformly throughout the volume of the mass. The larger outer jellies of *Ambystoma* eggs produce a wide, seemingly ova-free zone peripherally (white lines in Fig. 17A). If one tries to pick up a mass, it usually collapses and flows between the fingers, especially if the embryos are nearing hatching. However, masses of *Ambystoma gracile* and *A. maculatum* retain their configuration because the voluminous matrix is surprisingly dense, and these jellies often persist for more than a month after the embryos hatch. A mass often is vertically oblong and usually attached to vegetation. The surface of a mass may be smooth to irregular but never regularly lobate as in an old clump. A female may deposit a clutch in several masses.

Films. Films of eggs (Figs. 80A—hylid, 98A–B—microhylid, 105A—ranid) occur only in anurans and comprise a single-layered sheet of either adherent (e.g., hylids and ranids) or coherent (e.g., microhylids) eggs that float at the surface. Films sink easily when rained on, wetted eggs usually do not float again, and films may sink when the embryos approach hatching. Hylid and microhylid clutches are oviposited as several small films (= rafts) resulting from individual ovipositional bouts, while ranid films are usually laid as a single unit.

Foam nests. Leptodactylus fragilis in southern Texas is the only taxon in the scope of this book to oviposit eggs dispersed in a foam nest (Fig. 97A). The foam, placed in a subterranean burrow near a temporary puddle, is formed from oviductal secretions released by the female, and the ova are pale yellow.

Choices of appropriate geographic zones as defined before each key and ovipositional mode take one to other keys for more details. Correct familial and generic identifications are generally not difficult, but the paucity of accurate data, difficulty of observation, and variation with age of the eggs often make species identifications tenuous. The presence or absence of a capsular chamber that can distinguish salamander from frog eggs is used sparingly because of the difficulty of seeing the feature properly. If an answer does not seem logical, be sure to try the other parts of the key. The general lack of data is immediately obvious, but the construction of the key allows one to see patterns of geographical and morphological diversity.

Group names, formed by placing the first one listed in the species accounts in quotation marks (e.g., *Hyla "versicolor"* = *H. versicolor* and *H. chrysoscelis; Pseudacris "triseriata"* group = *P. triseriata, P. feriarum, P. fouquettei, P. kalmi, P. maculata,* and *P. nigrita*), are used in various places in the remainder of the text when a group of closely related taxa are encountered. Species that have totally insufficient data are listed within brackets below a presumed close relative. Each key is divided into geographical sectors that vary because of the species and the morphological diversity of the taxa involved. The key to eggs is divided into five geographic sectors: Hawaii, Pacific, Middle, Central, and Eastern. Oviposition mode is the next major character, and other characteristics of the clutch, ova, and range provide subsequent choices.

KEY TO EGGS

HAWAII

I. Oviposited as strings

Neotropical exotic on all major islands; nonflowing water and slow streams; sympatric: none of similar mode .... ***Rhinella marina* (Cane Toad)**

II. Oviposited as clumps

A. Neotropical exotic on Maui and Oahu; small terrestrial clump hidden in debris on forest floor; jelly thin and turgid; sympatric: none of similar development, but *Eleutherodactylus* oviposit similar arboreal eggs that develop directly .... ***Dendrobates auratus* (Green and Black Dart-poison) Frog**
Fig. 73B, photo of similar species, *D. leucomelas*

B. Oriental exotic on all major islands; small, aquatic, melded clump with flimsy jelly; nonflowing water and slow streams; sympatric: none of similar mode and habitat .... ***Glandirana rugosa* (Japanese Wrinkled Frog)**

III. Oviposited as films

Mainland exotic on all major islands; nonflowing water and slow streams; film greater than 150 mm diameter; 1 jelly layer; OD 1.2–1.7 mm; ED 6.4–10.4 mm; clutch 10,000–20,000; sympatric: none of similar mode ....
.... ***Lithobates catesbeianus* (American Bullfrog)**
Fig. 105A

PACIFIC COAST EAST TO INCLUDE SIERRA
NEVADA–CASCADE RANGES

I. Oviposited as independent eggs .... **Key Pacific Independent (p. 18)**

II. Oviposited in linear arrangements

    A. Oviposited as rosaries

        Most of British Columbia coast south in Cascade and Coast ranges to just north of San Francisco Bay; under stones and talus of mountain streams; ova nonpigmented, 2 jelly layers; ED about 8.0 mm diameter; OD 4.0–5.0 mm; clutch to about 68; sympatric: none of similar mode ....
        .... *Ascaphus truei* **(Coastal Tailed Frog)**
                                                        Fig. 96F

    B. Oviposited as strings

        1. Southeastern Alaska south into Mexico; nonflowing water, sometimes slow streams; OD 1.5–1.8 mm; jelly tube bilayered, 4.0–5.3 mm diameter; partitions present between uni- or biserial ova; clutch up to 16,500; 5–21 ova/cm; sympatric: *Anaxyrus californicus, A. canorus* ....
           .... *Anaxyrus boreas* **(Western Toad)**
                                                       Fig. 63G

        2. Coast Range of California from near Monterey south into Baja California, Mexico; slow streams or adjacent nonflowing water; OD 1.2–1.9 mm; ED 5.6–6.1 mm; 4–17 ova/cm; jelly tube unilayered; sympatric: *Anaxyrus boreas* .... *Anaxyrus californicus* **(Arroyo Toad)**

        3. Eldorado to Fresno cos., California in central Sierra Nevada above 1460 m; pools in mountain meadows and streams; small-diameter jelly tube bilayered, tube somewhat scalloped, partitions present between eggs; sympatric: *Anaxyrus boreas* .... *Anaxyrus canorus* **(Yosemite Toad)**
                                                        Fig. 64A

    C. Oviposited in three-dimensional arrangements

        1. Oviposited as small melded clumps with turgid jelly

            a. Humboldt Co. south into Sonoma and Lake cos., California; flattened clumps on lower surfaces of rocks in streams; OD 2.7 mm; ED 8.5 mm; clumps of 5–15 eggs; sympatric: *Taricha torosa* ....
            .... *Taricha rivularis* **(Red-bellied Newt)**
                                                        Fig. 55A

            b. Northern California south into Baja California; usually in ponds in coastal areas, streams in Coast and Sierra Nevada ranges; globular clumps attached to rocks and vegetation; OD 1.9–2.8 mm; ED 7.0–8.0 mm; sympatric: *Taricha rivularis* ....
            .... *Taricha torosa* **(California Newt)**
                                                        Fig. 56A

        2. Oviposited as large open (fresh clutch) or melded (older clutch) clumps, lobate surface visible even in older clumps; nonflowing water and slow streams .... **Key Pacific Clump (p. 20)**

        3. Oviposited as masses .... **Key Pacific Mass (p. 21)**

D. Oviposited as films; nonflowing water or slow streams

    1. Exotic in much of designated area; 1 jelly layer; OD 1.2–1.7 mm; ED 6.4–10.4 mm; clutch 10,000–20,000; sympatric: *Lithobates clamitans* ....
.... ***Lithobates catesbeianus*** **(American Bullfrog)**
Fig. 105A

    2. Exotic in southwestern British Columbia and adjacent Vancouver Island, Canada; OD 1.2–1.8 mm; ED 2.8–4.0 mm; 2 jelly layers; clutch 1000–5000; sympatric: *Lithobates catesbeianus* ....
.... ***Lithobates clamitans*** **(Green Frog)**
Fig. 107C

Key Pacific Independent

I. Oviposited as arrays

Coast Range and Cascades from southwestern British Columbia to northwestern California; nonpigmented ova under stones and among talus of mountain streams; OD 6.0 mm or greater; 5 jelly layers, 2 easily visible; clutch about 70; sympatric: *Dicamptodon copei* ..... ***Dicamptodon tenebrosus*** **(Coastal Giant Salamander)**
Fig. 20A, C

[Olympic Peninsula south to northwestern Oregon; sympatric: *Dicamptodon tenebrosus* .... ***Dicamptodon copei*** **(Cope's Giant Salamander)**]

[Coast Range from just south of San Francisco Bay to northwestern California; sympatric: no congeners ....
.... ***Dicamptodon ensatus*** **(California Giant Salamander)**]

II. Oviposited as grouped, unattached, nonpigmented singles; small streams and seeps; OD 5.6 mm or less; 3 jelly layers

Olympic Peninsula of Washington; sympatric: no congeners ....
.... ***Rhyacotriton olympicus*** **(Olympic Torrent Salamander)**
Fig. 50A

[Cascade Mountains from just north of Mt. St. Helens, Washington south to Lane Co., Oregon; sympatric: no congeners...
.... ***Rhyacotriton cascadae*** **(Cascades Torrent Salamander)**]

[Coast Range of Oregon and Washington south of Chehalis River to Yamhill Co., Oregon; sympatric: no congeners ....
.... ***Rhyacotriton kezeri*** **(Columbia Torrent Salamander)**]

[Coast Range from southern Yamhill Co., Oregon south to Mendocino Co., California; sympatric: no congeners ....
.... ***Rhyacotriton variegatus*** **(Southern Torrent Salamander)**]

III. Oviposited as single pigmented ova haphazardly attached to vegetation or scattered on bottom; usually in nonflowing water

A. Outer jelly layer round

1. West-central California; ephemeral nonflowing water; jelly layers watery; presence of outliers signal grouped singles instead of clumps; OD 2.0–4.0 mm; ED 4.5–10.0 mm; 3–4 jelly layers; sympatric: *Ambystoma macrodactylum, A. mavortium, Taricha granulosa* ....
.... ***Ambystoma californiense*** (**California Tiger Salamander**)
Similar to Fig. 19D

2. Southeastern Alaska south to northeastern California, disjunct area south of San Francisco Bay; nonflowing water; usually oviposit masses, but clutches of small numbers of eggs or clutches oviposited without structures for oviposition sites may be judged as singles; jelly watery and voluminous; OD 1.8–2.5 mm; ED 3.8–5.0 expanding to 12.0–17.0 mm; 2 jelly layers; sympatric: *Ambystoma californiense, A. mavortium, Taricha granulosa* ....
.... ***Ambystoma macrodactylum*** (**Long-toed Salamander**)
Fig. 14A–C

3. Scattered localities throughout designated area; nonflowing water; scattered singles attached to vegetation; jelly watery; sympatric: *Ambystoma macrodactylum, Taricha granulosa* ....
.... ***Ambystoma mavortium*** (**Western Tiger Salamander**)
Fig. 19D

4. Southeastern Alaska to south of San Francisco Bay; usually nonflowing water; singles attached to vegetation; OD 2.0 mm; ED 3.5 mm; jelly turgid; sympatric: *Ambystoma macrodactylum, A. mavortium* ....
.... ***Taricha granulosa*** (**Rough-skinned Newt**)
Fig. 54A–C

5. Southern Sierra Nevada Range of California; usually small streams; jelly turgid; ED 7.0–8.0 mm; OD 1.9–2.8 mm; sympatric: none in group..........
.... ***Taricha sierra*** (**Sierra Newt**)

6. Southwestern California, flowing and adjacent nonflowing water; scattered singles attached to plants, sometimes on the bottom; jelly watery; OD 1.7–2.4 mm; ED 3.7–5.0 mm; 1 jelly layer; sympatric: *Xenopus laevis* ....
.... ***Pseudacris cadaverina*** (**California Chorus Frog**)

7. African exotic in southwestern California; ephemeral and permanent nonflowing water and slow streams; singles attached to vegetation; OD about 2.0 mm; ED 3.0–4.0 mm; 2 or 3 jelly layers; sympatric: *Pseudacris cadaverina* .... ***Xenopus laevis*** (**African Clawed Frog**)

B. Outer jelly layer elongate, deposited in large, closely spaced arrays attached to plants or sticks in ephemeral, nonflowing water

West-central California; OD 1.0–1.6 mm; ED 3.2–4.4 mm; 2 jelly layers; clutch 300–500; sympatric: none with elongate outer jelly layer ....
.... ***Spea hammondii*** (**Western Spadefoot**)
Fig. 126A

Key Pacific Clump

I. Oviposited in streams and nearby nonflowing water

    A. Coast Range and Cascade Mountains of central Oregon south to southern California exclusive of Central Valley; OD 1.9–2.5 mm; ED 3.8–4.5 mm; 3 jelly layers; clump diameter 40 mm or less; clutch 100–1000; sympatric: none in habitat .... ***Rana boylii* (Foothill Yellow-legged Frog)**

    Figs. 118B, 120A, left

    B. Transverse ranges in southern California and central Sierra Nevada, 1370–3660 m; OD 1.8–2.3 mm; ED 6.4–14.0 mm; 3 jelly layers; clump diameter 28–40 mm; sympatric: *Rana boylii* ....

    .... ***Rana muscosa* (Southern Mountain Yellow-legged Frog)**

    Fig. 120A, right

    [Sierra Nevada of California and extreme western Nevada; sympatric: none in group .... ***Rana sierrae* (Sierra Nevada Yellow-legged Frog)**]

II. Oviposited in nonflowing water, including ponds and lake margins

    A. Southwestern British Columbia south to Elk Creek, Mendocino Co., California; OD 2.3–3.6 mm; ED 10.0–14.0 mm; 3 jelly layers; jelly layers thick so ova appear widely spaced; clump diameter 65–250 mm; clutch 750–4000; sympatric: *Lithobates pipiens*, *Rana cascadae*, *R. pretiosa* ....

    .... ***Rana aurora* (Northern Red-legged Frog)**

    Fig. 117C

    [South of Elk Creek, Mendocino Co., California into northern Baja California; sometimes in slow streams; sympatric: *Rana boylii* . ...

    ....*Rana draytoni* (**California Red-legged Frog**)]

    B. Olympic and Cascade mountains of Washington and Oregon south to Mt. Lassen, Lassen Co., California; OD 2.3 mm; ED greater than 6.0 mm; 3 jelly layers; jelly layers thick so ova appear widely spaced; clump diameter 75–125 mm; sympatric: *Rana aurora*, *R. pretiosa* ...***Rana cascadae* (Cascades Frog)**

    Fig. 119A, clump detached and floating

    C. Central Cascade Mountains; OD 1.8–2.3 mm; ED 5.0–15.0 mm; 1 or 2 jelly layers; jelly layers thin so ova appear narrowly spaced; clump diameter 75–150 mm; clutch 1100–2400; sympatric: *Rana aurora*, *R. cascadae* ....

    .... ***Rana pretiosa* (Oregon Spotted Frog)**

    Fig. 121A

    [Southeastern Alaska to southeastern British Columbia; nonflowing water or small streams; sympatric: *Lithobates pipiens*, *L. sylvaticus* ....

    .... ***Rana luteiventri* (Columbia Spotted Frog)**]

    Fig. 121B

    D. Few scattered populations throughout designated area; usually nonflowing water; OD 1.6 mm; ED 5.0 mm; 2 jelly layers; jelly thin so ova appear narrowly spaced; OD 1.0–2.0 mm; ED 2.5–5.6 mm; clump diameter 75–150 mm;

clutch 3000–6000; sympatric: *R. aurora*, *R. pretiosa* ....
.... *Lithobates pipiens* (**Northern Leopard Frog**)
Fig. 113A–C, F

E. Most of Alaska southeast through British Columbia to central Alberta; non-flowing water, often wooded and ephemeral; OD 1.8–2.4 mm; ED 4.2–9.4 mm; 2 jelly layers; jelly thin so ova appear narrowly spaced; clump diameter 60–100 mm; clutch 2000–3000; sympatric: *Rana luteiventris* ....
.... *Lithobates sylvaticus* (**Wood Frog**)
Fig. 114A

Key Pacific Mass

I. Capsular chamber present, so centers of ova are positioned below centers of inner jelly layer (Fig. 3A, D); masses usually greater than 60 mm diameter, less if deposited along a plant stem.. **Salamanders**

A. West-central British Columbia south to near San Francisco Bay; voluminous jelly very stiff, mass does not collapse out of water; inner jelly layers commonly with green algae; OD 2.0–3.0 mm; ED 6.0–7.0 mm; 1 jelly layer; mass diameter greater than 80 mm; sympatric: *Ambystoma macrodactylum* .... *Ambystoma gracile* (**Northwestern Salamander**)
Fig. 11C

B. Southeastern Alaska south to northeastern California, disjunct area south of San Francisco Bay; jelly voluminous and watery; nonflowing water; usually oviposits masses, small numbers of eggs may be judged to be singles or clumps; OD 1.8–2.5 mm; ED 3.8–5.0 expanding to 12.0–17.0 mm; 2 jelly layers; mass diameter 30–60 mm; sympatric: *Ambystoma gracile* ....
.... *Ambystoma macrodactylum* (**Long-toed Salamander**)
Fig. 14A–C

II. Capsular chamber absent, so ova centered in inner jelly layer (Fig. 3F); masses usually less than 60 mm diameter, jelly flimsy, usually distributed along a stem....
.... **Frogs**

Northern California, adjacent Oregon, Washington, and British Columbia to southern Alaska; OD 1.2–1.4 mm; ED 4.7–6.7 mm; 2 jelly layers; sympatric: none in group but could be confused with *Ambystoma macrodactylum* ....
.... *Pseudacris regilla* (**Northern Pacific Treefrog**)
Fig. 92C–D

[Southern California, Nevada, and adjacent Arizona south into Baja California; sympatric: none in group ....
.... *Pseudacris hypochondriaca* (**Baja California Treefrog**)]

[Central California, eastern Oregon and Idaho, and western Montana; sympatric: none in group .... *Pseudacris sierra* (**Sierran Treefrog**)]

EAST OF SIERRA NEVADA–CASCADE RANGES
TO INCLUDE ROCKY MOUNTAINS

I. Oviposited as independent eggs .... **Key Middle Independent (p. 23)**

II. Oviposited in linear arrangements

    A. Oviposited as rosaries

    Northern Rocky Mountains and Blue, Seven Devil, and Wallowa mountains in Idaho and Oregon; under rocks and talus of mountain streams; ova nonpigmented; 2 jelly layers, about 8 mm diameter; OD 4.0–5.0 mm; clutch to about 68 (data for *A. truei*); sympatric: none of similar mode ....
.... *Ascaphus montanus* **(Rocky Mountain Tailed Frog)**
See Fig. 96F

    B. Oviposited as wrapped rosaries

    Southeastern California across Rocky Mountains and south into Mexico; attached to vegetation in ephemeral, xeric sites, may be mistaken for a clump or mass; OD 1.4–1.6 mm; ED 6.0 mm; 1 jelly layer; clutch 350–500; ova darkly pigmented; jelly turgid, hatching ports remain open; sympatric: none of similar mode .... *Scaphiopus couchii* **(Couch's Spadefoot)**
Fig. 123A

    C. Oviposited as strings .... **Key Middle String (p. 24)**

III. Oviposited in three-dimensional arrangements

    A. Oviposited as clumps (also see II-B above) .... **Key Middle Clump (p. 26)**

    B. Oviposited as masses

        1. Capsular chamber present, so centers of ova below centers of inner jelly layer (Fig. 3A, D); masses usually greater than 60 mm diameter, less if deposited along the axis of a plant stem ....
.... *Ambystoma macrodactylum* **(Long-toed Salamander)**

        2. Capsular chamber absent, so ova centered in inner jelly layers (Fig. 3F); masses usually less than 60 mm diameter, jelly flimsy; usually distributed along a stem; Pacific Northwest from northern California, adjacent Oregon, Washington, and British Columbia to southern Alaska; OD 1.2–1.4 mm; ED 4.7–6.7 mm; 2 jelly layers; sympatric: none in group but could be confused with *Ambystoma macrodactylum*...
.... *Pseudacris regilla* **(Northern Pacific Treefrog)**
Fig. 92C–D

        [Southern Nevada and adjacent Arizona south into Baja California; sympatric: none in group ....
.... *Pseudacris hypochondriaca* **(Baja California Treefrog)**]

        [Central California, eastern Oregon and Idaho and western Montana; sympatric: none in group.............. *Pseudacris sierra* **(Sierran Treefrog)**]

3. Zone from northwestern to southeastern Arizona and adjacent New Mexico; mountain ephemeral sites and lake margins; capsular chamber absent, thus ova centered in inner jelly layer (Fig. 3F); OD 1.0–1.4 mm; sympatric: *Pseudacris maculata* ....
.... *Hyla wrightorum* (**Mountain Treefrog**)

4. Southeastern Yukon Territory south to central New Mexico; ephemeral nonflowing water; capsular chamber absent, thus ova centered in inner jelly layer (Fig. 3F); 1 jelly layer; sympatric: *Ambystoma macrodactylum, Hyla wrightorum, Pseudacris sierra* ....
.... *Pseudacris maculata* (**Boreal Chorus Frog**)

IV. Oviposited as coherent films, individual eggs easily pushed apart; nonflowing ephemeral water

South-central Arizona; ephemeral, nonflowing water; upper hemisphere of jelly of recently deposited eggs projects above water surface; 1 jelly layer wide relative to black ovum; OD 1.2–1.4 mm; ED 4.0 mm; sympatric: none of similar mode .... *Gastrophryne olivacea* (**Great Plains Narrow-mouthed Toad**)

V. Oviposited as adherent films, outer jellies stuck to each other so that individual eggs not easily pulled apart; top of outer jelly at water surface

A. Film diameter less than 150 mm

South-central Arizona; ephemeral desert pools; placed here on circumstantial evidence; sympatric: none of similar mode ....
.... *Smilisca fodiens* (**Lowland Burrowing Treefrog**)

B. Film diameter greater than 200 mm

1. Exotic at scattered sites throughout the designated area; 1 jelly layer; OD 1.2–1.7 mm; ED 6.4–10.4 mm; clutch 10,000–20,000; sympatric: *Lithobates clamitans* .... *Lithobates catesbeianus* (**American Bullfrog**)
Fig. 105A

2. Exotic near Salt Lake City, Utah, northeastern Washington, and northwestern Montana; OD 1.2–1.8 mm; ED 2.8–4.0 mm; 2 jelly layers; clutch 1000–5000; sympatric: *Lithobates catesbeianus* ....
.... *Lithobates clamitans* (**Green Frog**)
Fig. 107C

Key Middle Independent

I. Eggs oviposited as arrays

A. Central and northern Idaho and adjacent Montana; arrays of nonpigmented ova under stones and within talus of mountain streams; OD 6.0 mm or greater; 5 jelly layers, 2 easily visible ....
.... *Dicamptodon aterrimus* (**Idaho Giant Salamander**)
See Fig. 20A, C

B. Western Arizona to southwestern Colorado, east across Rocky Mountains and south into Mexico; ephemeral nonflowing water; arrays deposited in closely spaced groups attached to plants or sticks, may be difficult to visualize as an array; outer jelly elongate; OD 1.0–1.6 mm; ED 3.2–4.4 mm; 2 jelly layers; clutch 300–500; sympatric: none with elongate outer jelly layers .... *Spea multiplicata* (**Mexican Spadefoot**)

Fig. 128B

II. Eggs attached haphazardly as singles to aquatic substrates or free on the bottom; outer jelly round

A. Scattered localities throughout the designated area; nonflowing water; watery jelly layers 3, not voluminous, less than 12 mm; sympatric: *Ambystoma macrodactylum, Anaxyrus punctatus, Spea bombifrons* .... *Ambystoma mavortium* (**Western Tiger Salamander**)

Fig. 19D

B. Central British Columbia southeast to southwestern Alberta and south to central Idaho; nonflowing water; 2 jelly layers, voluminous and watery, swell to more than 12 mm diameter, usually oviposited as masses; sympatric: *Ambystoma mavortium* ..... *Ambystoma macrodactylum* (**Long-toed Salamander**)

Fig. 14A–C

C. West-central Idaho; usually nonflowing water; singles attached to vegetation; jelly rubbery; OD 2.0 mm; ED 3.5 mm; sympatric: *Ambystoma macrodactylum, A. mavortium* .... *Taricha granulosa* (**Rough-skinned Newt**)

Fig. 54A–C

D. Southeastern California to central Utah, east across Rocky Mountains to central Texas and south into Mexico; ephemeral pools and slow streams; OD 1.0–1.3 mm; ED 3.2–3.6 mm; 1 jelly layer; sympatric: *Ambystoma mavortium, Hyla arenicolor, Spea bombifrons* .... *Anaxyrus punctatus* (**Red-spotted Toad**)

E. Southeastern Nevada south into Mexico and east across Rocky Mountains; slow areas of rocky streams; OD 1.8–2.4 mm; ED 3.9–5.0 mm; 1 jelly layer; sympatric: *Ambystoma mavortium, Anaxyrus punctatus, Spea bombifrons* ... . *Hyla arenicolor* (**Canyon Treefrog**)

F. Southeastern Arizona north to southeastern Alberta and east across Rocky Mountains; ephemeral nonflowing water; 2 jelly layers not voluminous; ED less than 4 mm; OD 1.5 mm; clutch about 250; sympatric: *Ambystoma mavortium, Anaxyrus punctatus* .... *Spea bombifrons* (**Plains Spadefoot**)

Fig. 125A

Key Middle String

I. Small relictual ranges in east-central California and west-central Nevada; jelly tube obvious, strings long; OD 1.5–1.8 mm; jelly tube bilayered, 4.0–5.3 mm diameter; partitions present between uni- or biserial ova; clutch up to 16,500; 5–21 ova/cm based on *Anaxyrus boreas*; sympatric: no congeners

[Deep Springs Valley, east of Big Pine, Inyo Co., California; spring runs and adjacent sloughs .... *Anaxyrus exsul* (**Black Toad**)]

[Amargosa River drainage near Beatty, Nevada; spring runs and backwaters of river... *Anaxyrus nelsoni* (**Amargosa Toad**)]

Fig. 63F

II. Larger ranges; unilayered jelly tube diaphanous and short, difficult to see; ephemeral pools; sympatric: none of similar mode

[Southeastern Arizona and southern New Mexico and south into Mexico .... *Anaxyrus debilis* (**Green Toad**)]

[South-central Arizona and adjacent Mexico .... *Anaxyrus retiformis* (**Sonoran Green Toad**)]

III. Larger ranges, bilayered jelly tube obvious, string long

A. Much of area west of crest of Rocky Mountains from southern Alaska to northen Baja California except most of Arizona and New Mexico and parts of Utah and western Colorado; permanent and ephemeral nonflowing water, sometimes slow streams; OD 1.5–1.8 mm; jelly tube bilayered, 4.0–5.3 mm diameter; partitions present between uni- or biserial ova; clutch up to 16,500; 5–21 ova/cm; sympatric: *Anaxyrus cognatus, A. woodhousii* .... *Anaxyrus boreas* (**Western Toad**)

Fig. 63G

B. Southeastern California and adjacent Nevada to New Mexico; ephemeral, usually xeric sites; OD 1.2 mm; ED 2.0–2.7 mm; OD 1.2 mm; ova densely distributed; partitions between uniserial ova; scalloped tube margin commonly stated as diagnostic for this species is either not present or inconspicuous; clutch about 20,000; sympatric: *Anaxyrus boreas, A. woodhousii* .... *Anaxyrus cognatus* (**Great Plains Toad**)

C. Southwestern Utah and adjacent Nevada and a band from northwestern Arizona southeast into New Mexico; often in slow streams; ED 5.6–6.1 mm; OD 1.2–1.9 mm; 4–17 ova/cm; jelly tube unilayered based on *A. californicus*; sympatric: *Anaxyrus woodhousii, Incilius alvaria* .... *Anaxyrus microscaphus* (**Arizona Toad**)

D. Disjunct areas in southeastern Washington, southwestern Idaho south into southeastern California and east across Rocky Mountains; lake margins and ephemeral sites; OD 1.0–1.5 mm, tube diameter 2.6–4.6 mm, ova uni- or biserial, 6–10 ova/cm; clutch greater than 25,000; sympatric: *Anaxyrus boreas, A. cognatus, A. microscaphus, Incilius alvaria* .... *Anaxyrus woodhousii* (**Woodhouse's Toad**)

Fig. 70A

[Western half of Alberta, Canada; sympatric: *Anaxyrus cognatus*....................... .... *Anaxyrus hemiophrys* (**Canadian Toad**)]

E. Extreme southeastern California and southern half of Arizona; ephemeral nonflowing water; ED 2.1–2.2 mm; OD 1.1–1.7 mm; tube diameter 2.1–2.7 mm; clutch 7500–8000; 4–11 ova/cm; sympatric: *Anaxyrus microscaphus, A. woodhousii* .... *Incilius alvaria* (**Sonoran Desert Toad**)

Key Middle Clump

I. Ranges north of latitude of continuous southern borders of Utah, Colorado, and Kansas

A. Clumps less than 75 mm diameter; jelly quite flimsy

Central British Columbia south to west-central California and east to central Wyoming; ephemeral, nonflowing water; clump less than 40 mm diameter, jelly very flimsy, usually melded to appear like a mass, attached to vegetation; sympatric: no congeners..
.... ***Spea intermontana* (Great Basin Spadefoot)**
Fig. 127B

B. Clumps 100 mm diameter or greater; jelly reasonably firm

1. Throughout much of designated area, several disjunct ranges in western region; nonflowing water; OD 1.0–2.0 mm; ED 2.5–5.6 mm; clump diameter 75–150 mm; clutch 3000–6000 or greater; 2 jelly layers; jelly thin, so ova appear narrowly spaced; sympatric: *Lithobates sylvaticus, Rana luteiventris*.......................... ***Lithobates pipiens* (Northern Leopard Frog)**
Fig. 113A–C, F

2. Disjunct population in northern Idaho and central Alberta southeast to Mississippi River, mostly Minnesota and north; nonflowing water, often wooded sites; OD 1.8–2.4 mm; ED 4.2–9.4 mm; 2 jelly layers; jelly thin, so ova appear narrowly spaced; clump diameter 60–100 mm; clutch 2000–3000; sympatric: *Lithobates pipiens, Rana luteiventris*...
.... ***Lithobates sylvaticus* (Wood Frog)**
Fig. 114A

3. Southeastern Alaska to southeastern British Columbia and Cascade Mountains in Yakima Co., Washington; nonflowing water and small streams; OD 1.8–2.2 mm; ED 5.0–7.1 mm; 1 jelly layer; clump diameter 75–150 mm; clump sometimes at or extending above water surface; sympatric: *Lithobates pipiens, L. sylvaticus* ....
.... ***Rana luteiventris* (Columbia Spotted Frog)**
Fig. 121B

II. Ranges south of latitude of continuous southern borders of Utah, Colorado, and Kansas

A. Central and southeastern Arizona and adjacent New Mexico; presumably similar to other southwestern members of *Lithobates pipiens* group; nonflowing water and slow streams; sympatric: *Lithobates blairi, L. pipiens, L. yavapaiensis*
.... ***Lithobates chiricahuensis* (Chiricahua Leopard Frog)**
Fig. 106A

[Exotic in Gila River drainage, Arizona; nonflowing water and slow streams; sympatric: *Lithobates yavapaiensis* ....
.... ***Lithobates berlandieri* (Rio Grande Leopard Frog)**]

[Small area in southeastern Arizona; sympatric: *Lithobates chiricahuensis*.........
.... *Lithobates blairi* (**Plains Leopard Frog**)]

[Extinct in Vegas Valley, Clarke Co., Nevada, but other populations occur along the Mogollon Rim of north-central Arizona; nonflowing water and slow streams; sympatric: *Lithobates chiricahuensis*.......................................
.... *Lithobates fisheri* (**Vegas Valley Leopard Frog**)]

[Northwestern Arizona and adjacent California southeast to southwestern Arizona; nonflowing water and slow streams; sympatric: *Lithobates berlandieri, L. chiricahuensis* ....
.... *Lithobates yavapaiensis* (**Lowland Leopard Frog**)]

B. Southern Nevada and adjacent Arizona; nonflowing water and slow streams; mean clump diameter 45 mm; sympatric: none in group ....
.... *Lithobates onca* (**Relict Leopard Frog**)
Fig. 111D

C. Widespread north of central Arizona; nonflowing water, often ephemeral; OD 1.0–2.0 mm; ED 2.5–5.6 mm; clump diameter 75–150 mm; clutch 3000–6000 or greater; sympatric: *Rana luteiventris*..
.... *Lithobates pipiens* (**Northern Leopard Frog**)
Fig. 113A–C, F

D. South-central Arizona; rocky streams; OD 3.7–5.0 mm; ED 2.0–2.2 mm; clutch diameter about 50 mm; 2 jelly envelopes; clutch about 2200; sympatric: *Lithobates chiricahuensis, L. yavapaiensis* ....
.... *Lithobates tarahumarae* (**Tarahumara Frog**)
Fig. 115B

E. Much of area north of central Nevada with many disjunct populations; small streams and lakes; OD 1.8–2.2 mm; ED 5.0–7.1 mm; 1 jelly layer; clump diameter 75–150 mm; often in nonflowing water; clump often at or extending above water surface; sympatric: *Lithobates pipiens* ....
.... *Rana luteiventris* (**Columbia Spotted Frog**)
Fig. 121B

EAST OF ROCKY MOUNTAINS TO MISSISSIPPI RIVER

I. Oviposited as independent eggs

A. Oviposited as small arrays of 120 mm diameter or less, nonpigmented in streams and caves; Edwards Plateau, Texas .... **Key Central Array (p. 30)**

B. Oviposited as small arrays of 120 mm diameter or less, nonpigmented ova in streams and caves; Ozark region (**note**: all *Eurycea* are presumed to oviposit arrays, but when few eggs are deposited, one may interpret them as singles; data insufficient in most cases)

1. Northern half of Arkansas and adjacent Oklahoma; OD 2.5–3.0 mm; 2 jelly layers; sympatric: *Eurycea lucifuga, E. multiplicata, E. spelaea, E. tynerensis* .... *Eurycea longicauda* (**Long-tailed Salamander**)

2. Southern half of Missouri and adjacent Arkansas and Oklahoma; OD 2.5–3.2 mm; 2 jelly layers; sympatric: *Eurycea longicauda, E. multiplicata, E. spelaea, E. tynerensis* ..... ***Eurycea lucifuga* (Cave Salamander)**

3. Central Arkansas to south-central Missouri and adjacent Oklahoma; OD 2.0–2.6 mm; 2 jelly layers; sympatric: *Eurycea longicauda, E. spelaea, E. tynerensis* .... ***Eurycea multiplicata* (Many-ribbed Salamander)**

4. Southern Missouri and adjacent Arkansas and Oklahoma; probably attached to cave walls at water line; OD 2.0–2.2 mm; 3 jelly layers; sympatric: *Eurycea longicauda, E. lucifuga, E. multiplicata, E. tynerensis* .... ***Eurycea spelaea* (Grotto Salamander)**
Fig. 36A

5. Southwestern Missouri and adjacent Oklahoma and Arkansas; OD 3.5 mm; sympatric: *Eurycea longicauda, E. lucifuga, E. multiplicata, E. spelaea* .. ..*Eurycea tynerensis* **(Oklahoma Salamander)**

C. Oviposited in large arrays, much greater than 120 mm diameter; larger nonflowing water and slow streams

1. West-central Louisiana and adjacent Texas; sympatric: no congeners .... ***Necturus beyeri* (Gulf Coast Waterdog)**

2. Arkansas and areas of adjacent states and drainages of Mississippi and Missouri rivers north into southern Manitoba; sympatric: no congeners .... ***Necturus maculosus* (Mudpuppy)**
Fig. 48A

II. Oviposited as independent arrangements other than arrays or probable arrays that are difficult to recognize because of the small number of eggs or ovipositional substrate not uniform .... **Key Central Other (p. 30)**

III. Oviposited as linear arrangements .... **Key Central Linear (p. 32)**

IV. Oviposited as three-dimensional arrangements

A. Oviposited as clumps of nonpigmented ova; nonflowing water....................... .... **Key Central Clump (p. 34)**

B. Oviposited as clusters of nonpigmented ova; flowing water

[Eastern Texas through southern Louisiana .... ***Desmognathus auriculatus* (Southern Dusky Salamander)**]
Similar to Fig. 23A–B

[West-central Arkansas and adjacent Oklahoma; sympatric: no congeners.......... .... ***Desmognathus brimleyorum* (Ouachita Dusky Salamander)**]

[Northeastern Missouri and north-central Louisiana and adjacent southern Arkansas........................ ***Desmognathus fuscus* (Northern Dusky Salamander)**]

C. Oviposited as masses, because capsular chamber present, centers of ova below centers of inner jelly layer (Fig. 12A); nonflowing water

1. Eastern Oklahoma northeast to east-central Missouri; ephemeral nonflowing water; OD about 2.0 mm; 2 flimsy jelly layers; sympatric: *Ambystoma
   maculatum, A. talpoideum, A. texanum* ....
   .... ***Ambystoma annulatum* (Ringed Salamander)**
   Fig. 7A

2. Most of Louisiana, Arkansas, and Missouri and adjacent Oklahoma and
   Texas; ephemeral nonflowing water; OD 2.5–3.0 mm; ED 6.0–8.0 mm;
   1 jelly layer; jelly voluminous, dense and stiff, sometimes milky white;
   inner jelly layers commonly with green algae; sympatric: *Ambystoma annulatum, A. talpoideum, A. texanum* ....
   .... ***Ambystoma maculatum* (Spotted Salamander)**
   Figs. 15A, 113A

3. Northern two-thirds of Louisiana and adjacent Texas and Oklahoma; nonflowing water, wooded sites; OD 0.9–1.2 mm; ED 2.3–3.6 mm; 1, rarely 2
   flimsy jelly layers; sympatric: *Ambystoma annulatum, A. maculatum, A.
   texanum* .... ***Ambystoma talpoideum* (Mole Salamander)**
   Fig. 17A

4. East of a line connecting central Texas and southeastern Iowa; ephemeral
   nonflowing water; OD 1.6–2.5 mm; ED 6.0–6.5 mm; 2 flimsy jelly layers;
   sympatric: *Ambystoma annulatum, A. maculatum, A. talpoideum* ....
   .... ***Ambystoma texanum* (Small-mouthed Salamander)**
   Fig. 18A

D. Oviposited as masses, because capsular chamber absent, ova centered in inner
   jelly layer (Fig. 3E–F); masses usually less than 60 mm diameter, eggs with
   flimsy jelly, usually distributed along a stem

   1. Southern tip of Texas north to central Kansas, west to New Mexico border
      and east to Missouri border; grassy nonflowing water; sympatric: *Pseuda-
      cris streckeri, P. "triseriata"* ....
      .... ***Pseudacris clarkii* (Spotted Chorus Frog)**

   2. Southern Texas north to south-central Kansas and east to central Arkan-
      sas; ephemeral nonflowing water; sympatric: *Pseudacris clarkii, P. "trise-
      riata"* .... ***Pseudacris streckeri* (Strecker's Chorus Frog)**

   3. Northern Alberta south and east to southeastern Texas and east to Mississippi
      River; ephemeral sites, usually grassland; sympatric: *Pseudacris clarkii,
      P. streckeri* .... ***Pseudacris "triseriata"* (Trilling Chorus Frogs)**
      Fig. 94A

E. Oviposited as nonpigmented ova in subterranean foam nest

   Southern Texas; nest near ephemeral nonflowing water; small ova yellow;
   sympatric: none of similar mode ....
   .... ***Leptodactylus fragilis* (Mexican White-lipped Frog)**
   Fig. 97A

F. Oviposited as films; usually nonflowing water, sometimes ephemeral sites, or larger nonflowing water or slow streams .... **Key Central Film (p. 35)**

Key Central Array

I. Range north of Colorado River; sympatric: no congeners (**note**: data lacking)

[Near Salado, Bell Co., Texas ... *Eurycea chisholmensis* (**Salado Salamander**)]

[Near Georgetown, Williamson Co., Texas ....
.... *Eurycea naufragia* (**Georgetown Salamander**)]

[Jollyville Plateau, Travis and Williamson Cos., Texas ....
.... *Eurycea tonkawae* (**Jollyville Plateau Salamander**)]

II. Range south of Colorado River

A. Cascade Caverns, Kendall Co., Texas; sympatric: no congeners ....
.... *Eurycea latitans* (**Cascade Caverns Salamander**)
Fig. 29A

[Comal Co., Texas........... *Eurycea tridentifera* (**Comal Blind Salamander**)]

[Medina Co., Texas..... *Eurycea troglodytes* (**Valdina Farms Salamander**)]

B. San Marcos Springs, Hays Co., Texas; sympatric: *Eurycea rathbuni*..............
.... *Eurycea nana* (**San Marcos Salamander**)
Fig. 32C–F

C. Bexar Co., Texas; sympatric: no congeners ....
.... *Eurycea neotenes* (**Texas Salamander**)

[Near Wimberley, Hays Co., Texas in Blanco River drainage; sympatric: no congeners .... *Eurycea pterophila* (**Fern Bank Salamander**)]

D. San Marcos, Hays Co., Texas; sympatric: *Eurycea nana*.................................
.... *Eurycea rathbuni* (**Texas Blind Salamander**)
Fig. 34B–C

[Aquifer beneath Blanco River, Hays Co., Texas; sympatric: no congeners ....
.... *Eurycea robusta* (**Blanco Blind Salamander**)]

[Travis Co., Texas; sympatric: *Eurycea sosorum* ....
.... *Eurycea waterlooensis* (**Austin Blind Salamander**)]

E. Travis Co., Texas; sympatric: *Eurycea waterlooensis*.....................................
.... *Eurycea sosorum* (**Barton Springs Salamander**)
Fig. 35A

Key Central Other

I. Southeastern Texas coast north and east to southeastern Missouri and east to Mississippi River; eggs oviposited as grouped singles, weakly adherent eggs, guarded by parent at semiterrestrial site before autumnal rains form pools; jelly

layers tough and resilient; sympatric: none of similar mode and habitat ....
.... *Ambystoma opacum* **(Marbled Salamander)**
Fig. 16A–B

II. Eggs oviposited as scattered singles among or attached to semiterrestrial or aquatic vegetation or free on substrate; usually nonflowing water

   A. Outer jelly round

      1. Single eggs attached haphazardly to vegetation

         a. Most of designated area, absent from most of Louisiana and Arkansas; nonflowing water; jelly layers 3, not voluminous, less than 12 mm diameter; sympatric: *Eurycea quadridigitata, Hemidactylium scutatum, Acris blanchardi, Anaxyrus punctatus, Hyla arenicolor, Rhinophrynus dorsalis*.
.... *Ambystoma mavortium* **(Western Tiger Salamander)**
Fig. 19D

         b. Southern Arkansas, eastern Texas, and most of Louisiana; attached to vegetation in swampy areas, a probable array not easily recognized because of nonuniform substrate; OD about 2.0 mm; ED about 3.0 mm; 2 jelly layers; sympatric: none in habitat ....
.... *Eurycea quadridigitata* **(Dwarf Salamander)**

         c. East of a line extending from southeastern Texas to northeastern Iowa; usually wooded ephemeral sites; singles attached to twigs; OD 0.6–0.9 mm; ED 2.2–2.4 mm; 1–2 jelly layers; sympatric: none in habitat or breedings season .... *Pseudacris crucifer* **(Spring Peeper)**
Fig. 89A

         d. South Texas; ephemeral nonflowing water; OD 2.0 mm; ED 3.0–4.0 mm; ova pale; sympatric: *Ambystoma mavortium, Acris blanchardi* .....
.... *Rhinophrynus dorsalis* **(Burrowing Toad)**

         e. Rocky Mountains from southeastern Alberta south into Mexico and east to eastern Missouri; ephemeral, nonflowing water; 2 jelly layers not voluminous; OD 1.0–1.6 mm; ED 3.2–4.4 mm; clutch 300–500; sympatric: *Acris blanchardi, Anaxyrus punctatus, Pseudacris crucifer, Rhinophrynus dorsalis* .... *Spea bombifrons* **(Plains Spadefoot)**
Fig. 125A

      2. Single eggs scattered on bottom, sometimes laid as flimsy clumps that soon fall apart

         a. Eastern two-thirds of Texas north to southern South Dakota and east to Mississippi River except for range of Northern Cricket Frog in southeastern Texas and Louisiana; usually in larger nonflowing waters or slow streams; sympatric: *Ambystoma mavortium, Anaxyrus punctatus, Hyla arenicolor, Rhinophrynus dorsalis* ....
.... *Acris blanchardi* **(Blanchard's Cricket Frog)**

         [Southeastern Texas northeast to southeastern Missouri and east to Mississippi River; sympatric: *Ambystoma mavortium* ....
.... *Acris crepitans* **(Northern Cricket Frog)**]

    b. Southeastern California to central Utah, east across Rocky Mountains to central Texas and south into Mexico; usually free on bottom of ephemeral pools and slow-moving streams as scattered single eggs; OD 1.0–1.3 mm; ED 3.2–3.6 mm; 1 jelly layer; sympatric: *Ambystoma mavortium, Acris blanchardi, Hyla arenicolor* ....
.... ***Anaxyrus punctatus*** (**Red-spotted Toad**)

    c. Trans-Pecos, Texas; usually free as single eggs on bottoms of rocky streams; OD 1.8–2.4 mm; ED 3.9–5.0 mm; 1 jelly layer; sympatric: *Ambystoma mavortium, Anaxyrus punctatus*...
.... ***Hyla arenicolor*** (**Canyon Treefrog**)

B. Outer jelly oval or elongate, and turgid; attached singly to vegetation; non-flowing water

    1. Southeastern Texas coast south into Mexico; outer jelly oval and turgid, attached singly; sympatric: *Notophthalmus viridescens* ....
.... ***Notophthalmus meridionalis*** (**Black-spotted Newt**)
Fig. 51A

    2. East of a line connecting south-central Texas coast and central Minnesota; outer jelly oval and turgid, attached singly; OD about 1.5 mm; ED 2.4–3.6 mm; 3 jelly layers; sympatric: *Notophthalmus meridionalis* ....
.... ***Notophthalmus viridescens*** (**Eastern Newt**)

    3. Eastern New Mexico south into Mexico and east to central Texas; ephemeral, nonflowing water; outer jelly elongate, attached in tight groups to vegetation; OD 1.0–1.6 mm; ED 3.2–4.4 mm; 2 jelly layers; clutch 300–500; sympatric: none with elongate outer jelly layers ....
.... ***Spea multiplicata*** (**Mexican Spadefoot**)
Fig. 128B

C. Oviposited in groups in nest, may appear as a clump

    1. Disjunct areas in Arkansas, Missouri, and Louisiana; semiterrestrial, often among *Sphagnum*, guarded by female, communal nests common, a probable array not easily recognized because of nonuniform substrate; sympatric: none of similar mode and habitat ....
.... ***Hemidactylium scutatum*** (**Four-toed Salamander**)
Fig. 40A

    2. Oviposited in a nest on the bottom of nonflowing water; crystalline inclusions in opaque outer jelly layer; ova pale; OD 2.5–3.0 mm; ED 4.4–7.5 mm; southern quarter of Texas north and east to northeastern Missouri and east to Mississippi River (**note**: some authorities consider sirens in southern Texas a distinct taxon, *S. texana*, more closely related to *S. lacertina*); sympatric: none of similar mode and habitat ....
.... ***Siren intermedia*** (**Lesser Siren**)

Key Central Linear

I. Oviposited as rosaries of large nonpigmented ova

    A. Southeastern Texas coast north and east to southeastern Missouri and east to Mississippi River; in burrow near nonflowing water, guarded by female;

sympatric: none of similar mode, ovum size, or habitat ....
.... *Amphiuma tridactylum* (**Three-toed Amphiuma**)

B. Ozark Region; larger, cool streams and rivers; eggs in nest beneath large rocks, guarded by male; OD about 6.0 mm; ED 18.0–20.0 mm; sympatric: none of similar mode, ovum size, or habitat ....
.... *Cryptobranchus alleganiensis* (**Hellbender**)
Fig. 22B

II. Oviposited as wrapped rosaries of small, darkly pigmented ova; may be mistaken for a clump or mass

A. Eastern third of New Mexico east to southeastern Oklahoma and south into Mexico; wrapped around vegetation in ephemeral, xeric sites; OD 1.4–1.6 mm; ED 6.0 mm; 1 jelly layer; clutch 350–500; jelly turgid, hatching ports remain open; sympatric: *Scaphiopus hurterii* ....
.... *Scaphiopus couchii* (**Couch's Spadefoot**)
Fig. 123A

B. Southeastern Missouri and adjacent Arkansas; wound around vegetation in ephemeral sites; jelly watery; OD 1.4–2.0 mm; ED 4.0–5.6 mm; sympatric: none in group .... *Scaphiopus holbrookii* (**Eastern Spadefoot**)
Fig. 124A

[Eastern half of Oklahoma south to southern Texas; jelly watery; sympatric: *Scaphiopus couchii* .... *Scaphiopus hurterii* (**Hurter's Spadefoot**)]

III. Oviposited as strings

A. Short egg tubes unilayered and diaphanous, difficult to see

Rocky Mountains of New Mexico and Colorado east to eastern third of Texas and south into Mexico; sympatric: none of similar mode......................................
.... *Anaxyrus debilis* (**Green Toad**)

B. Long egg tube unilayered and obvious

1. East of a line connecting eastern Texas and southeastern Iowa; edges of pools and ephemeral nonflowing water; tubes long, obvious; OD 1.0–1.4 mm; tube diameter 2.6–4.6 mm; ova uniserial or staggered; sympatric: *Anaxyrus woodhousii*...........................*Anaxyrus fowleri* (**Fowler's Toad**)
Fig. 65A

2. South-central New Mexico north to central Montana and east to central Missouri except Minnesota and most of Iowa; ephemeral, nonflowing water and lake margins; OD 1.0–1.5 mm, tube diameter 2.6–4.6 mm, ova uni- or biserial, 6–10 ova/cm; clutch greater than 25,000; sympatric: *Anaxyrus americanus, A. baxteri, A. cognatus, A. hemiophrys, A. houstonensis, A. speciosus, Incilius nebulifer, Rhinella marina* ....
.... *Anaxyrus woodhousii* (**Woodhouse's Toad**)
Fig. 70A

C. Egg tube bilayered and long

1. Most of area except Coastal Plain; OD 1.0–1.4 mm; tube diameter 3.4–4.0 mm; partitions present between uniserial ova; sympatric: *Anaxyrus*

cognatus, A. hemiophrys, A. woodhousii, Incilius nebulifer ....
.... *Anaxyrus americanus* (**American Toad**)
Fig. 62A

[Southeastern Wyoming; sympatric: *Anaxyrus woodhousii* ....
.... *Anaxyrus baxteri* (**Wyoming Toad**)]
Fig. 66A

[North-central Alberta southeast to northeastern South Dakota; sympatric: *Anaxyrus americanus, A. cognatus, A. woodhousii* ....
.... *Anaxyrus hemiophrys* (**Canadian Toad**)]

[Bastrop Co., east-central Texas; sympatric: *Anaxyrus speciosus, A. woodhousii, Incilius nebulifer* .... *Anaxyrus houstonensis* (**Houston Toad**)]
Fig. 62E

2. Much of Great Plains from southwestern Saskatchewan to Texas panhandle; ephemeral, usually xeric sites; OD 1.2 mm; ED 2.0–2.7 mm; partitions present between uniserial ova; indentations in tube wall between successive eggs commonly stated as diagnostic for this species are either not present or inconspicuous; sympatric: *Anaxyrus americanus, A. hemiophrys, A. speciosus, A. woodhousii, Incilius nebulifer*................................
.... *Anaxyrus cognatus* (**Great Plains Toad**)

3. Central New Mexico east to eastern Texas; nonflowing water, usually ephemeral; OD 1.2–1.6 mm; ED 1.8–2.4 mm; 4–7 ova/cm; partitions absent between uniserial ova; sympatric: *Anaxyrus cognatus, A. houstonensis, A. woodhousii, Incilius nebulifer*....................................... *Anaxyrus speciosus* (**Texas Toad**)

4. Southeastern half of Texas across southern Louisiana to Mississippi River; ephemeral, nonflowing water; OD 1.2 mm; tube diameter 3.0 mm with straight margins; partitions absent between uniserial or staggered ova; sympatric: *Anaxyrus americanus, A. cognatus, A. houstonensis, A. speciosus, A. woodhousii, Rhinella marina*..
.... *Incilius nebulifer* (**Gulf Coast Toad**)

5. Southern Texas; sympatric: *Anaxyrus woodhousii, Incilius nebulifer*.........
.... *Rhinella marina* (**Cane Toad**)

Key Central Clump

I. Oviposited as small clumps, surface not obviously lobate, often fall apart as single eggs

A. West Texas north to northern Nebraska and east to Mississippi River minus parts of eastern Texas, Arkansas, and southern Missouri; slow streams and nonflowing water; sympatric: *Hyla cinerea, H. squirella*................................
....*Acris blanchardi* (**Blanchard's Cricket Frog**)

[Eastern Texas east to Mississippi River; sympatric: *Hyla cinerea, H. squirella*
....*Acris crepitans* (**Northern Cricket Frogs**)]

B. Eastern half of Texas north and east to southeastern Missouri and east to Mississippi River; permanent and ephemeral sites with emergent vegetation; OD 0.8–1.6 mm; ED 3.6–4.0 mm; 2 jelly layers; sympatric: *Acris crepitans, Hyla squirella* .... ***Hyla cinerea* (Green Treefrog)**

C. Southeastern quarter of Texas east to Mississippi River; ephemeral, nonflowing water; OD 0.8–1.0 mm; ED 1.4–2.0 mm; 2 jelly layers; sympatric: *Acris crepitans, Hyla cinerea* .... ***Hyla squirella* (Squirrel Treefrog)**

II. Eggs oviposited as large clump, surface obviously lobate even after melded in older clumps, structure persists until hatching

A. Irregular distribution from eastern Texas northeast to southeastern Iowa; OD 2.4–2.6 mm; ED 3.6–4.4 mm; 2 jelly layers; clump diameter 90–140 mm; large ova widely spaced because of thick jelly layers; sympatric: *Lithobates blairi, L. palustris, L. "pipiens"* .... ***Lithobates areolatus* (Crawfish Frog)**
Fig. 113F

B. Southwestern three-quarters of Texas; nonflowing water and slow streams; ova darkly pigmented; sympatric: *Lithobates blairi, L. sphenocephalus*.........
.... ***Lithobates berlandieri* (Rio Grande Leopard Frog)**

C. Southeastern Texas north and east to northeastern Missouri and east to Mississippi River; nonflowing water, often wooded; ova pale to yellowish; sympatric: *Lithobates sphenocephalus* ...... ***Lithobates palustris* (Pickerel Frog)**

D. Central New Mexico north to northern Manitoba and east to Mississippi River; nonflowing water, often ephemeral; OD 1.0–2.0 mm; ED 2.5–5.6 mm; clump diameter 75–150 mm; clutch 3000–6000 or greater; sympatric: *Lithobates blairi* .... ***Lithobates pipiens* (Northern Leopard Frogs)**
Fig. 113A–C, F

[Much of Great Plains north of central Texas to southern Iowa; sympatric: *Lithobates pipiens* .... ***Lithobates blairi* (Plains Leopard Frog)**]

[Eastern third of Texas north and east to northern Missouri and east to Mississippi River; sympatric: *Lithobates berlandieri, L. palustris*..................................
.... ***Lithobates sphenocephalus* (Southern Leopard Frog)**]
Fig. 113B

E. Wide band from central Alberta southeast to Mississippi River, disjunct populations in northwestern Arkansas; nonflowing water, often wooded; OD 1.8–2.4 mm; ED 4.2–9.4 mm; 2 jelly layers; clump diameter 60–100 mm; clutch 2000–3000; sympatric: *Lithobates palustris, L. pipiens, L. sphenocephalus* .... ***Lithobates sylvaticus* (Wood Frog)**
Fig. 114A

Key Central Film

I. Oviposited in coherent films, individual eggs easily pushed apart; upper hemisphere of jelly of recently deposited eggs projects above water surface, film diameter less than 150 mm; ovum intensely black and jelly diameter large relative to ovum size

A. Eastern third of Texas north and east to central Missouri and east to Mississippi River; ephemeral, nonflowing water; sympatric: *Gastrophryne olivacea, Hypopachus variolosus* ....
.... *Gastrophryne carolinensis* (**Eastern Narrow-mouthed Toad**)
Fig. 98A–B

[Trans-Pecos, Texas north and east to central Missouri and east to western Arkansas; sympatric: *Gastrophryne carolinensis, Hypopachus variolosus* ....
.... *Gastrophryne olivacea* (**Western Narrow-mouthed Toad**)]

B. South Texas; ephemeral, nonflowing water; sympatric: *Gastrophryne carolinensis, G. olivacea* .... *Hypopachus variolosus* (**Sheep Frog**)

II. Oviposited in adherent films, individual eggs not easily pulled from film, top of outer jelly lies at water surface

A. Film diameter less than 200 mm

1. East of a line from southern tip of Texas north and east to southern Illinois and east to Mississippi River; permanent and ephemeral nonflowing water with emergent vegetation; OD 0.8–1.6 mm; ED 3.6–4.0 mm; 2 jelly layers; most often lays clumps; sympatric: *Hyla avivoca, H. "versicolor"* ....
.... *Hyla cinerea* (**Green Treefrog**)

2. Eastern half of Texas north to southern Manitoba and east to Mississippi River; permanent and ephemeral nonflowing water; sympatric: *Hyla avivoca*
.... *Hyla "versicolor"* (**Gray Treefrogs**)
Fig. 83H

[Scattered populations in central and western Arkansas and central Louisiana; swampy sites with emergent woody vegetation; probably oviposits clumps that sometimes break up and float at the surface; sympatric: *Hyla cinerea, H. "versicolor"* .... *Hyla avivoca* (**Bird-voiced Treefrog**)]

3. South Texas; OD 1.3 mm; ED 1.5 mm; 1 jelly layer; ephemeral, nonflowing water; sympatric: *Hyla cinerea* ...... *Smilisca baudinii* (**Mexican Treefrog**)

B. Film diameter greater than 200 mm

1. Throughout much of designated area but more disjunct populations to the west; larger, permanent nonflowing water: 1 jelly layer; OD 1.2–1.7 mm; ED 6.4–10.4 mm; clutch 10,000–20,000; sympatric: *Lithobates clamitans, L. grylio* .... *Lithobates catesbeianus* (**American Bullfrog**)
Fig. 105A

2. Eastern third of Texas north to southern Ontario and east to Mississippi River; permanent, often swampy, nonflowing water; OD 1.2–1.8 mm; ED 2.8–4.0 mm; 2 jelly layers; clutch 1000–5000; sympatric: *Lithobates catesbeianus, L. grylio* .... *Lithobates clamitans* (**Green Frog**)
Fig. 107C

3. Coastal Plain in southeastern corner of Texas east to Mississippi River; larger permanent nonflowing water; 2 jelly layers; sympatric: *Lithobates catesbeianus, L. clamitans* .... *Lithobates grylio* (**Pig Frog**)

MISSISSIPPI RIVER TO ATLANTIC COAST

I. Oviposited as independent eggs

    A. Oviposited as small arrays of a few eggs, 120 mm or less diameter, in springs, streams, and caves .... **Key Eastern Array I (p. 37)**

    B. Oviposited in large arrays, greater than 120 mm diameter in lakes and rivers .... **Key Eastern Array II (p. 39)**

    C. Oviposited as independent arrangements other than arrays or probable arrays that are difficult to recognize because of the small number of eggs or nonuniform substrate; most often in nonflowing water.... **Key Eastern Other (p. 39)**

II. Oviposited as linear arrangements.......................... **Key Eastern Linear (p. 41)**

III. Oviposited as three-dimensional arrangements

    A. Oviposited as clumps of pigmented ova; nonflowing water .... **Key Eastern Clump (p. 42)**

    B. Oviposited as clusters of nonpigmented ova; flowing water; sympatric: none within group... **Key Eastern Cluster (p. 44)**

    C. Oviposited as masses

    Because capsular chamber present, centers of ova below center of inner jelly layer .... **Key Eastern Mass (p. 46)**
Fig. 3A, D

    D. Oviposited as films of pigmented eggs; usually nonflowing water, sometimes ephemeral or larger nonflowing water and slow streams................................. .... **Key Eastern Film (p. 48)**

Key Eastern Array I

I. North-central Kentucky and adjacent Ohio and Indiana, disjunct in eastern and western Kentucky and adjacent West Virginia; arrays of pigmented ova on the undersides of rocks in small streams; sympatric: none of similar mode .... *Ambystoma barbouri* **(Streamside Salamander)**
Fig. 8A–B

II. Oviposited in dark zones of caves; ova nonpigmented (**note**: eggs of *Plethodon* are nonpigmented and direct developing; *Eurycea, Desmognathus, Gyrinophilus porphyriticus*, and *Pseudotriton* may also occur in cave entrances; most data lacking)

    [Northern Florida and adjacent Georgia; ovarian ED 2.0–2.2 mm; sympatric: no confamilial taxa .... *Eurycea wallacei* **(Georgia Blind Salamander)**]

    [Roane Co., Tennessee; sympatric: no congeners .... *Gyrinophilus gulolineatus* **(Berry Cave Salamander)**]

    [South-central Tennessee and adjacent Alabama; sympatric: no congeners .... *Gyrinophilus palleucus* **(Tennessee Cave Salamander)**]

[Greenbrier Co., West Virginia; sympatric: *Gyrinophilus porphyriticus* ....
.... ***Gyrinophilus subterraneus* (West Virginia Spring Salamander)**]

III. Oviposited among cobble or under rocks in springs and streams or among debris and vegetation at lowland sites (array structure not immediately obvious in latter case); ova nonpigmented

A. Egg diameter 6.0 mm or greater, 1 jelly layer involved in pedicel

    1. Wide band from southwestern Maine to northeastern Mississippi; sympatric: *Pseudotriton ruber, P. montanus* ....
.... ***Gyrinophilus porphyriticus* (Spring Salamander)**

    2. Coastal Plain from southeastern Mississippi to New Jersey and large area in central Kentucky and Tennessee; sympatric: *Gyrinophilus porphyriticus, Pseudotriton ruber* ..... ***Pseudotriton montanus* (Mud Salamander)**

          Fig. 41A

    3. Wide band from southeastern New York to southeastern Mississippi; sympatric: *Gyrinophilus porphyriticus, Pseudotriton montanus* ....
.... ***Pseudotriton ruber* (Red Salamander)**

          Fig. 42C

B. Egg diameter less than 6 mm; 2 jelly layers involved in pedicel

    1. Labrador and Quebec south to northeastern Ohio and northern Virginia; sympatric: *Eurycea longicauda, E. lucifuga*...
.... ***Eurycea bislineata* (Northern Two-lined Salamander)**

          Fig. 27A–B

    [North-central Alabama near Birmingham northeast to northwestern Georgia; sympatric: *Eurycea cirrigera, E. longicauda, E. lucifuga* ....
.... ***Eurycea aquatica* (Brown-backed Salamander)**]

    [Eastern Illinois to central Virginia south exclusive of Florida Peninsula and the range of *E. wilderae*; sympatric: *Eurycea guttolineata, E. lucifuga* ....***Eurycea cirrigera* (Southern Two-lined Salamander)**]

    [Small area in Graham Co., North Carolina and Sevier and Monroe cos., Tennessee; sympatric: *Eurycea cirrigera, E. longicauda* ....
.... ***Eurycea junaluska* (Junaluska Salamander)**]

    [Appalachian Mountains from southwestern Virginia to northern Georgia; sympatric: *Eurycea longicauda, E. lucifuga* ....
.... ***Eurycea wilderae* (Blue-ridge Two-lined Salamander)**]

    [Stephens Co., Georgia; sympatric: *Eurycea wilderae*...............................
.... ***Urspelerpes brucei* (Patch-nosed Salamander)**]

    2. Mississippi River east from southern Illinois to western Virginia and southeastern New York; OD 2.5–3.0 mm; sympatric: *Eurycea bislineata, E. lucifuga, E. wilderae* ....
....***Eurycea longicauda* (Long-tailed Salamander)**

[South of the range of *E. longicauda* from northeastern Virginia south to Gulf of Mexico and west to Mississippi River; sympatric: *Eurycea cirrigera, E. lucifuga* .... ***Eurycea guttolineata* (Three-lined Salamander)**]

3. Mississippi River in southern Illinois east to northern Virginia and south to central Alabama; OD 2.5–3.2 mm; sympatric: *Eurycea bislineata, E. cirrigera, E. guttolineata, E. longicauda, E. wilderae*............................
.... ***Eurycea lucifuga* (Cave Salamander)**

Key Eastern Array II

IV. Arrays large, lower surfaces of submerged objects in lakes and rivers

A. Much of north-central North America from west-central Mississippi to western North Carolina and north to southern Ontario, exotic at various northeastern sites; sympatric: perhaps *N. beyeri* since opening of the Tennessee-Tombigbee Waterway .... ***Necturus maculosus* (Mudpuppy)**
Fig. 48A

[Upper part of Black Warrior River, Alabama; sympatric: *Necturus beyeri* ....
.... ***Necturus alabamensis* (Blackwarrior Waterdog)**]

[Southeastern Louisiana and adjacent Mississippi east to base of Florida panhandle; sympatric: *Necturus alabamensis*...
.... ***Necturus beyeri* (Gulf Coast Waterdog)**]

B. Neuse and Tar rivers, North Carolina; sympatric: *Necturus punctatus*............
....***Necturus lewisi* (Neuse River Waterdog)**

C. Coastal Plain from southeastern Georgia north to southern Virginia; sympatric: *Necturus lewisi* .... ***Necturus punctatus* (Dwarf Waterdog)**

Key Eastern Other

I. Most of designated area south of central Indiana and excluding Florida Peninsula; grouped singles, weakly adherent eggs, guarded by parent at semiterrestrial sites before autumnal rains fill pools; jelly layers tough and resilient; sympatric: none of similar mode and habitat .... ***Ambystoma opacum* (Marbled Salamander)**
Fig. 16A–B

II. Eggs haphazardly attached to plants or free on substrate

A. Outer jelly layer distorted with some indication of being suspended, not perfectly round

1. Coastal Plain from Mississippi River to southeastern North Carolina; swampy areas and small lowland streams; appear as scattered singles attached to vegetation; sympatric: *Eurycea chamberlaini, Stereochilus marginatus*.............................. ***Eurycea quadridigitata* (Dwarf Salamander)**

[Coastal Plain of North and South Carolina to central Georgia; sympatric: *Eurycea quadridigitata, Hemidactylium scutatum, Sterochilus marginatus* .... ***Eurycea chamberlaini* (Chamberlain's Dwarf Salamander)**]

2. Many disjunct areas from southwestern Mississippi north to northern Wisconsin and east to southeastern New York; scattered in *Sphagnum* clumps, semiterrestrial boggy sites, guarded by female, communal nests common; sympatric: none of similar ovipositional mode ....
.... *Hemidactylium scutatum* (**Four-toed Salamander**)
Fig. 40A

3. Coastal Plain from northwestern Florida to southeastern Virginia; among vegetation or dead leaves in swampy areas; ova yellowish; ED 3.0–3.5 mm; outer 2 layers involved in short pedicel; sympatric: *Eurycea quadridigitata, E. chamberlaini* ....
.... *Stereochilus marginatus* (**Many-lined Salamander**)
Fig. 43A–B

B. Outer jelly layer round

1. Coastal Plain from southeastern South Carolina to southeastern Virginia; jelly flimsy; sympatric: *Ambystoma talpoideum, Pseudacris crucifer* ....
.... *Ambystoma mabeei* (**Mabee's Salamander**)

2. Mississippi Embayment and Coastal Plain from Mississippi River to east-central South Carolina and disjunct areas in Alabama to Virginia; ephemeral, forested pools, typically oviposits masses; OD 0.9–1.2 mm; ED 2.3–3.6 mm; 1, rarely 2 jelly layers; jelly flimsy; temporary woodland pools; sympatric: *Ambystoma mabeei, Pseudacris crucifer* ....
.... *Ambystoma talpoideum* (**Mole Salamander**)
Fig. 17B

3. Almost entire designated area except Florida Peninsula; attached to twigs in ephemeral, often forested, pools; OD 0.6–0.9 mm; ED 2.2–2.4 mm; 1–2 layers; jelly flimsy; sympatric: *Ambystoma talpoideum, A. mabeei* ....
.... *Pseudacris crucifer* (**Spring Peeper**)
Fig. 89A

4. Northern two-thirds of Florida Peninsula and adjacent Georgia and South Carolina; jelly somewhat turgid; sympatric: none of similar mode ....
.... *Pseudobranchus striatus* (**Northern Dwarf Siren**)
Fig. 57A

[Central two-thirds of peninsular Florida ....
................................*Pseudobranchus axanthus* (**Southern Dwarf Siren**)]

C. Outer jelly layer oval; jelly turgid and somewhat opaque

1. Southern third of Georgia and northern third of Florida Peninsula; usually ephemeral pools; sympatric: *Notophthalmus viridescens* ....
.... *Notophthalmus perstriatus* (**Striped Newt**)

2. Most of designated area except prairie regions in Illinois to Ohio; ephemeral and permanent nonflowing water; sympatric: *Notophthalmus perstriatus* ....
.... *Notophthalmus viridescens* (**Eastern Newt**)

III. Eggs grouped in nest on bottom; jelly tough and opaque; nonflowing water, often swampy, sometimes ephemeral

A. Coastal Plain from southern Texas to Mississippi River, north in Mississippi Embayment to southwestern Michigan and Coastal Plain from Mississippi River to southeastern Virginia; OD 2.5–3.0 mm; ED 4.4–7.5 mm; sympatric: *Siren lacertina* .... ***Siren intermedia* (Lesser Siren)**

B. Coastal Plain from southwestern Alabama to southeastern Virginia; ED about 4.0 mm; sympatric: *Siren intermedia* ....
.... ***Siren lacertian* (Greater Siren)**
Fig. 59A

Key Eastern Linear

I. Oviposited as rosaries

A. Coastal Plain from Mississippi River east and north to southeastern Virginia and entire Florida Peninsula; in burrow near nonflowing water, guarded by parent; sympatric: *Amphiuma tridactylum, A. pholeter*....................................
....***Amphiuma means* (Two-toed Amphiuma)**
Fig. 21A

[Southern Mississippi north to southwestern Tennessee and east into southwestern Alabama; sympatric: *Amphiuma means* ....
....*Amphiuma tridactylum* **(Three-toed Amphiuma)**]

[Disjunct populations from southeastern Mississippi to northwestern Florida panhandle; sympatric: *Amphiuma means* ....
....*Amphiuma pholeter* **(One-toed Amphiuma)**]

B. Greater Ohio River drainage; nest below large rocks in streams and rivers; guarded by parent; OD about 6.0 mm; ED 18.0–20.0 mm, ova nonpigmented; sympatric: none of similar mode, size, or habitat ....
.... ***Cryptobranchus alleganiensis* (Hellbender)**
Fig. 22B

II. Oviposited as wrapped rosaries

Mississippi River east to Atlantic Coast in Massachusetts and most of Florida Peninsula; ephemeral, nonflowing water, can be mistaken for a clump or mass; OD 1.4–2.0 mm; ED 4.0–5.6 mm; jelly flimsy; sympatric: none of similar mode
.... ***Scaphiopus holbrookii* (Eastern Spadefoot)**
Fig. 124A

III. Eggs oviposited as strings

A. Jelly tube unilayered and diaphanous, difficult to see; string short

Coastal Plain from Mississippi River to southeastern Virginia; ephemeral, nonflowing water; OD about 0.8–1.0 mm; tube diameter 1.2–1.4 mm; partitions between uniserial ova; sympatric: none of similar mode...
.... ***Anaxyrus quercicus* (Oak Toad)**

B. Jelly tube unilayered and obvious; string long

Most of area except southeastern Coastal Plain and Florida and north of northern Ohio; ephemeral, nonflowing water; OD 1.0–1.4 mm; tube diameter 2.6–4.6 mm;

tube with straight margins; partitions absent between uniserial or staggered eggs; sympatric: *Anaxyrus americanus, A. terrestris* ....
.... *Anaxyrus fowleri* (**Fowler's Toad**)

Fig. 65A

C. Jelly tube bilayered and obvious, string long

   1. Most of area except Coastal Plain; nonflowing water and lake margins; OD 1.0–1.4 mm; tube diameter 3.4–4.0 mm; partitions present between uniserial ova; sympatric: *Anaxyrus terrestris, Incilius nebulifer* ....
.... *Anaxyrus americanus* (**American Toad**)

Fig. 62A

   2. Coastal Plain from Mississippi River to southeastern Virginia; ephemeral sites and lake margins; OD 1.0–1.5 mm; tube diameter 2.2–4.6 mm; partitions absent between uniserial, widely spaced ova; sympatric: *Anaxyrus americanus, Incilius nebulifer* .... *Anaxyrus terrestris* (**Southern Toad**)

   3. Southwestern Mississippi; ephemeral, nonflowing water; OD 1.2 mm, tube diameter 3.0 mm, partitions absent between uniserial or staggered ova; sympatric: *Anaxyrus americanus, A. terrestris* ....
.... *Incilius nebulifer* (**Gulf Coast Toad**)

   4. Exotic in west-central and southeastern Florida; sympatric: *Anaxyrus terrestris* .... *Rhinella marina* (**Cane Toad**)

Key Eastern Clump

  I. Oviposited as small clumps, not well formed, surface irregular and not distinctly lobate, eggs often fall apart as singles soon after oviposition

    A. Southeastern Louisiana north to southeastern Tennessee and east to southeastern Virginia including all of Florida Peninsula; permanent, nonflowing water; OD 0.9–1.2 mm; ED 2.3–3.7 mm; 2 jelly layers; clutch about 300; sympatric: *Acris crepitans, Hyla avivoca, H. gratiosa, H. cinerea, H. squirella* .... *Acris gryllus* (**Southern Cricket Frog**)

Fig. 75A

[Southeastern Louisiana north to southern Tennessee and east to southeastern New York except all of Florida Peninsula and most of Atlantic Coastal Plain; sympatric: *Acris gryllus, Hyla avivoca, H. gratiosa, H. cinerea, H. squirella* ......
.... *Acris crepitans* (**Northern Cricket Frog**)]

[Southern Tennessee north to central Wisconsin and east to eastern Ohio; sympatric: *Hyla avivoca, H. cinerea* ....
.... *Acris blanchardi* (**Blanchard's Cricket Frog**)]

    B. Southeastern Louisiana north to southern Illinois and southeast to southern South Carolina; swampy sites with emergent woody vegetation; OD 0.8–1.4 mm; ED 4.0–5.8 mm; 2 jelly layers; clumps sometimes fall apart as floating singles; sympatric: *Acris crepitans, A. gryllus, Hyla cinerea, H. gratiosa, H. squirella* ....
.... *Hyla avivoca* (**Bird-voiced Treefrog**)

C. Southeastern Louisiana north to southern Illinois and east and north in the Coastal Plain to New Jersey, including all of Florida Peninsula; permanent and ephemeral nonflowing water with emergent vegetation; OD 0.8–1.6 mm; ED 3.6–4.0 mm; 2 layers; sympatric: *Acris crepitans, A. gryllus, Hyla avivoca, H. gratiosa, H. squirella* ....*Hyla cinerea* (**Green Treefrog**)

D. Southeastern Louisiana east and north in the Coastal Plain to New Jersey including most of Florida Peninsula and disjunct populations in western Kentucky and northwestern Mississippi; permanent and ephemeral nonflowing water; sympatric: *Acris crepitans, A. gryllus, Hyla avivoca, H. cinerea, H. squirella* ....*Hyla gratiosa* (**Barking Treefrog**)

E. Coastal Plain from Mississippi River east and north to southeastern Virginia including all of Florida Peninsula; ephemeral, nonflowing water; OD 0.8–1.0 mm; ED 1.4–2.0 mm; 2 jelly layers; sympatric: *Acris crepitans, A. gryllus, Hyla avivoca, H. cinerea, H. gratiosa* .... *Hyla squirella* (**Squirrel Treefrog**)

II. Oviposited as large clump, surface lobate even in old, melded clumps, structure maintained until hatching

A. Central Illinois to southeastern Mississippi east to western Indiana; usually in ephemeral nonflowing water; OD 3.6–4.4 mm; ED 2.0–2.6 mm; 2 jelly layers; large ova widely spaced; clump diameter 90–140 mm; sympatric: *Lithobates blairi, L. palustris, L. "pipiens," L. sylvaticus, L. virgatipes* ....
.... *Lithobates areolatus* (**Crawfish Frog**)
Fig. 113F

[Coastal Plain from Mississippi River to east-central North Carolina except southern Florida; sympatric: none in group..
....*Lithobates capito* (**Gopher Frog**)]

[Southwestern Mississippi; sympatric: none in group ....
.... *Lithobates sevosus* (**Dusky Gopher Frog**)]
Fig. 102C

B. Much of area north of central Alabama except for prairie areas in Illinois and Indiana; nonflowing water, often wooded; OD 1.6–1.9 mm; ED 2.8 mm; ova pale to yellowish; 2 jelly layers; clump diameter 87–100 mm; sympatric: *Lithobates areolatus, L. blairi, L. "pipiens," L. septentrionalis, L. sylvaticus, L. virgatipes* .... *Lithobates palustris* (**Pickerel Frog**)

C. Widespread in designated area with large gaps in northern part of range; nonflowing water, often ephemeral; OD 1.0–2.0 mm; ED 2.5–5.6 mm; clump diameter 75–150 mm; clutch 3000–6000 or greater; sympatric: *Lithobates "areolatus," L. blairi, L. palustris, L. septentrionalis, L. sylvaticus*...............
.... *Lithobates pipiens* (**Northern Leopard Frogs**)
Fig. 113A–C, F

[Southwestern to northeastern Illinois to west-central Indiana; sympatric: *Lithobates areolatus, L. palustris, L. "pipiens," L. sylvaticus* ....
.... *Lithobates blairi* (**Plains Leopard Frog**)]

[Most of area south of central Illinois except for Appalachian Mountains; sympatric: Lithobates "areolatus," L. palustris, L. sylvaticus, L. virgatipes ....
.... *Lithobates sphenocephalus* (**Southern Leopard Frog**)]

[Great Swamp, New Jersey, and Staten Island and Orange and Putnam cos., New York.. *Lithobates* **sp. nov. (NCN)**]

D. North into Canada from central Minnesota and central New York; streams and lake margins; OD 1.3–1.6 mm; ED 6.0–7.0 mm; 2 jelly layers; clump diameter 75–125 mm; sympatric: *Lithobates palustris, L. pipiens, L. sylvaticus* .... *Lithobates septentrionalis* (**Mink Frog**)

E. Most of area north of southeastern Tennessee and adjacent northeastern North Carolina except for prairie regions in central Illinois and adjacent Indiana north to northern Quebec, disjunct population in east-central Alabama; wooded nonflowing water; OD 1.8–2.4 mm; ED 4.2–9.4 mm; 2 jelly layers; clump diameter 60–100 mm; clutch 2000–3000; sympatric: *Lithobates* "*areolatus*," *L. blairi, L. palustris, L. pipiens, L. septentrionalis* .... *Lithobates sylvaticus* (**Wood Frog**)
Fig. 114A

F. Coastal Plain from northeastern Florida to Maryland; swampy nonflowing water; OD 1.4–1.8 mm; ED 3.8–6.9 mm; 1 jelly layer; clump diameter 75–100 mm; sympatric: *Lithobates capito, L. palustris, L. pipiens, L. sphenocephalus* .... *Lithobates virgatipes* (**Carpenter Frog**)

Key Eastern Cluster

I. Egg suspension cords short, eggs appear tightly attached to substrate

Southern Appalachian Mountains; lower surfaces of stones in small streams; eggs tightly bunched, can be misinterpreted as a clump; usually remain attached to substrate during development; sympatric: several *Desmognathus* spp. .... *Desmognathus quadramaculatus* (**Black-bellied Salamander**)
Fig. 26A–B

[Wolf and Helton creeks, Union Co., Georgia and Clay Co., North Carolina; sympatric: *Desmognathus quadramaculatus*... .... *Desmognathus folkertsi* (**Dwarf Black-bellied Salamander**)]

[Northeastern Georgia to southwestern Virginia above 300 m; sympatric: *Desmognathus quadramaculatus*... .... *Desmognathus marmoratus* (**Shovel-nosed Salamander**)]
Fig. 24C

II. Egg suspension cords longer, attachment points often visible; eggs often detach from suspension points, look for stubs of attachment cords; often among leaves or seepy areas, talus, under objects, or among organic debris along small streams, in crevices of wet rock faces; many cases of sympatry but insufficient data to distinguish among species; one should try to capture the attending parent

A. Ranges in Appalachian Mountains

   1. Appalachian Mountains from Pennsylvania southwest to southwestern
      Alabama; sympatric: *Desmognathus abditus, D. carolinensis, D. fuscus,
      D. imitator, D. ochrophaeus, D. orestes, D. santeetlah, D. welteri..............
      ....* **Desmognathus monticola (Seal Salamander)**

   2. Adirondack Mountains to southern Quebec south to southwestern Virginia
      and southeastern Kentucky; sympatric: *Desmognathus monticola* ....
      .... ***Desmognathus ochrophaeus* (Allegheny Mountain Dusky
      Salamander)**

      [Cumberland Plateau, Tennessee from Wartburg, Morgan Co. to Tracy City,
      Grundy Co.; sympatric: *Desmognathus monticola* ....
      .... ***Desmognathus abditus* (Cumberland Dusky Salamander)**]

      [Between Linville Falls and McKinley Gap on Blue Ridge Divide and Iron
      Mountain Gap on North Carolina-Tennessee border to Pigeon River valley
      (= Blue Ridge, Black, Bald, and Unaka mountains); sympatric: *Desmog-
      nathus monticola* ....
      .... ***Desmognathus carolinensis* (Carolina Mountain Dusky Salamander)**]

      [Central Smoky Mountains above 900 m; sympatric: *Desmognathus ochro-
      phaeus* ....***Desmognathus imitator* (Imitator Salamander)**]

      [Disjunct regions on Appalachian Plateau of northeastern Alabama and
      southwestern Blue Ridge Physiographic Province south of the Pigeon River;
      sympatric: *Desmognathus monticola*...
      .... ***Desmognathus ocoee* (Ocoee Salamander)**]
      Fig. 25C

      [Southwestern Virginia to near Linville Falls on the Blue Ridge Divide;
      sympatric: *Desmognathus monticola*..
      .... ***Desmognathus orestes* (Blue Ridge Dusky Salamander)**]

   3. Small area of higher elevations on border between Tennessee and North
      Carolina; sympatric: *Desmognathus fuscus, D. monticola, D. ochro-
      phaeus* ....*Desmognathus santeetlah* **(Santeetlah Dusky Salamander)**

   4. Western Kentucky and adjacent North Carolina and Tennessee; sympatric:
      Desmognathus monticola, D. ochrophaeus ....
      .... ***Desmognathus welteri* (Black Mountain Salamander)**

B. Ranges on Coastal Plain and non-Appalachian interior

   1. Wide band from eastern Nova Scotia southwest to southwestern Missis-
      sippi, exclusive of southeastern Coastal Plain; under objects and debris at
      edges of streams, creeks, and seeps; sympatric: *Desmognathus auricula-
      tus*..................................... ***Desmognathus fuscus* (Dusky Salamander)**
      Fig. 23B, see Plate 2L, similar taxon

      [Drainages of Choctawhatchee, Chattahoochee, and Apalachicola rivers in
      southeastern Alabama and adjacent Georgia and Florida; sympatric: no

congeners ....
.... ***Desmognathus apalachicolae* (Apalachicola Dusky Salamander)**]

[Coastal Plain from Mississippi River to southeastern Virginia; sympatric: *Desmognathus conanti*...
.... ***Desmognathus auriculatus* (Southern Dusky Salamander)**]

Fig. 23A

[Northeastern Missouri and north-central Louisiana and adjacent southern Arkansas; sympatric: *Desmognathus auriculatus*..
.... ***Desmognathus conanti* (Spotted Dusky Salamander)**]

Key Eastern Mass I

I. Capsular chambers present, centers of ova below centers of outer jelly layer (Fig. 3D); ova pigmented; peripheral ova-free zone because jelly layers considerably larger than ovum diameter

A. Mass diameter 45 mm or less

  1. Coastal Plain from Apalachicola River east to southeastern South Carolina; OD 2.0–2.6 mm; temporary woodland pools; sympatric: *Ambystoma mabeei, A. talpoideum, A. tigrinum*...
.... ***Ambystoma cingulatum* (Frosted Flatwoods Salamanders)**

    [Coastal Plain from Apalachicola River to southwestern Alabama; sympatric: *A. talpoideum, A. tigrinum* ....
.... ***Ambystoma bishopi* (Reticulated Flatwoods Salamander)**]

  2. Coastal Plain from southeastern South Carolina to southeastern Virginia; sympatric: *Ambystoma cingulatum, A. talpoideum, A. tigrinum*...............
.... ***Ambystoma mabeei* (Mabee's Salamander)**

  3. Mississippi Embayment and Coastal Plain from Mississippi River to east-central South Carolina and disjunct areas in Alabama to Virginia; temporary woodland pools; OD 0.9–1.2 mm; ED 2.3–3.6 mm; 1, rarely 2 jelly layers; sympatric: *Ambystoma cingulatum, A. mabeei, A. texanum, A. tigrinum* .... ***Ambystoma talpoideum* (Mole Salamander)**

Fig. 17A

  4. Mississippi north to northern Illinois and east through Ohio; temporary woodland pools; OD 1.6–2.5 mm; ED 6.0–6.5 mm; 2 jelly layers; sympatric: *Ambystoma "cingulatum," A. talpoideum, A. tigrinum*.........................
.... ***Ambystoma texanum* (Small-mouthed Salamander)**

Fig. 18A

  5. Coastal Plain from southern Mississippi to Delaware exclusive of Florida Peninsula and much of Illinois and Wisconsin east to central Ohio; temporary, often grassland pools; OD 2.0–4.0 mm; ED 4.5–10.0 mm; 3 jelly layers; sympatric: *Ambystoma "cingulatum," A. mabeei, A. talpoideum, A. texanum* .... ***Ambystoma tigrinum* (Eastern Tiger Salamander)**

Fig. 19F

B. Mass diameter 45 mm or greater

1. Central Wisconsin east to Quebec and south to latitude of central Kentucky; temporary wooded pools; OD 2.0–2.5 mm; 3 jelly layers; jelly matrix watery; sympatric: *Ambystoma laterale, A. maculatum, A. opacum, A. texanum, A. tigrinum* .... ***Ambystoma jeffersonianum*** (**Jefferson Salamander**)
Fig. 12A

[Eastern Ontario east to Labrador and south to central Indiana; sympatric: *Ambystoma jeffersonianum, A. maculatum, A. opacum, A. texanum, A. tigrinum* .... ***Ambystoma laterale*** (**Blue-Spotted Salamander**)]

2. Most of area south of northern Ohio except Florida Peninsula; temporary woodland pools; OD 6.0–8.0 mm; ED 2.5–3.0 mm; 2 jelly layers; jelly very dense, sometimes milky white, mass can be removed from water without collapsing; inner jelly layers commonly with green algae; sympatric: *Ambystoma "jeffersonianum," A. tigrinum* ....
.... ***Ambystoma maculatum*** (**Spotted Salamander**)
Figs. 15A, 113A

3. Coastal Plain from southern Mississippi to Maryland north through Alabama to northeastern Kentucky and throughout central Midwest; temporary, often grassland pools; OD 2.0–4.0 mm; ED 4.5–10.0 mm; 3 jelly layers; jelly watery, mass cannot be picked up; sympatric: *Ambystoma "jeffersonianum," A. maculatum* .... ***Ambystoma tigrinum*** (**Eastern Tiger Salamander**)
Fig. 19F

II. Capsular chambers absent, so ova centered in inner jelly layers (Fig. 3F); mass usually 45 mm diameter or less (**note**: particularly difficult group with minimal data)

A. Ranges on Coastal Plain

1. Coastal Plain from southeastern Georgia to southeastern Virginia; OD 1.3–1.7 mm; ED 6.8–8.6 mm; sympatric: *Pseudacris nigrita, P. ornata, P. ocularis, P. "triseriata"* .... ***Pseudacris brimleyi*** (**Brimley's Chorus Frog**)

2. Coastal Plain from southeastern Mississippi to east-central North Carolina; OD 0.9–1.0 mm; ED 2.6–2.8 mm; 1 jelly layer; sympatric: *Pseudacris brimleyi, P. ornata, P. ocularis, P. "triseriata"* ....
.... ***Pseudacris nigrita*** (**Southern Chorus Frog**)

3. Central Florida Panhandle to southeastern Virginia; OD 0.6–0.8 mm; ED 1.2–2.0 mm; sympatric: *Pseudacris brimleyi, P. nigrita, P. ornata, P. "triseriata"* .... ***Pseudacris ocularis*** (**Little Grass Frog**)

4. Coastal Plain from southeastern Louisiana to southeastern North Carolina; OD 1.6 mm; ED 3.2–4.2 mm; sympatric: *Pseudacris brimleyi, P. nigrita, P. ocularis, P. "triseriata"* ..... ***Pseudacris ornata*** (**Ornate Chorus Frog**)

5. Most of Coastal Plain except for southeastern area and central Appalachian Highlands; OD 0.9–1.1 mm; ED 3.2–4.0 mm; 1 jelly layer; sympatric: *Pseudacris brimleyi, P. nigrita, P. ocularis, P. ornata, P. streckeri* .....
.... ***Pseudacris "triseriata"*** (**Trilling Chorus Frogs**)
Fig. 94A

[Northeastern half of Mississippi east and north to central Pennsylvania; sympatric: *Pseudacris brachyphona* .... ***Pseudacris feriarum*** (**Upland Chorus Frog**)]

[Eastern third of Texas north to northern Oklahoma, east to southern Missouri and south to coastal Mississippi; sympatric: *Pseudacris feriarum, P. nigrita* .... ***Pseudacris fouquettei*** (**Cajun Chorus Frog**)]

[Staten Island, New York, south throughout Delmarva Peninsula; sympatric: *Pseudacris feriarum* .... ***Pseudacris kalmi*** (**New Jersey Chorus Frog**)]

B. Ranges in interior regions

    1. Appalachian area from eastern Pennsylvania to central Alabama; OD 1.2–2.0 mm; ED 6.0–8.5 mm; sympatric: *Pseudacris feriarum* .... ***Pseudacris brachyphona*** (**Mountain Chorus Frog**)

    2. Disjunct sandy areas in southern, central, and northern Illinois; OD 1.2–1.3 mm; ED 3.0–7.2 mm; sympatric: *Pseudacris* "*triseriata*" .... ***Pseudacris illinoensis*** (**Illinois Chorus Frog**)

    3. Most of area except southeastern Coastal Plain and central Appalachian Highlands; OD 0.9–1.1 mm; ED 3.2–4.0 mm; sympatric: *Pseudacris brachyphona, P. streckeri* .... ***Pseudacris* "*triseriata*"** (**Trilling Chorus Frogs**) Fig. 94A

[Northeastern half of Mississippi east and north to central Pennsylvania; sympatric: *Pseudacris fouquettei* .... ***Pseudacris feriarum*** (**Upland Chorus Frog**)]

[Eastern third of Texas north to northern Oklahoma, east to southern Missouri, and south to coastal Mississippi; sympatric: *Pseudacris feriarum* .... ***Pseudacris fouquettei*** (**Cajun Chorus Frog**)]

Key Eastern Film

  I. Southeastern Louisiana north to southern Illinois and east to eastern Virginia; ephemeral nonflowing water; eggs in coherent films, individual eggs easily pushed apart; upper hemisphere of jelly of recently deposited eggs projects above water surface, film diameter less than 150 mm; ovum intensely black and jelly diameter large relative to ovum size; sympatric: none of similar mode .... ***Gastrophryne carolinensis*** (**Eastern Narrow-mouthed Toad**) Fig. 98A–B

  II. Eggs oviposited as adherent films, individual eggs not easily removed from film, top of outer jelly lies at water surface; jelly diameter not large relative to ovum size

    A. Film diameter less than 150 mm, ova pale to dark brown

       1. Disjunct areas in New Jersey, the Carolinas, and the Florida Panhandle; permanent or ephemeral boggy sites, usually with emergent vegetation; OD 1.2–1.4 mm; ED 3.5–4.0 mm; 2 jelly layers; often in small pockets of usually boggy water and often just a few eggs; sympatric: *Hyla avivoca, H.*

cinerea, H. femoralis, H. gratiosa, H. "versicolor" ....
.... ***Hyla andersonii*** (**Anderson's Treefrog**)

2. Southeastern Louisiana north to southern Illinois and east and north in the Coastal Plain to New Jersey, including all of Florida Peninsula; usually oviposits clumps; permanent and ephemeral nonflowing water; OD 0.8–1.6 mm; ED 3.6–4.0 mm; 2 jelly layers; sympatric: *Hyla andersonii, H. avivoca, H. femoralis, H. gratiosa, H. "versicolor"* ....
.... ***Hyla cinerea*** (**Green Treefrog**)

3. Southeastern Louisiana east and north in the Coastal Plain to New Jersey including most of Florida Peninsula and disjunct populations in western Kentucky; permanent and ephemeral nonflowing water; sympatric: *Hyla andersonii, H. avivoca, H. cinerea, H. "versicolor"* ....
.... ***Hyla gratiosa*** (**Barking Treefrog**)

4. Coastal Plain from Mississippi River to southeastern Virginia and most of Florida Peninsula; ephemeral nonflowing water; OD 0.8–1.2 mm; 2 jelly layers; sympatric: *Hyla andersonii, H. cinerea, H. gratiosa, H. "versicolor"* .... ***Hyla femoralis*** (**Pine Woods Treefrog**)
Fig. 80A

5. All of designated area north to southern Ontario except most of Florida Peninsula; permanent and ephemeral nonflowing water; sympatric: *Hyla cinerea, H. femoralis, H. gratiosa, Osteopilus septentrionalis*...................
.... ***Hyla "versicolor"*** (**Gray Treefrogs**)
Fig. 83H

6. Southeastern Louisiana north to southern Illinois and southeast to southern South Carolina; swampy sites with emergent woody vegetation; probably oviposits clumps that sometimes break free and float at the surface; OD 0.8–1.4 mm; ED 4.0–5.8 mm; 2 jelly layers; sympatric: *Hyla cinerea, H. femoralis, H. gratiosa, H. "versicolor"* ....
.... ***Hyla avivoca*** (**Bird-voiced Treefrog**)

7. Exotic in Atlantic coastal areas from Key West to at least Jacksonville, Florida, Gulf Coast north to at least Tampa Bay; ephemeral nonflowing water; outer jelly layer particularly sticky; sympatric: *Hyla cinerea, H. femoralis, H. gratiosa* ....
....*Osteopilus septentrionalis* (**Cuban Treefrog**)

B. Film diameter greater than 200 mm; ovum black and jelly diameter not large relative to ovum size

1. All of designated area north to northern New York except southern two-thirds of Florida Peninsula; larger nonflowing water; film greater than 150 mm diameter; 1 jelly layer; OD 1.2–1.7 mm; ED 6.4–10.4 mm; clutch 10,000–20,000; sympatric: *Lithobates clamitans, L. grylio, L. heckscheri, L. okaloosae* .... ***Lithobates catesbeianus*** (**American Bullfrog**)
Fig. 105A

2. All of designated area north to central Quebec except southern two-thirds of Florida Peninsula; larger, often swampy, nonflowing water; OD 1.2–1.8 mm;

ED 2.8–4.0 mm; 2 jelly layers; clutch 1000–5000; sympatric: *Lithobates catesbeianus, L. grylio, L. heckscheri, L. okaloosae*....................................
.... ***Lithobates clamitans* (Green Frog)**
Fig. 107C

[Okaloosa and Santa Rosa cos., Florida; side pools of swampy creeks; sympatric: *Lithobates catesbeianus, L. clamitans, L. heckscheri, L. grylio* ....
.... ***Lithobates okaloosae* (Florida Bog Frog)**]
Fig. 110B

3. Coastal Plain from Mississippi River east and north to eastern South Carolina and all of Florida Peninsula; larger nonflowing water; 2 jelly layers; sympatric: *Lithobates catesbeianus, L. clamitans, L. heckscheri, L. okaloosae* .... ***Lithobates grylio* (Pig Frog)**

[Coastal Plain from south-central Mississippi east and north to southeastern North Carolina and all of Florida Peninsula; sympatric: *Lithobates catesbeianus, L. clamitans, L. grylio, L. okaloosae*..
.... ***Lithobates heckscheri* (River Frog)**]

EMBRYOS AND HATCHLINGS

Embryos and hatchlings have unique transitory features and have not yet developed all larval features. Embryos are poorly known, and the paucity of easily distinguishable morphological characters and the rates at which they change adds to the identification problems. Associating data from eggs and embryos and working with living specimens within a local fauna always helps with the identifications.

Salamander embryos are distinctive from frog embryos early in development, and the salamander morphological theme is apparent from early development. The external gills associated with a free gular fold are usually prominent, and most *Ambystoma* and salamandrids have a transient balancer extending from each side of the head below the eye. Front limbs usually develop before the hind limbs.

Anuran embryos and hatchlings have several transient features of interest. External gills of various shapes during stages 19–23 (Figs. 65C, 83F, 124C; reviewed by Nokhbatolfoghahai and Downie 2008) are eventually overgrown by the operculum and atrophy. The operculum closes in different patterns among taxa to form the spiracle(s). Variations of the transient adhesive glands (Nokhbatolfoghahai and Downie 2005) posterior to the presumptive mouth are also useful in identifying hatchlings.

Most frog embryos encountered are bufonids, hylids, ranids, or scaphiopodids. Bufonid embryos are small, uniformly and often darkly colored, and have small external gills. Hylid embryos are long and narrow, usually become vertically curved head-to-tail while still in the egg jelly, and usually are patterned. Microhylid embryos are small, uniformly dark, and distinctly dumbbell-shaped in dorsal view from late neurula to advanced tailbud stages. Ranids are uniformly, usually darkly colored, and large relative to other embryos. Scaphiopodid embryos are uniformly colored either pale or dark and have a distinctive hammer-headed appearance in lateral view because of the immense V-shaped adhesive gland that persists much longer than in other taxa as a dark, heart-shaped pigmentary scar posterior to the oral disc.

We describe some of the features of embryos and hatchlings and present a key to genera. The best scenario would involve a well-documented clutch of eggs, sampled and examined over short time periods. Unfortunately, detailed documentation of morphological shifts over short periods of developmental time are few and far between. Formalin (5–10%) preserves embryos quite well even though the egg jellies often deteriorate with time. Even though the diversity of color and patterns is relatively small, color photographs taken at various magnifications to assess general morphology and specific features have been helpful. Because the data are so highly incomplete, we have often had to extrapolate a character state for a genus based on knowledge of one or a few species within the group. The key is divided into western and eastern sectors for salamanders and frogs.

KEY TO GENERA OF EMBRYONIC AND HATCHLING AMPHIBIANS

I. Approximate stages 30–45: front limb buds present from about stage 37; hind limbs develop either posthatching or with front buds; gill rami, usually with some pigment, obvious in later stages except short and moundlike in *Desmognathus* and *Rhyacotriton*; rami branched only in amphiumids and sirenids; gill fimbriae usually lamellar on trailing edge of rami; jaw angle at or posterior to plane of eye; balancer of variable size present on side of head in salamandrids and *Ambystoma* exclusive of the *Ambystoma tigrinum* group; yolk mass always long relative to body; coloration varies from unicolored to patterned (R.G. Harrison 1969; also Duellman and Trueb 1970:130) .... **Salamanders**

A. Pacific Coast East to include Rocky Mountains

    1. Pacific Coast east to Montana and south to Mexican border; ova pigmented, attached as singles (2 spp.) or in masses (2 spp.) attached to plants in nonflowing water. Embryos and hatchlings (Figs. 11A, 15B–D): body robust; tail fin originates on back; gill rami long, fimbriae plumose; eyes small and dorsal; balancers present in *A. gracile* and *A. macrodactylum*, absent in *A. californiense* and *A. mavortium*; toes relatively long, cylindrical, and blunt (*A. gracile* and *A. macrodactylum*) or flattened and pointed in *A. californiense* and *A. mavortium*; body usually striped at hatching, dorsum of tail muscle sometimes banded .... ***Ambystoma* (Mole Salamanders)**

    2. Southern British Columbia to San Francisco Bay and northern Idaho; large, nonpigmented ova in arrays in talus or under stones of mountain streams. Embryos and hatchlings: body particularly robust; tail fin originates near tail-body junction; gill rami medium, fimbriae short; eyes small and dorsal; balancers absent; short toes cylindrical and blunt; uniformly dark, hint of iridophore pigmentation sometimes visible on head .... ***Dicamptodon* (Giant Salamanders)**

3. Northwestern Washington to northwestern California; large, nonpigmented ova as unattached singles grouped under rocks and talus of streams and seeps. Embryos and hatchlings: body slender; low tail fin originates posterior to tail-body junction; gill rami particularly short and moundlike, fimbriae few; eyes large, protuberant and lateral; balancers absent; short toes cylindrical and blunt; yellowish to golden ground color with or without small to medium-sized dark contrasting spots, specks, and blotches .... *Rhyacotriton* (**Torrent Salamanders**)

4. Southern Alaska to southern California and central Sierra Nevada range, isolated population in Idaho; single pigmented ova with turgid jelly attached to plants in nonflowing water or rounded (nonflowing or flowing water) or flattened clumps (flowing water). Embryos and hatchlings (Figs. 54B–C, 55B, 56B): body slender, not robust; tail fin originates on body of hatchling; gill rami long, fimbriae appear ragged; eyes medium and lateral; balancers present; toes cylindrical, narrow, and pointed; striped in 3 species, unicolored in *T. rivularis* .... *Taricha* (**Newts**)

B. East of Rocky Mountains to Atlantic Coast

1. Cylindrical body disproportionately elongate relative to diameter; limbs small to tiny; gill rami branched; gular fold absent

a. Eastern third of Texas north to southeastern Missouri and east and north in the Coastal Plain to southeastern Virginia; large, nonpigmented ova in rosaries in semiterrestrial burrows near nonflowing water. Embryos and hatchlings (Fig. 21B–C, Plate 2F): balancers absent; gills with few fimbriae; gill slit 1; tail fin barely noticeable; tiny digits 1/1, 2/2, or 3/3 with blunt tips on tiny limbs; uniformly dark .... *Amphiuma* (**Amphiumas**)

b. Northwestern Florida, throughout Florida Peninsula and north in Coastal Plain to southeastern South Carolina; pale ova as scattered singles attached to plants. Embryos and hatchlings (Fig. 59B): balancers absent; gills with abundant fimbriae; gill slit 1; transparent tail fin originates well forward on body; 3 fingers pointed; only small front limbs present throughout ontogeny; intensely black ground color with bright contrasting markings in red or yellow .... *Pseudobranchus* (**Dwarf Sirens**)

c. Southern Texas northeast to southwestern Michigan and east in Coastal Plain to northeastern Virginia; pale ova with crystalline inclusions in jelly, grouped in nest on the bottom, Embryos and hatchlings (Fig. 59B): crystalline inclusions in opaque outer jelly layer; gill slits 3; transparent tail fin originates well forward on body; 4 fingers pointed; only small front limbs present throughout ontogeny; intensely black ground color with bright contrasting markings in red or yellow .... *Siren* (**Sirens**)

2. Body and four limbs normally proportioned; gill rami not branched; gular fold present

a. Throughout designated area; pigmented ova in masses in nonflowing water. Embryos and hatchlings (Figs. 8C, 15B–D): body, and especially head, robust; balancers present except in *A. mavortium* and *A. tigrinum*; gill rami long with numerous fimbriae; tail fin originates on back; mandibular grooves of hatchlings do not bisect lower lip anteriorly; toes relatively long, cylindrical and blunt, except flattened and pointed in *A. mavortium* and *A. tigrinum*; digits 4/5; body usually striped at hatching, dorsum of tail often banded ....

.... ***Ambystoma*** **(Mole Salamanders)**

b. Greater Ohio River drainage; large, nonpigmented ova in rosaries in nest beneath large rocks in streams. Embryos and hatchlings (Fig. 22A): body, and especially head, robust and depressed; balancers absent; gill rami medium with numerous fimbriae; tail fin originates at tail-body junction; mandibular grooves of hatchlings do not bisect lower lip anteriorly; toes robust with blunt tips; digits 4/5; uniformly dark to faintly mottled .... ***Cryptobranchus*** **(Hellbender)**

c. Throughout most of designated area; large, nonpigmented ova in clusters hidden in talus or under stones in or near small streams and seeps. Embryos and hatchlings (Figs. 23C, E–F, 26B): balancers absent; gill rami very short and moundlike, fimbriae few and filiform; low tail fin originates at tail-body junction; mandibular grooves of hatchlings bisect lower lip anteriorly; toes narrow, cylindrical and pointed; digits 4/5; usually with dorsolateral series of pale dots on darker ground color ......

.... ***Desmognathus*** **(Dusky Salamanders)**

d. Throughout most of designated area; large nonpigmented ova in arrays in small streams, seeps, and caves, rarely in nonflowing water; if eggs are few or scattered on plants, can be misinterpreted as singles. Embryos and hatchlings (Figs. 27C–D, 33A, 40A inset): balancers absent; gill rami medium, fimbriae few and filiform; low tail fin originates at tail-body junction; mandibular grooves of hatchlings do not bisect lower lip anteriorly; toes cylindrical and blunt; digits usually 4/5, 4/4 in 3 taxa; often with dorsolateral series of pale dots ....

....................***Eurycea, Gyrinophilus, Hemidactylium, Pseudotriton,***
Stereochilus*, and *Urspelerpes **(Brook Salamanders and relatives)**

e. Eastern Arkansas north to southern Manitoba and east to Atlantic Coastal Plain; large, nonpigmented ova in arrays on bottom of submerged surfaces in larger nonflowing and flowing water. Embryos and hatchlings (Fig. 47A–B): balancers absent; gill rami long, fimbriae numerous; low tail fin originates at tail-body junction; mandibular grooves of hatchlings do not bisect lower lip anteriorly; toes cylindrical and blunt; digits 4/4; coloration uniform, striped, or speckled ....

.... ***Necturus*** **(Mudpuppy and Waterdogs)**

f. Southern Texas north to central Minnesota and east to Atlantic Coast; single pale ova with turgid, oval jelly attached singly to plants in nonflowing water. Embryos and hatchlings (Figs. 51A–B): body slender and not robust; tail fin originates on body; mandibular grooves of hatchlings do not bisect lower lip anteriorly; gill rami long, fimbriae appear ragged; eyes medium and lateral; balancers present; toes cylindrical, narrow, and pointed; hatchlings boldly striped .... *Notophthalmus* (**Newts**)

II. Stages 21–24: limb buds absent; transient external gills short and variable; mouth small with jaw angle well anterior of eye; body somewhat globular at early stages, somewhat elongate at later stages; yolk mass long relative to body at early stages, short relative to body once tail bud forms; adhesive glands of variable morphology usually present; hatching glands on top of head variable but not easily seen..**Frogs**

A. Hawaii and east on mainland to include Rocky Mountains

   1. Ova large and nonpigmented

      a. Small terrestrial clumps hidden on forest floor. Embryos and hatchlings: uniformly dark; moved to a phytotelmon by parent after hatching; neotropical exotic on Oahu and Maui, Hawaii ....
             .... *Dendrobates* (**Green and Black Dart-poison Frog**)

      b. Rosaries of large, nonpigmented ova under stones and talus in mountain streams. Embryos and hatchlings (Fig. 96G–H): gills and adhesive gland absent; body bulky, large suctorial oral disc noticeable early in development; eyes small and dorsal; nonpigmented to pale well after hatching; Pacific Northwest including much of Idaho and adjacent Montana ....
             .... *Ascaphus* (**Tailed Frogs**)

   2. Pigmented ova oviposited below the water surface

      a. Pigmented ova oviposited in strings with obvious outer tube, tubes short and diaphanous in 2 taxa, typically in nonflowing water. Embryos and hatchlings (Fig. 65C): gills small; adhesive gland U-shaped at beginning and 2 longitudinally oval patches while functional; body stocky; most commonly uniformly dark; throughout much of designated area ....
             .... *Anaxyrus* and *Incilius* (**Toads**)

      b. Small, pigmented ova oviposited as either singles (*Hyla*), small clumps (*Acris*, *Hyla*), or masses (*Pseudacris*), usually in nonflowing water. Embryos and hatchlings (Figs. 82A, 83E–F, 94B): gills with a central ramus and presumptive fimbriae emerging mostly ventrally; adhesive gland U-shaped at beginning and 2 longitudinally oriented, oval patches while functional; often with iridophore pigment pattern near hatching; body noticeably long and narrow in lateral view, often curves vertically head toward tail in stages 17–20; throughout much of designated area............
             .... *Acris*, *Hyla*, and *Pseudacris* (**Treefrogs and Chorus Frogs**)

      c. Large clumps of pigmented ova in nonflowing water and streams. Embryos and hatchlings (Figs. 104B, 107A, 121D): gills usually with 2 rami visible, fimbriae few but distinct; adhesive gland U-shaped at beginning and 2 posteriorly convergent oval patches while functional; body stocky; usually uniformly dark; throughout much of designated area..................
             .... *Lithobates* and *Rana* (**True Frogs**)

d. Pigmented ova, in nonflowing water, typically ephemeral, xeric sites. *Scaphiopus*: wrapped rosaries with turgid jelly; *Spea*: haphazard arrays with stalked outer jelly of each egg, flimsy clumps, or groups of singles attached to vegetation. Embryos and hatchlings (Fig. 124B–D): gills with confluent base and fimbriae in a single fan; adhesive gland large, V-shaped, and projects noticeably from head to present a hammer-headed lateral silhouette while functional, atrophies anterior-to-posterior and densely pigmented gland scar well posterior of oral disc remains well after oral apparatus forms; body robust; uniformly dark or pale; throughout much of designated area .... ***Scaphiopus* and *Spea* (Spadefoots)**

e. Groups of pale, single ova attached to plants in nonflowing water. Embryos and hatchlings: gills small, with each fimbria emerging independently from the base; adhesive gland single, medial, and circular; body notably long and narrow in lateral view; coloration almost uniformly pale; African exotic in southern Arizona .... ***Xenopus* (African Clawed Frog)**

3. Pigmented ova oviposited as films in nonflowing water

   a. Small, pigmented ova oviposited as adherent surface films in nonflowing water, often in multiple rafts. Embryos and hatchlings (Figs. 82A, 83E–F, 94B): gills with a central ramus and presumptive fimbriae emerging mostly ventrally; adhesive gland U-shaped at beginning and 2 longitudinally oriented, oval patches while functional; often with iridophore pigment pattern after stage 22; body notably long and narrow in lateral view, often curves vertically head toward tail in stages 17–20; southern Arizona .... ***Smilisca* (Treefrogs)**

   b. Small, pigmented ova oviposited as coherent surface films in nonflowing water. Embryos and hatchlings: gills quite small; adhesive gland U-shaped at beginning and 2 separate conical mounds ventrolateral to presumptive mouth while functional; body dumbbell-shaped in dorsal view in tail bud stages; uniformly dark coloration; southern Arizona .... ***Gastrophryne* (Narrow-mouthed Toads)**

   c. Large clumps of pigmented ova in nonflowing water and streams, usually as a single unit. Embryos and hatchlings (Figs. 104B, 107A): gills usually with 2 rami visible, fimbriae few but distinct; adhesive gland U-shaped at beginning and two posteriorly convergent oval patches while functional; body stocky; usually uniformly dark; Hawaii and much of mainland .... ***Lithobates* (True Frogs)**

B. East of Rocky Mountains to Atlantic Coast

   1. Small, pale yellow ova in a subterranean foam nest at ephemeral site. Embryos and hatchlings: uniformly dark; gills large; adhesive gland likely U-shaped; southern Texas .... ***Leptodactylus* (Mexican White-lipped Frog)**

   2. Pigmented ova deposited below the water surface in various ovipositional modes, typically in nonflowing water

a. Pigmented ova oviposited in long strings, short strings in 1 species, most often in ephemeral nonflowing water. Embryos and hatchlings (Fig. 65C): gills small; adhesive gland U-shaped at beginning and two longitudinally oval patches while functional; body robust; usually uniformly dark; throughout designated area ...... ***Anaxyrus, Incilius,* and *Rhinella*** **(Toads)**

b. Pigmented ova oviposited as singles, clumps, or masses in nonflowing water. Embryos and hatchlings (Figs. 82A, 83E–F, 94B): gills with a central ramus and presumptive fimbriae emerging mostly ventrally; adhesive gland U-shaped at beginning and two longitudinally oriented, oval patches while functional; often with iridophore pigment pattern near hatching; body noticeably long and narrow in lateral view, often curves vertically head toward tail in stages 17–20; throughout designated area ....
.... *Acris, Hyla,* **and** *Pseudacris* **(Cricket Frogs, Treefrogs, and Chorus Frogs)**

c. Pigmented eggs oviposited in large clumps, usually in nonflowing water. Embryos and hatchlings (Figs. 104B, 107A): gills usually with two rami visible, fimbriae few but distinct; adhesive gland U-shaped at beginning and two posteriorly convergent oval patches while functional; body stocky; usually uniformly dark; throughout designated area ....
.... *Lithobates* **(True Frogs)**

d. Pale ova oviposited as groups of single eggs attached to vegetation in ephemeral, nonflowing water. Embryos and hatchlings (Fig. 122A, C): adhesive gland transversely linear; body shape globular, large yolk mass much wider anteriorly than posteriorly; obvious from stage 22 that eyes will be positioned laterally; uniformly pale; southern Texas ....
.... *Rhinophrynus* **(Burrowing Toad)**

e. Pigmented ova oviposited as wrapped rosaries, groups of singles (sometimes with elongate outer jelly of each ovum), or arrays in ephemeral, xeric sites. Embryos and hatchlings (Fig. 124B–D): gills with confluent base and fimbriae in a single fan; adhesive gland large, V-shaped, and projects noticeably from head to present a hammer-headed lateral silhouette, atrophies anterior to posterior, and densely pigmented scar well posterior of oral disc remains well after oral apparatus forms; body robust; uniformly dark or pale; throughout much of designated area ....
.... *Scaphiopus* **and** *Spea* **(Spadefoots)**

3. Pigmented ova oviposited as films

a. Pigmented ova oviposited as adherent films, often divided into multiple rafts, in nonflowing water. Embryos and hatchlings (Figs. 82A, 83E–F, 94B): gills with a central ramus and presumptive fimbriae emerging mostly ventrally; adhesive gland U-shaped at beginning and two longitudinally oriented, oval patches while functional; body noticeably long and narrow in lateral view, often curves vertically head toward tail in stages

17–20, often with iridophore pigment pattern near hatching; throughout much of designated area .... *Hyla* and *Smilisca* (**Treefrogs**)

b. Pigmented ova oviposited as coherent films in nonflowing water, typically ephemeral. Embryos and hatchlings: gills quite small; adhesive gland U-shaped at beginning and two separate conical mounds ventrolateral to stomodeum while functional; body dumbbell-shaped in dorsal view in tail bud stages; uniformly dark; western Texas east to Atlantic Coastal Plain ....

.... *Gastrophryne* and *Hypopachus* (**Narrow-mouthed Toads and Sheep Frog**)

c. Pigmented ova oviposited in large adherent surface films in nonflowing water. Embryos and hatchlings (Figs. 104B, 107A): gills usually with two rami visible, fimbriae few but distinct; adhesive gland U-shaped at beginning and two posteriorly convergent oval patches while functional; body robust; uniformly dark; throughout designated area ....

.... *Lithobates* (**True Frogs**)

LARVAE

To identify larval amphibians, one needs to be familiar with a new, sometimes difficult and often subjective set of variable morphological traits. Recent verification of variations in size, shape, and coloration of tadpoles in response to coinhabiting competitors and predators (e.g., Relyea 2001) adds to the variations. The following text and figures serve as a primer for identifications of larvae (see Altig 2007 and Altig and McDiarmid 1999 for additional information on tadpoles). Because circumstances usually do not allow microscopic examination in the field, biologists must learn to associate subtle characters of coloration, shape, size, and habitat with the morphology learned from preserved specimens. Acceptable views of the mouthparts can usually be made by placing a tadpole in a test tube with a bit of water and examining the oral apparatus with a loupe. Mouthparts of large tadpoles can be examined with a loupe while holding the tadpole upside-down between one's index finger and applying slight forward and downward pressure with the thumb. Most point-and-shoot digital cameras and some camera phones can be used to take close-ups of mouthparts that are adequate for identification.

ORDER CAUDATA

Salamanders and Relatives

Relative to anuran tadpoles, the morphology of salamander larvae is quite conservative, and a few external characters are usually sufficient to allow identifications to family and genus. More subtle morphologies and coloration must be employed for species identifications, and ontogenetic and geographic variations are often not well documented.

Relevant morphological traits of salamander larvae (Fig. 4) include the number of costal grooves, shape of the fins, head shape, eye size and position, and limb and digit morphology. With more detailed information, the morphology of labial folds (e.g., Altig and Ireland 1984) could be expanded and used for identification. When the jaws of a heavily narcotized larva are opened widely (Fig. 17E), the labial folds appear as an intricately pleated sheet that spans the lateral part of the gape. The mandibular groove (Fig. 24A) extends almost parallel to the lower jaw curvature and separates the jaw from the labial fold. In most cases, the mandibular grooves stop short of bisecting the anterior margin of the lower lip, but in larval *Desmognathus* each groove bisects the jaw margin slightly lateral to the midline and isolates a small, medial lip segment (compare Figs. 24A and 42A).

The number of costal grooves does not change at metamorphosis, but postmetamorphic allometric changes require separate evaluations of leg length for larvae. The number of costal folds between adpressed limbs reflects the relationship between body length and limb length (e.g., many folds between adpressed limbs = short legs or long body; few folds between adpressed limbs or limb overlap = long legs or short body). Subjective evaluations of toe lengths and shapes are often useful.

Gill structure (e.g., Valentine and Dennis 1964) could be quite informative if more detailed features of the rami and fimbriae were known. Pond forms usually have larger and longer gills than stream inhabitants (Fig. 6), but the gills of a pond larva in well-oxygenated water may superficially resemble those of a stream form (Bond 1960). The gills of sirenids reduce to small stubs during aestivation, and the gills of all larvae are among the first features to show metamorphic reduction. The various configurations and positions of the nostrils are useful in anuran larvae, but this feature has seldom been evaluated in salamanders.

Differences in the *coloration* (i.e., visual impression of color plus pattern) of the iris, although lost in preservative, were obvious to us while photographing larvae for this book. The presence, shape, and mode of attachment of the gular fold also need further examination. Body coloration, provided by complex arrangements of

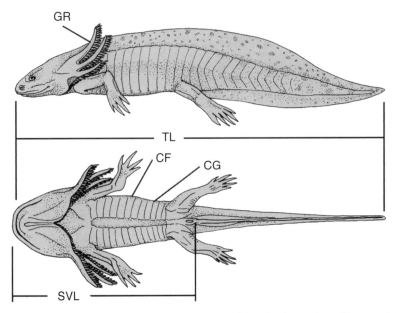

Figure 4. **Standard measurements and morphology of larval salamanders.** CF = costal fold bounded by costal grooves on each side, CG = costal groove, GR = gill ramus, SVL = snout vent length, TL = total length (DK).

various kinds of chromatophores, is quite important for identification, is ontogenetically and geographically variable, and has never been well documented. Amphibian larvae from turbid water are more pale, sometimes patternless, and appear almost albinistic compared with the brighter or more contrasty colorations of individuals from clear water. Field-collected larvae generally have more intense coloration than those held or reared in the laboratory, so photographs and notes should be taken immediately after capture. Cave-adapted salamanders appear silvery white or pinkish in life and tan, gray, or white in preservative. Hatchling patterns are sometimes retained into later larval stages or postmetamorphically (Olsson 1993, Olsson and Löfberg 1992). Tails of many hatchling *Ambystoma* are banded dorsally, and this pattern may or may not be accompanied by a pale, lateral stripe (Brandon 1961, 1964), or a lateral stripe may occur without the dorsal bands (e.g., *A. gracile*). Larvae of *A. talpoideum*, and to a lesser extent those of *A.* "*cingulatum*" and *A. mabeei*, retain early larval patterns throughout larval ontogeny. Larval patterns of *Rhyacotriton* and *Stereochilus* are retained in the adults. Metamorphs of *A. mavortium* and *A. tigrinum* have at least a semblance of the eventual adult pattern regardless of size at metamorphosis, while metamorphs of all other *Ambystoma* do not exhibit the adult coloration until they are larger. Patterns modified by the distribution of neuromasts are common in at least desmognathines and various *Eurycea*. Few data are available on the actual functions of the color

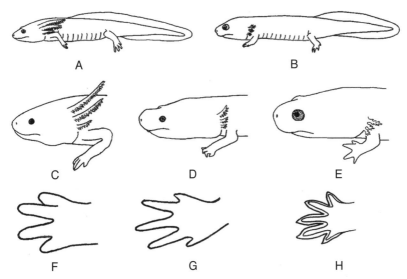

Figure 5. **Larval salamander morphology.** Examples of larvae that live in (A) nonflowing water (e.g., *Ambystoma*; see also Fig. 10A) or (B) flowing water (e.g., *Dicamptodon*; see also Fig. 20B, D, E), most plethodontines, and *Rhyacotriton* (see also Fig. 50B–C). Examples of gills: (C) pond larva with gill rami much longer than the numerous fimbriae (e.g., *Ambystoma*; see also Fig. 13A), (D) stream-inhabiting larva with short rami and fewer, shorter fimbriae (e.g., *Eurycea*; see also Fig. 36C), (E) stream-inhabiting larva with very short rami and few fimbriae (e.g., *Desmognathus* and *Rhyacotriton*; see also Fig. 26C, D; JD). Examples of digits: (F) short and rounded with blunt tips (e.g., plethodontids; see also Fig. 27F), (G) long and narrow with pointed tips (e.g., salamandrids; see also Fig. 56C), (H) flattened with lateral flanges (e.g., *Ambystoma tigrinum* group; see also Fig. 19B).

patterns of larval amphibians (e.g., J. P. Caldwell 1982, Altig and Channing 1993), but most of them are assumed to function in some form of crypsis.

Gaining a general perspective of the morphological features of specific ecological and taxonomic morphotypes aids identifications, and because dorsal views are a common observational aspect, we present a guide (Fig. 6) to both salamander larvae and tadpoles. The common resemblance between salamander metamorphs and their respective adults enhances the chance of their identifications. Larval paedomorphs usually look like the larviform adults. Attempts to identify larval *Ambystoma*, *Desmognathus*, and *Eurycea* are particularly frustrating.

Salamander larvae can be distinguished from anuran tadpoles (traits in parentheses) as follows: at least front limb buds or limbs present at all posthatching stages (front limb buds hidden under operculum, hind limb buds externally visible after stage 26), external gills usually present throughout larval ontogeny (external gills absent after stage 24), and tail and body about same width in dorsal view (body notably wider than tail). Salamander larvae occur in many habitats, but the morphological diversity, feeding niche, and habitat uses are notably

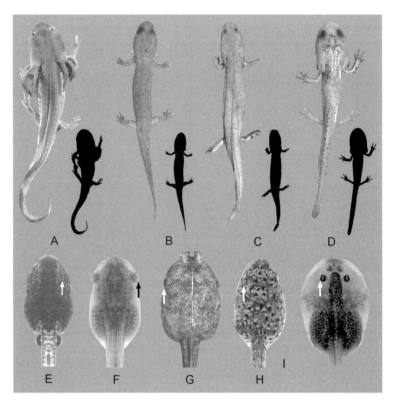

Figure 6. **Ecological types of larvae.** Comparisons should be made of general proportions and shapes of larvae and their silhouettes in dorsal view without concern for species-specific traits. Salamander larvae: (A) ambystomatid—*Ambystoma* (JPC): nonflowing, benthic with bulky body, large, depressed head with small eyes and rounded snout, large gills, dorsal fin extends onto back, and small, dorsal eyes, (B) *Desmognathus* (DBM): flowing water, benthic inhabitant with streamlined body, small gills with very short rami, low tail fin originating near tail:body junction, pointed toe tips sometimes keratinized, and eyes near lateral margin of head, (C) plethodontine plethodontid: benthic in flowing water with streamlined body, medium-sized gills, low tail fin originating near tail:body junction, and eyes near lateral margin of head, (D) salamandrid: usually nonflowing water with an angular head, dorsal fin originating on body, and toes long, narrow and usually pointed. Tadpoles (vertical white arrows = lateral margins of eyes): (E) bufonid (CCC): nonflowing water with depressed body, dorsal eyes, large nares, and a rounded tail tip with fin origin near tail:body junction, (F) most hylids (CCC): nonflowing water, nektonic with a head that is flat on top, body globular and usually compressed, eyes usually distinctly lateral, and tail fin usually originating on body, (G) microhylid (DLM): nonflowing water, nektonic with rounded, depressed body, eyes extremely lateral, and dorsal fin originating at tail:body junction, (H) ranid (CCC): usually nonflowing water with a depressed to globular body, eyes dorsal, and low to medium dorsal fin originating at tail:body junction, (I) scaphiopodid omnivore (CCC): nonflowing water with a depressed body, small, very dorsal eyes, and tail fin originating near tail:body junction.

narrower than those of tadpoles. Various sorts of paedomorphs and carnivorous morphotypes exist. Also, larval variations are poorly documented, and caution is advised in evaluating limbs; for example, the full complement of digits in *Amphiuma* is 1, 2, or 3, but other species with 4–5 digits will have similar counts at select points during early ontogeny. The absence of hind limbs throughout life in sirenids is different from that of typical four-legged larvae of all other species at stages near hatching. Hatchling salamanders (forelimbs incompletely differentiated with fewer than 4 fingers; hind limbs absent or incompletely developed, and hind limbs always absent in sirenids) are not included in this key.

The key, which emphasizes free-living, feeding larvae in the middle of ontogeny, is initially divided into those larviform taxa that lack or have reduced gills, amphiumids and sirenids with elongate bodies, those species with 4/4 digits, and then those taxa with digits 4/5, which are divided into three geographic units (i.e., Pacific Coast east to include the Rocky Mountains, east of Rocky Mountains to Mississippi River, and east of Mississippi River to Atlantic Coast).

KEY TO LARVAL AND LARVIFORM SALAMANDERS

Gills Absent or Reduced...**p. 63**

Cylindrical, Elongate Body, Limbs Small, Gular Fold Absent, Gill Rami ....
Branched.. **p. 63**

Body and Limbs Normally Proportioned, Digits 4/4 .... **p. 64**

Body and Limbs Normally Proportioned, Digits 4/5 .... **p. 66**

GILLS ABSENT OR REDUCED

I. Postlarva; body cylindrical and elongate, limbs small to tiny

    A. Four tiny limbs; digits 1/1, 2/2, or 3/3; gill slit 1; gills atrophy for life soon after hatching; east of Rocky Mountains to Atlantic Coast (see next heading) .... **Amphiumids**

    B. Small front limbs only; digits 3–4/0; gill slits 1–3; gill fimbriae and part of rami atrophy during aestivation; east of Rocky Mountains to Atlantic Coast (see next heading) .... **Sirenids**

II. Postlarva; about 180 mm TL or greater; south-central Missouri and adjacent Arkansas west of the Mississippi River and greater Ohio River drainage east of Mississippi River; rocky streams and rivers; depressed body and limbs normally proportioned; digits 4/5; gill slit 1; about 180–740 mm TL ....
.... *Cryptobranchus alleganiensis* (**Hellbender**)

CYLINDRICAL, ELONGATE BODY, LIMBS SMALL, GULAR FOLD ABSENT, GILL RAMI BRANCHED

I. Four tiny limbs, digits 1/1, 2/2, or 3/3 without keratinized tips; gills with few fimbriae, atrophy soon after hatching; keratinized jaw sheaths absent; gill slit 1; small ridgelike tail fin atrophies soon after hatching; nearly uniformly dark at all sizes .... **Amphiumas**

[Eastern quarter of Texas northeast to southeastern Missouri; digits 3/3 ....
....*Amphiuma tridactylum* (**Three-toed Amphiuma**)]
Fig. 21B–C, Plate 2F

[Coastal Plain from southeastern Louisiana to southeastern Virginia; digits 2/2; sympatric: *Amphiuma pholeter, A. tridactylum* ....
.... ***Amphiuma means*** (**Two-toed Amphiuma**)]

[Coastal Plain from southeastern Mississippi to northwestern Florida; digits 1/1; sympatric: *Amphiuma means, A. tridactylum* ....
.... ***Amphiuma pholeter*** (**One-toed Amphiuma**)]

II. Hind limbs absent throughout life; gilled rami with abundant fimbriae present throughout life but atrophy during aestivation; keratinized jaw sheaths present; transparent tail fins in larvae extend anteriorly almost to the head, opaque fins of postlarvae restricted to tail; larvae brightly patterned, paedomorphs drab

A. Digits 3/0; gill slit 1 .... **Sirens and Dwarf Sirens**

1. Southern two-thirds of Florida Peninsula; larva: uniformly dark with bright silver lateral stripe; paedomorph: nearly unicolored dorsally; to about 250 mm TL; sympatric: *Pseudobranchus striatus* ....
.... ***Pseudobranchus axanthus*** (**Southern Dwarf Siren**)

2. Florida panhandle and adjacent Georgia north to southeastern South Carolina; larva: as for *P. axanthus*; paedomorph: body with multiple stripes; to about 155 mm TL; sympatric: *Pseudobranchus axanthus* ....
.... ***Pseudobranchus striatus*** (**Northern Dwarf Siren**)
Fig. 57B–C, Plate 6I

B. Digits 4/0; gill slits 3

1. Southern Michigan south to southern Illinois and the Gulf Coast and along the Coastal Plain to southeastern Virginia and northern half of the Florida Peninsula; costal grooves 31–36; larvae with prominent red markings across tip of snout and along side of body, remainder dark, shiny black; adults uniformly gray with yellowish or greenish tinge, nose band may persist; hatchling 10 mm TL, paedomorph to 690 mm TL [**note**: populations from Rio Grande Valley may be distinct (*S. texana*) and may be more closely related to *S. lacertina*]; sympatric: *Siren lacertina* ....
.... ***Siren intermedia*** (**Lesser Siren**)
Fig. 58A–C, Plate 6J–K

2. Florida Peninsula and southeastern Coastal Plain from southwestern Alabama to eastern Virginia; costal grooves 36–39; larvae with prominent yellowish markings on shiny, dark black body; adults uniformly gray to black ground color, sometimes with abundant iridophore pigmentation in life that imparts a greenish to golden tinge; hatchling about 60 mm TL, paedomorph to 980 mm TL; sympatric: *Siren intermedia* ....
.... ***Siren lacertian*** (**Greater Siren**)
Fig. 59B–C, Plate 6L

BODY AND LIMBS NORMALLY PROPORTIONED, DIGITS 4/4

I. Boggy, swampy, nonflowing water or slow streams; midline of gular fold not attached to adjacent throat tissue; gill slits 3; less than 40 mm TL ....
.... **Brook Salamanders and relatives**

A. Coastal Plain from southeastern Texas to southern Arkansas east to southeastern North Carolina, including most of Florida Peninsula; costal grooves 17–18; dorsal fin originates at tail-body junction; dark eye line or not; lateral pattern stippled; to 38 mm TL; sympatric: none of similar morphology and habitat .... ***Eurycea quadridigitata*** (**Dwarf Salamander**)
Fig. 33A–B, Plate 4E

B. Coastal Plain of North and South Carolina; costal grooves 15–16; small individuals with dorsolateral pale spots; sympatric: *Eurycea quadridigitata*, *Hemidactylium scutatum* ....
.... ***Eurycea chamberlaini*** (**Chamberlain's Dwarf Salamander**)

C. West-central Arkansas and southwestern Missouri and throughout much of area east of Mississippi River with many disjunct populations; costal grooves 12–15; dorsal fin extends well onto back at least at earlier stages; lateral pattern of dashes; to 28 mm TL; sympatric: none of similar morphology ....
.... ***Hemidactylium scutatum*** (**Four-toed Salamander**)
Fig. 40A insert, Plate 5C

II. Larger, permanent nonflowing water, streams and rivers; midline of gular fold attached to adjacent throat tissue; gill slits 2; lateral pattern uniform or mottled; costal grooves indistinct, especially in postlarvae; to 276 mm TL as paedomorph .... **Waterdogs**

A. West of Mississippi River

1. Eastern Texas into southwestern Louisiana; larva and paedomorph: spotted dorsally and ventrally; sympatric: no congeners ....
.... ***Necturus beyeri*** (**Gulf Coast Waterdog**)
Fig. 46A–C, Plate 5I

2. Central Louisiana and Arkansas and adjacent Oklahoma north to southeastern Manitoba along Mississippi River; larva: striped; paedomorph: spotted dorsally, center of belly clear; sympatric: no congeners ....
.... ***Necturus maculosus*** (**Mudpuppy**)
Fig. 48B–D, Plate 5K

B. Mississippi River east to Appalachian Mountains

1. Upper Black Warrior River, Alabama; streams and river; larva: middorsal dark stripe bordered laterally by white or cream stripes; paedomorph: maroon to brownish with variable spotting; sympatric: *Necturus beyeri* ...
.... ***Necturus alabamensis*** (**Blackwarrior Waterdog**)
Fig. 45A, Plate 5H

2. Coastal Plain from Mississippi River east to Mobile Bay; streams and rivers; larva and paedomorph: maroon to brownish with scattered small and sometimes large spots with diffuse edges; sympatric: *Necturus alabamensis* ....
.... ***Necturus beyeri*** (**Gulf Coast Waterdog**)
Fig. 46A–C, Plate 5I

3. Much of north-central North America from west-central Mississippi north; streams, rivers, and lakes; larva: middorsal dark stripe bordered dorsolaterally by white or cream stripes; paedomorph: uniformly gray to brownish to

boldly spotted; sympatric: none, unless *N. beyeri* since the opening of the Tennessee-Tombigbee Waterway .... ***Necturus maculosus*** **(Mudpuppy)**
Fig. 48B–D, Plate 5K

C. Appalachian Mountains east to Atlantic Coast

1. Neuse and Tar rivers, North Carolina; larva: striped; paedomorph: dorsum and venter spotted; sympatric: *Necturus punctatus* ....
.... ***Necturus lewisi*** **(Neuse River Waterdog)**
Fig. 47A–D, Plate 5J

2. Southeastern Virginia to south-central Georgia; larva and paedomorph: usually uniformly gray to black, sometimes brown to maroon with small spots, venter not spotted; sympatric: *Necturus lewisi* ....
.... ***Necturus punctatus*** **(Dwarf Waterdog)**
Fig. 49A–B, Plate 5L

BODY AND LIMBS NORMALLY PROPORTIONED, DIGITS 4/5

Pacific Coast East to Include Rocky Mountains

I. Nonflowing water, often ephemeral; body robust; gular fold fits loosely against adjacent throat; dorsal fin extends well onto back; gill slits 4, gill rami long, extend above plane of back; gill rakers obvious; eyes small and not protuberant; snout long and depressed, broadly rounded in dorsal view; jaw sheaths often present; ground color varies from pale to dark, with or without contrasting marks .... **Mole Salamanders**

A. West-central California; ephemeral ponds; more than 15 gill rakers on third gill arch; costal grooves 11–13; digits flat and pointed with lateral flanges; greenish gray mottled to uniformly silvery depending on water turbidity; to 180 mm TL; sympatric: *Ambystoma macrodactylum* ....
.... ***Ambystoma californiense*** **(California Tiger Salamander)**
Fig. 10A, Plate 1D

B. Cascade Mountains and west from southeastern Alaska to just north of San Francisco Bay; ephemeral and permanent ponds; 7–10 gill rakers on third gill arch; gill fimbriae about same length along length of ramus; costal grooves 11–12; digits round with blunt tips; dorsal dark pigmentation with contrasting darker marks; prominent lateral stripe in small specimens; to 210 mm TL as pedotypes, larvae smaller; sympatric: *Ambystoma macrodactylum* ....
.... ***Ambystoma gracile*** **(Northwestern Salamander)**
Fig. 11A–B, D, Plate 1E

C. Southeastern Alaska south and east to eastern Montana and southwest to disjunct populations just south of San Francisco Bay; usually ephemeral ponds; 9–13 gill rakers on third gill arch; gill fimbriae at tips of rami shorter than at base, ramus of first gill shorter than head length; costal grooves 12–13; digits round with blunt tips; small specimens uniformly dark, larger larvae brownish to greenish without large marks; to 80 mm TL; sympatric: *Ambystoma*

gracile, A. mavortium ....
.... ***Ambystoma macrodactylum*** (**Long-toed Salamander**)
Fig. 14D–E, Plate 1I

D. Disjunct populations scattered throughout western North America; nonflowing water, including lakes; digits with pointed tips and lateral flanges; ramus of first gill equal to or longer than head length; 15–24 gill rakers on third gill arch; to 250 mm TL; costal grooves 12–13; to 70–90 mm TL as first-year larva, 130 mm TL as second-year larva, 330 mm TL as pedotype; sympatric: *Ambystoma gracile, A. macrodactylum* ....
.... ***Ambystoma mavortium*** (**Western Tiger Salamander**)
Fig. 19E

II. Mountain streams; body robust; gular fold fits tightly against adjacent throat; gill slits 4, gill rami medium, usually do not extend above the back; gill rakers obvious; eyes small and not protuberant; snout bulbous, rounded and depressed; jaw sheaths absent; to 150 mm TL as larva, larger than 300 mm TL as pedotype; distinguished reliably only by range .... **Giant Salamanders**

[Northern Idaho and adjacent Montana; sympatric: no congeners ....
.... ***Dicamptodon aterrimus*** (**Idaho Giant Salamander**)]
Fig. 20D

[Olympic Peninsula south to northwestern Oregon; belly dark; sympatric: *D. tenebrosus* .... ***Dicamptodon copei*** (**Cope's Giant Salamander**)]
Plate 2C

[Coast Range from Sonoma and Mendocino counties to just south of San Francisco Bay; sympatric: *D. tenebrosus* ....
.... ***Dicamptodon ensatus*** (**California Giant Salamander**)]
Plate 2D

[Coast Range and Cascades from northwestern California to southwestern British Columbia; 5–7 gill rakers on third gill arch; belly pale; sympatric: *Dicamptodon ensatus, D. copei* .... ***Dicamptodon tenebrosus*** (**Coastal Giant Salamander**)]
Fig. 20B, E, Plate 2E

III. Seeps and small mountain streams; body streamlined but bulky; gill rami exceptionally short and moundlike, few, short fimbriae do not extend above plane of back; gill slits 4; gill rakers small to absent; eyes large, protuberant, and lateral; snout short and broadly rounded; jaw sheaths absent; yellowish to golden ground color with variable amounts of small dark markings; species not reliably distinguishable except by range .... **Torrent Salamanders**

[Olympic Peninsula of Washington; dorsum not spotted, venter spotted; sympatric: no congeners .... ***Rhyacotriton olympicus*** (**Olympic Torrent Salamander**)]
Fig. 50C

[Cascade Mountains from Skookumchuck River, Lewis Co., Washington to Lane Co., Oregon; dorsum blotched, venter not spotted; sympatric: no congeners ....
......................................***Rhyacotriton cascadae*** (**Cascades Torrent Salamander**)]
Fig. 50D, Plate 6A

[Coast ranges of Oregon and Washington south of Chehalis River to Yamhill Co., Oregon; dorsum and venter not spotted; sympatric: no congeners ....
.... *Rhyacotriton kezeri* (**Columbia Torrent Salamander**)]
Fig. 50B

[Coast Range from south of Yamhill Co., Oregon south to Mendocino Co., California; dorsum and venter spotted; sympatric: no congeners ....
.... *Rhyacotriton variegatus* (**Variegated Torrent Salamander**)]
Plate 6B

IV. Nonflowing water and small streams; body shape not robust; gill slits 4, gill rami long, fimbriae ragged; gill rakers small; eyes medium and lateral; dorsal fin originates well anterior to tail-body junction in early stages, near tail-body junction or more posteriorly at later stages; digits narrow and pointed, often with keratinized tips; snout often angular but not strongly depressed; jaw sheaths absent; to 75 mm TL, usually smaller .... **Newts**

 A. Cascade Mountains and west from southeastern Alaska to southern California, also northern Idaho; usually nonflowing water; mottled and blotched on gray to brown ground color; dorsal fin extends to shoulders in young specimens; sympatric: *Taricha torosa* ....
.... *Taricha granulosa* (**Rough-skinned Newt**)
Fig. 54D, Plate 6F

 B. Sonoma and Lake cos. north into Humboldt Co., California; small streams; uniformly brown to black; dorsal fin extends to midbody in smaller individuals and to near tail-body junction in larger ones; sympatric: *Taricha torosa* ..
.... *Taricha rivularis* (**Red-bellied Newt**)
Fig. 55B, C, Plate 6G

 C. Most of Coast Range of California; usually nonflowing water; uniform yellowish to cream ground color with black dorsolateral stripes extending onto tail; dorsal fin extends to shoulders in younger specimens; sympatric: *Taricha rivularis* .... *Taricha torosa* (**California Newt**)
Fig. 56B, E, Plate 6H

  [Sierra Nevada of California; often in small streams; sympatric: no congeners ...
.... *Taricha sierra* (**Sierra Newt**)]

East of Rocky Mountains to Mississippi River

 I. Nonflowing water; gill rami long; tail fin originates somewhere on the back; digits long; mandibular grooves do not bisect lower lip on each side of the midline

 A. Eyes small, dorsal, and not protuberant; body robust; digits rounded and blunt; snout long and depressed, rounded in dorsal view; keratinized jaw sheaths often present .... **Key Central Ambystoma (p. 69)**

 B. Eyes medium and lateral; body shape not robust; digits narrow and pointed, often with keratinized tips; snout often angular but not strongly depressed; jaw sheaths absent .... **Newts**

1. Southeastern Texas coast and adjacent Mexico; dorsum uniformly gray to brown; lower flanks pale and spotted; sympatric: *Notophthalmus viridescens* .... ***Notophthalmus meridionalis*** (**Black-spotted Newt**)

   Fig. 51B–D, Plate 6C

2. East of line connecting south-central Texas coast and central Minnesota; dorsum pale with tiny dark dots; sympatric: *Notophthalmus meridionalis* ... . ***Notophthalmus viridescens*** (**Eastern Newt**)

   Fig. 53A–C, Plate 6E

II. Flowing water; mandibular grooves bisect the lower lip on each side of the midline; gill fimbriae short, gill rami short and moundlike; toe tips pointed and often keratinized; sympatric: none within group .... **Dusky Salamanders**

[Eastern Texas through southern Louisiana.. ....*Desmognathus auriculatus* (**Southern Dusky Salamander**)]

   Plate 2I

[West-central Arkansas and adjacent Oklahoma; sympatric: no congeners .... *Desmognathus brimleyorum* (**Ouachita Dusky Salamander**)]

   Plate 2J

[Northeastern Missouri and north-central Louisiana and adjacent southern Arkansas .... *Desmognathus fuscus* (**Northern Dusky Salamander**)]

   Fig. 23E

III. Flowing water; mandibular grooves do not bisect the lower lip on each side of the midline; gill fimbriae and gill rami long; toe tips blunt and not keratinized; low tail fin originates at tail-body junction

   A. Rocky streams and rivers in south-central Missouri and adjacent Arkansas; gill slits 4; body bulky; eyes small and dorsal; snout depressed and broadly rounded in dorsal view; gular fold fits loosely against throat .... ***Cryptobranchus alleganiensis*** (**Hellbender**)

   Fig. 22C, Plate 2G

   B. Large eyes lateral; gill rami medium, usually extending above plane of back; toe tips rounded and not keratinized; gular fold fits tightly against throat .... **Brook Salamanders**

      1. Ozark Region of Arkansas, Missouri, and Oklahoma .... **Key Ozark Eurycea (p. 70)**

      2. Edwards Plateau of central Texas .... **Key Texas Eurycea (p. 71)**

Key Central Ambystoma

   I. Costal grooves 14–16
      A. East-central Oklahoma east and north to west-central Missouri; costal grooves 15–16; throat white; to 70 mm TL; sympatric: *Ambystoma texanum* .... ***Ambystoma annulatum*** (**Ringed Salamander**)

   Fig. 7B, Plate 1A

B. Eastern third of Texas north to southern Iowa and east to Mississippi River; costal grooves 14–15; throat pigmented; to 65 mm TL; sympatric: *Ambystoma annulatum* .... ***Ambystoma texanum*** **(Small-mouthed Salamander)**
Fig. 18B–D, Plate 2A

II. Costal grooves 11–13

A. Extreme eastern Texas and Oklahoma east to Mississippi River; costal grooves 11–13; throat white; dorsum uniformly pale or dark; digits round; to 62 mm TL; sympatric: *Ambystoma mavortium, A. opacum, A. talpoideum* ... . ***Ambystoma maculatum*** **(Spotted Salamander)**
Fig. 15E, Plate 1J

B. Most of designated area except for most of states bordering Mississippi River from Louisiana north to Nebraska; costal grooves 12–13; digits with pointed tips and lateral flanges; ramus of first gill equal to or longer than head length; 15–24 gill rakers on third gill arch; to 70–90 mm TL as first-year larva, 130 mm TL as second-year larva, 330 mm TL as pedotype; sympatric: *Ambystoma maculatum, A. opacum, A. talpoideum* .... ***Ambystoma mavortium*** **(Western Tiger Salamander)**
Fig. 19E

C. Southeastern Texas coast north and east to southeastern Missouri and east to Mississippi River; costal grooves 11–13; throat uniformly but sparsely pigmented; uniformly pale to dark, ventrolateral pale dots sometimes present; digits round; to 76 mm TL; sympatric: *Ambystoma maculatum, A. mavortium, A. talpoideum* .... ***Ambystoma opacum*** **(Marbled Salamander)**
Fig. 16D, Plate 1K

D. Northern two-thirds of Louisiana, adjacent Texas, and southeastern Oklahoma; costal grooves 10–11; throat pigmented; dark ground color with pale lateral and ventrolateral stripes; dorsum of tail banded throughout larval life; eye line present; digits round with blunt tips; to 80 mm TL, larger as pedotype; sympatric: *Ambystoma maculatum, A. mavortium, A. opacum* .... ***Ambystoma talpoideum*** **(Mole Salamander)**
Fig. 17D, F, Plate 1L

Key Ozark Eurycea

I. Costal grooves 15 or fewer

A. Southern two-thirds of Missouri, most of northern half of Arkansas and adjacent Oklahoma; costal grooves 13–14; throat, chin, and soles of feet nonpigmented; dorsum may be black when first collected but is usually uniform pale brown to tan with scattered darker specks and dots; large larvae with indications of bands on tail like adults; ventrolateral pigment extends to near level of limb insertions; sympatric: *Eurycea lucifuga* .... ***Eurycea longicauda*** **(Long-tailed Salamander)**
Fig. 30A, Plate 3L

B. Southern half of Missouri and adjacent Arkansas and Oklahoma; costal grooves 14–15; throat, chin, and soles of feet pigmented; dorsum may be black when first collected but is usually uniform pale brown to tan; ventrolateral pigment extends beyond level of limb insertions; sympatric: *Eurycea longicauda* .... ***Eurycea lucifuga* (Cave Salamander)**

Fig. 31A, Plate 4A

II. Costal grooves 16 or more

A. Central Arkansas and adjacent Oklahoma south of Arkansas River; costal grooves 19–20; 7 or more costal grooves between adpressed limbs; uniformly yellowish to pale brown; to 40 mm TL; sympatric: *Eurycea spelaea, E. tynerensis* .... ***Eurycea multiplicata* (Many-ribbed Salamander)**

Plate 4B

B. Southern Missouri and adjacent Arkansas and Oklahoma; costal grooves 16–17; 4–6 costal grooves between adpressed limbs; uniformly pale gray to pinkish; to 120 mm TL; sympatric: *Eurycea multiplicata, E. tynerensis* ....
.... ***Eurycea spelaea* (Grotto Salamander)**

Fig. 36B–C, Plate 4H

C. Southwestern Missouri and adjacent Oklahoma and Arkansas; costal grooves 19–20; 7 or more costal grooves between adpressed limbs; cream to gray, finely marked with small dots; to 80 mm TL; sympatric: *Eurycea multiplicata, E. spelaea* .... ***Eurycea tynerensis* (Oklahoma Salamander)**

Plate 4I

Key Texas Eurycea

I. Ranges north of Colorado River; epigean sites; costal grooves 15–16; snout uniformly rounded in profile; sympatric: no congeners

[Salado, Bell Co., Texas .... ***Eurycea chisholmensis* (Salado Salamander)**]

Plate 3I

[Georgetown, Williamson Co., Texas...
.... ***Eurycea naufragia* (Georgetown Salamander)**]

Fig. 28A

[Jollyville Plateau, Travis and Williamson cos., Texas ....
.... ***Eurycea tonkawae* (Jollyville Plateau Salamander)**]

Fig. 28B

II. Ranges south of Colorado River; hypogean sites

A. Costal grooves 11–12

1. San Marcos, Hays Co., Texas; sympatric: none in hypogean sites ....
.... ***Eurycea rathbuni* (Texas Blind Salamander)**

Plate 4F

[Aquifer beneath Blanco River, Hays Co., Texas; sympatric: none in hypogean sites........................ ***Eurycea robusta* (Blanco Blind Salamander)**]

Fig. 34A

[Honey Creek Cave, Comal Co., Texas; sympatric: none in hypogean sites
.... *Eurycea tridentifera* (**Comal Blind Salamander**)]

[Barton Springs, Travis Co., Texas; sympatric: none in hypogean sites .....
.... *Eurycea waterlooensis* (**Austin Blind Salamander**)]
Plate 4J

B. Costal grooves 13–17

Cascade Caverns, Kendall Co., Texas, sympatric: no congeners ....
.... *Eurycea latitans* (**Cascade Caverns Salamander**)

[Valdina Farms Sinkhole, Medina Co., Texas; sympatric: no congeners ....
.... *Eurycea troglodytes* (**Valdina Farms Salamander**)]
Fig. 293B

III. Ranges south of Colorado River; epigean sites; costal grooves 13–17

A. San Marcos Springs, Hays Co., Texas; sympatric: none in epigean sites..........
.... *Eurycea nana* (**San Marcos Salamander**)
Fig. 32A–B, Plate 4C

B. Bexar Co., Texas; sympatric: no congeners ....
.... *Eurycea neotenes* (**Texas Salamander**)

[Blanco River drainage, Hays, Co., Texas; sympatric: no congeners ....
.... *Eurycea pterophila* (**Fern Bank Salamander**)]
Plate 4D

C. Barton Springs, Travis Co., Texas; sympatric: none in epigean sites...............
.... *Eurycea sosorum* (**Barton Springs Salamander**)
Fig. 35B–C, Plate 4G

Mississippi River East to Atlantic Coast

I. Flowing water

A. Mandibular grooves bisect lower lip each side of midline. Toes pointed and often keratinized; gill fimbriae short, gill rami short and moundlike; dorsal fin originates near tail-body junction .... **Key Eastern Desmognathus (p. 73)**

B. Mandibular grooves do not bisect lower lip each side of midline; gill rami long; small streams and seeps and large streams and rivers, usually with rocky substrates

1. Greater Ohio River drainage; gills and gular fold present, gill slits 4; digits thick and fleshy; gular fold fits loosely against throat; sympatric: none of similar morphology ... *Cryptobranchus alleganiensis* (**Hellbender larva**)
Fig. 22C, Plate 2G

2. Flowing water; large eyes lateral; gill rami medium, usually extending above plane of back; toe tips rounded and not keratinized; gular fold fits tightly against throat .... **Key Eastern Plethodontid (p. 75)**

II. Nonflowing water

    A. Nonflowing water; eyes small, dorsal, and not protuberant; body robust; digits rounded with blunt tips; depressed snout rounded in dorsal view; keratinized jaw sheaths often present .... **Key Eastern Ambystoma (p. 78)**

    B. Nonflowing water; eyes medium and lateral; dorsal fin originates well anterior to tail-body junction in early stages, near tail-body junction or more posteriorly at later stages; digits narrow and pointed, often with keratinized tips; snout often angular but not strongly depressed, not rounded in dorsal view; jaw sheaths absent; to 75 total length, usually smaller .... **Newts**

        1. Southeastern third of Georgia and northern half of Florida Peninsula; eye diameter equal to nostril-eye distance; tail with dusky spots; sympatric: *Notophthalmus viridescens* .... ***Notophthalmus perstriatus*** **(Striped Newt)**
                Fig. 52A–B, Plate 6D

        2. Most of eastern United States north to southern Quebec except for prairie regions in Illinois, Indiana, and Ohio; eye diameter less than nostril-eye distance; tail with tiny dark dots; sympatric: *Notophthalmus perstriatus* .... ***Notophthalmus viridescens*** **(Eastern Newt)**
                Fig. 53A–C, Plate 6E

Key Eastern Desmognathus

I. Appalachian highlands; greater than 35 mm TL

    A. Southwestern Virginia south to northeastern Georgia; slender body and demarcation between dorsal and ventral coloration abrupt and distinct; dorsum uniform brown to black to variously mottled; head narrow and somewhat pointed in dorsal view; internal nostrils positioned far laterally; sympatric: *Desmognathus "quadramaculatus"* .... ***Desmognathus marmoratus*** **(Shovel-nosed Salamander)**
                Fig. 24D, Plate 3A

    B. Southern West Virginia to northeastern Georgia; robust body; demarcation between dorsal and ventral coloration diffuse; dorsum uniform brown to black or with small black markings; head broadly rounded in dorsal view; internal nostrils rounded and positioned adjacent to midline of roof of mouth; sympatric: *Desmognathus marmoratus* .... ***Desmognathus quadramaculatus*** **(Black-bellied Salamander)**
                Fig. 26D, Plate 3D

    [Wolf and Helton creeks, Union Co., Georgia and Clay Co., North Carolina; sympatric: *Desmognathus quadramaculatus* .... ***Desmognathus folkertsi*** **(Dwarf Black-bellied Salamander)**]

    C. Eastern Kentucky and adjacent Virginia and Tennessee; Cumberland Mountains of southwestern Virginia and adjacent Kentucky and Tennessee; robust body; demarcation between dorsal and ventral coloration diffuse; dorsum dark with paler mottling; head broadly rounded in dorsal view; internal nostrils rounded and positioned adjacent to midline of roof of mouth; sympatric: none of large size .... *Desmognathus welteri* (**Black Mountain Salamander**)

Plate 3F

II. East of Mississippi River—Appalachian highlands; less than 35 mm TL

   A. Southwestern Virginia to southeastern Kentucky through the Adirondack Mountains to southern Quebec; 2–7, commonly 4–5, dorsolateral circular spots between limb insertions, circles may be incomplete medially, sometimes with dark stripes laterally connecting spots, or only stripe present, or dorsum unicolored reddish to brown; sympatric: *Desmognathus "fuscus," D. marmoratus, D. monticola, D. "quadramaculatus"* ....

     ..... *Desmognathus ochrophaeus* (**Allegheny Mountain Dusky Salamander**)

Fig. 25A

     [Southeastern North Carolina and adjacent Georgia, South Carolina, and Tennessee; sympatric: *Desmognathus monticola* ....

     .... *Desmognathus carolinensis* (**Carolina Mountain Dusky Salamander**)]

Fig. 25E, Plate 2K

     [Central Smoky Mountains above 900 m; sympatric: *Desmognathus ochrophaeus*............................ *Desmognathus imitator* (**Imitator Salamander**)]

     [Disjunct regions on Appalachian Plateau of northeastern Alabama and southwestern Blue Ridge Physiographic Province south of Pigeon River; sympatric: *Desmognathus monticola*..

     .... *Desmognathus ocoee* (**Ocoee Salamander**)]

Fig. 25B

     [Floyd Co., Virginia to somewhere between Linville Falls and McKinney Gap and to headwaters of Toms and Clark creeks near Iron Mountain Gap; sympatric: *Desmognathus monticola*...

     .... *Desmognathus orestes* (**Blue Ridge Dusky Salamander**)]

Plate 3C

     [Southeastern Blue Ridge Mountains southward to Patrick County and east to the Piedmont .... *Desmognathus planiceps* (**Flat-headed Salamander**)]

   B. Wide band from eastern Nova Scotia southwest through Mississippi; 5–13 dorsolateral round spots between limb insertions, often with black stripe along lateral margins of spots; sympatric: *Desmognathus auriculatus, D. marmoratus, D. monticola, D. "quadramaculatus," D. welteri* ....

     .... *Desmognathus fuscus* (**Dusky Salamander**)

Fig. 23E, see Plate 2L

   C. Central Appalachian Mountains from southwestern Virginia to northeastern Georgia; dorsolateral round spots between limb insertions sometimes present in young individuals, varies from mottled and blotched to uniformly dark; gills usually different from body color; discrete boundary between dorsal and ventral coloration; sympatric: *Desmognathus monticola, D. "quadramaculatus,"* .... *Desmognathus marmoratus* (**Shovel-nosed Salamander**)

Fig. 24D, Plate 3A

D. Appalachian Mountains from southwestern Pennsylvania to southwestern Alabama; 4–5 dorsolateral round spots between limb insertions; gill fimbriae/side 16–17; sympatric: *Desmognathus "fuscus," D. marmoratus, D. "quadramaculatus"* .... ***Desmognathus monticola* (Seal Salamander)**
Plate 3B

E. Central Appalachian Mountains from southern West Virginia to northeastern Georgia; dorsolateral round spots between limb insertions of young individuals; sympatric: *Desmognathus marmoratus, D. welteri* ....
.... ***Desmognathus quadramaculatus* (Black-bellied Salamander)**
Fig. 26D, Plate 3D

[Wolf and Helton creeks, Union Co., Georgia and Clay Co., North Carolina; sympatric: *Desmognathus quadramaculatus* ....
.... ***Desmognathus folkertsi* (Dwarf Black-bellied Salamander)**]

F. Central Appalachian Mountains on Tennessee-North Carolina border; dorsolateral round spots between limb insertions; sympatric: *Desmognathus "fuscus," D. marmoratus, D. monticola, D. "quadramaculatus"* ....
.... ***Desmognathus santeetlah* (Santeetlah Dusky Salamander)**
Fig. 23F, Plate 3E

G. Cumberland Mountains of southwestern Virginia and adjacent Kentucky and Tennessee; 5–8 dorsolateral round spots between limb insertions; gill fimbriae/side 17–27; sympatric: *Desmognathus "fuscus," D. monticola* ....
.... ***Desmognathus welteri* (Black Mountain Salamander)**
Plate 3F

III. Coastal Plain and interior; 35 mm TL or less
A. Choctawhatchee, Chattahoochee, and Apalachicola river drainages in southeastern Alabama and adjacent Georgia and Florida; 5–7 dorsolateral round spots between limb insertions; sympatric: no congeners ....
.... ***Desmognathus apalachicolae* (Apalachicola Dusky Salamander)**
Fig. 25D, Plate 2H

B. Coastal Plain from Mississippi River to southeastern Virginia; round dorsolateral spots between limb insertions, although often uniformly dark; sympatric: *Desmognathus "fuscus"* ....
.... ***Desmognathus auriculatus* (Southern Dusky Salamander)**
Fig. 23C, Plate 2I

C. Wide band from eastern Nova Scotia southwest through Mississippi; 5–13 (mode = 6) dorsolateral round spots between limb insertions, often with black stripe along lateral margins of spots; sympatric: *Desmognathus auriculatus* .... ***Desmognathus fuscus* (Dusky Salamander)**

[Southeastern Louisiana north to southern Illinois and east to southeastern Georgia; sympatric: *Desmognathus auriculatus, D. fuscus*..
.... ***Desmognathus conanti* (Spotted Dusky Salamander)**]

Key Eastern Plethodontid

I. Hypogean sites; snout variably depressed in lateral profile; costal grooves 13–19

A. North-central Florida and adjacent Georgia; costal grooves 12–13; eyes tiny pigmented specks; silvery white with few scattered melanophores; to 78 mm TL; sympatric: none of similar morphology ....
.... *Eurycea wallacei* (**Georgia Blind Salamander**)
Fig. 37A–B, Plate 4L

B. Costal grooves 17–19; to 230 mm TL; eyes small but obvious; general coloration pinkish with or without darker markings
   1. Roane Co., Tennessee; sympatric: none of similar morphology ....
   .... *Gyrinophilus gulolineatus* (**Berry Cave Salamander**)
Fig. 38A

   2. South-central Tennessee and adjacent Alabama; sympatric: none of similar morphology...
   .... *Gyrinophilus palleucus* (**Tennessee Cave Salamander**)
Fig. 38B, Plate 5A

   3. Greenbrier Co., West Virginia; sympatric: *Gyrinophilus porphyriticus* .....
   .... *Gyrinophilus subterraneous* (**West Virginia Spring Salamander**)
Fig. 38C

II. Epigean sites; costal grooves 13–16; dorsum with series of paired spots or dots, sides of tail uniformly colored

A. Northeastern Alabama northeast to northwestern Georgia; costal grooves 13–14; throat and soles of feet white; dorsal pigmentation does not extend ventrally below a line connecting ventral limb insertions; lateral surfaces of body darker than dorsum; sympatric: *Eurycea cirrigera, E. guttolineata, E. lucifuga* .... *Eurycea aquatica* (**Brown-backed Salamander**)
Plate 3G

B. Most of eastern United States south of central Indiana and Ohio except for range of *E. wilderae* (see below) and Florida Peninsula; costal grooves 14–16; throat and soles of feet white; dorsal pigmentation does not extend ventrally below a line connecting ventral limb insertions; sympatric: *Eurycea aquatica, E. "longicauda," E. lucifuga* ....
.... *Eurycea bislineata* (**Northern Two-lined Salamander**)
Fig. 27C, Plate 3H

[Eastern Illinois to central Virginia and southwest to Mississippi River exclusive of peninsular Florida and range of *E. wilderae* at higher elevations in Appalachian Mountains; sympatric: *Eurycea aquatica, E. guttolineata, E. junaluska, E. longicauda, E. lucifuga*..
.... *Eurycea cirrigera* (**Southern Two-lined Salamander**)]
Plate 3J

[Appalachian Mountains from southwestern Virginia to northern Georgia; sympatric: *Eurycea junaluska, E. longicauda, Urspelerpes brucei* ....
.... *Eurycea wilderae* (**Blue Ridge Two-lined Salamander**)]
Plate 4K

C. Graham Co., North Carolina and Sevier and Monroe cos., Tennessee; costal grooves 14; throat and soles of feet white; dorsal pigmentation not extending

beyond a line connecting ventral limb insertions; sympatric: *Eurycea longi-cauda, E. lucifuga, E. wilderae* ....
.... ***Eurycea junaluska*** (**Junaluska Salamander**)
Plate 3K

D. Western Kentucky east and north to southeastern New York; costal grooves
13–14; throat lightly pigmented anteriorly; soles of feet not or slightly pigmented; dorsal pigmentation not extending ventrally below line connecting
ventral limb insertions; sympatric: *Eurycea aquatica, E. bislineata, E. cirrigera, E. junaluska, E. lucifuga, E. wilderae* ....
.... ***Eurycea longicauda*** (**Long-tailed Salamander**)
Fig. 30A, Plate 3L

[Most of Mississippi east and north to southeastern Virginia; sympatric: *Eu-
rycea cirrigera, E. lucifuga* ....
.... ***Eurycea guttolineata*** (**Three-lined Salamander**)]
Fig. 30B

E. Central part of eastern United States from southern Illinois to northeastern
West Virginia and south to central Alabama; costal grooves 14–15; throat
well pigmented and pattern extends medially at first gill; soles of feet pigmented; dorsal pigmentation extends ventrally beyond a line connecting ventral limb insertions; sympatric: *Eurycea bislineata, E. cirrigera, E. junaluska,
E. "longicauda"* .... ***Eurycea lucifuga*** (**Cave Salamander**)
Fig. 31A, Plate 4A

F. Stephens Co., Georgia; costal grooves 15; distinct pale patch on snout; sympatric: *Eurycea wilderae* .... ***Urspelerpes brucei*** (**Patch-nosed Salamander**)
Fig. 44A–B, Plate 5G

III. Epigean sites; costal grooves 16–18
A. Appalachian Mountains and adjacent areas from northeastern Mississippi
northeast to south-central Maine; costal grooves 17–19; usually more than 5.5
costal grooves between adpressed limbs; squared, slightly upturned snout in
larger individuals; ground color dark lavender-gray, pale tan, various shades
of brown, or russet; dorsal markings small and scattered, sometimes a reticulated pattern; side of head behind jaws not spotted; chin often pigmented;
neuromasts above otic region of skull arranged in an ellipse; to 100+ mm TL;
sympatric: *Pseudotriton montanus, P. ruber* ....
.... ***Gyrinophilus porphyriticus*** (**Spring Salamander**)
Fig. 39A–B, Plate 5B

B. Coastal Plain from Mississippi River to southern New Jersey except for Mississippi Embayment, southern two-thirds of Florida Peninsula, and Appalachian
Mountains, and Ohio River valley; costal grooves 16–18; usually 5.5 or fewer costal grooves between adpressed limbs; snout rounded; ground color usually some
shade of pale to medium brown with scattered dorsal markings; neuromasts above
otic region of skull arranged in a circle; to about 55 mm TL; sympatric: *Gyrinophilus porphyriticus, P. ruber* .... ***Pseudotriton montanus*** (**Mud Salamander**)
Fig. 41B–C, Plate 5D

C. Eastern United States east of Mississippi and Ohio rivers and south of southern New York except Atlantic Coastal Plain; costal grooves 16–17; usually 5.5 or fewer costal grooves between adpressed limbs; snout rounded; ground color usually some shade of pale to medium brown, dorsal markings small and scattered; neuromasts above otic region of skull arranged in a circle; to 86 mm TL; sympatric: *Gyrinophilus porphyriticus, P. montanus*........................
.... *Pseudotriton ruber* (**Red Salamander**)
Fig. 42B, D, Plate 5E

D. Coastal Plain from northeastern Florida to southeastern Virginia; costal grooves 18; neuromasts obvious on snout; slender body with wedge-shaped snout in lateral profile; dark brown to maroon dorsally and yellowish to white ventrally with ventrolateral pale stripe, sides streaked; small individuals with dorsal fin extending nearly to head, to tail-body junction in larger individuals; to 60 mm TL; sympatric: none of similar morphology ....
.... *Stereochilus marginatus* (**Many-lined Salamander**)
Fig. 43F, Plate 5F

Key Eastern Ambystoma

I. Small rocky streams; eastern Kentucky and adjacent Indiana, Tennessee, Ohio, and West Virginia; costal grooves 14–15 throat pigmented; dusky gray to brown with pale bands on tail and body, becoming faint with size: to 60 mm TL; sympatric: *Ambystoma texanum* at some sites ....
.... *Ambystoma barbouri* (**Streamside Salamander**)
Fig. 8C–E, Plate 1B

II. Nonflowing water; costal grooves 13–15

A. Coastal Plain from Apalachicola River east to central coastal South Carolina; costal grooves 13–14; gaudy striped pattern distinct among all salamander larvae; dorsal fin unmarked; eye stripe prominent; throat not pigmented; to 70 mm TL; sympatric: none with similar coloration ....
.... *Ambystoma cingulatum* (**Frosted Flatwoods Salamander**)
See Fig. 9A–B, Plate 1C

[Apalachicola River west to southwestern Alabama; sympatric: none of similar coloration .... *Ambystoma bishopi* (**Reticulated Flatwoods Salamander**)]
Fig. 9A–B, Plate 1C

B. Mississippi north to northern Illinois and east through Ohio; costal grooves 14–15; throat pigmented; to 50 mm TL; sympatric: *Ambystoma "cingulatum"*
.... *Ambystoma texanum* (**Small-mouthed Salamander**)
Fig. 18B–D, Plate 2A

III. Nonflowing water; costal grooves 10–13

A. Coastal Plain from Apalachicola River east to central coastal South Carolina; costal grooves 13–14; gaudy striped pattern distinct from all salamander larvae; dorsal fin unmarked; eye stripe prominent; throat nonpigmented; to

70 mm TL; sympatric: none of similar coloration ....
.... ***Ambystoma cingulatum*** (**Frosted Flatwoods Salamander**)
See Fig. 9A–B, Plate 1C

[Apalachicola River west to southwestern Alabama; as for *A. cingulatum*; sympatric: none of similar coloration ....
.... ***Ambystoma bishopi*** (**Reticulated Flatwoods Salamander**)]
Fig. 9A–B, Plate 1C

B. Central Wisconsin east to Quebec and south to latitude of central Kentucky; costal grooves 12–13; back banded; throat white; usually dark ground color with conspicuous lateral stripe with ragged edges; digits round with blunt tips; to 70 mm TL; sympatric: *Ambystoma maculatum*, *A. opacum*, *A. tigrinum*, and various triploids .... ***Ambystoma jeffersonianum*** (**Jefferson Salamander**)
Plate 1F

[Eastern Ontario east to Labrador and south to central Indiana; sympatric: *Ambystoma jeffersonianum*, *A. maculatum*, *A. opacum*, *A. tigrinum* ....
.... ***Ambystoma laterale*** (**Blue-spotted Salamander**)]
Fig. 12B, Plate 1G

C. Coastal Plain from southeastern South Carolina to southeastern Virginia; costal grooves 13; throat white; dark ground color; usually with distinct pale, lateral stripe; to 60 mm TL; sympatric: *Ambystoma maculatum*, *A. opacum*, *A. talpoideum*, *A. tigrinum* .... ***Ambystoma mabeei*** (**Mabee's Salamander**)
Fig. 13A–B, Plate 1H

D. Most of eastern United States except southeastern Coastal Plain, Florida, and prairie regions of Illinois and Wisconsin; costal grooves 11–13; throat white; dorsum uniformly pale or dark; digits round; to 62 mm TL; sympatric: *Ambystoma "jeffersonianum," A. mabeei*, *A. opacum*, *A. talpoideum*, *A. tigrinum* .... ***Ambystoma maculatum*** (**Spotted Salamander**)
Fig. 15E, Plate 1J

E. Most of eastern United States south of northern Indiana except for southern Appalachian Mountains and Florida Peninsula; costal grooves 11–13; throat uniformly but sparsely pigmented; dorsum uniformly pale to dark brownish, pale ventrolateral dots usually present; digits round; to 76 mm TL; sympatric: *Ambystoma mabeei*, *A. maculatum*, *A. talpoideum*, *A. tigrinum* ....
.... ***Ambystoma opacum*** (**Marbled Salamander**)
Fig. 16D, Plate 1K

F. Mississippi Embayment and Coastal Plain to central South Carolina, several disjunct populations north to Kentucky and Virginia; costal grooves 10–11; throat pigmented; dark ground color with pale lateral and ventrolateral stripes; dorsum of tail banded throughout larval life; eye line present; digits round with blunt tips; to 80 mm TL, larger as pedotype; sympatric: *Ambystoma mabeei*, *A. maculatum*, *A. opacum*, *A. tigrinum* ....
.... ***Ambystoma talpoideum*** (**Mole Salamander**)
Fig. 17D, F, Plate 1L

G. Coastal Plain from southern Mississippi to Maryland north through Alabama to northeastern Kentucky and throughout the central Midwest; costal grooves 12–13; throat white; uniformly gray to black or with various mottled patterns; digits flattened with pointed tips and lateral flanges; to 90 mm TL, larger as pedotype; sympatric: *Ambystoma maculatum, A. opacum, A. talpoideum* ......
...................................*Ambystoma tigrinum* (**Eastern Tiger Salamander**)

Fig. 19A, Plate 2B

TAXONOMIC ACCOUNTS

The taxonomic accounts summarize the salient points of morphology and natural history, list the pertinent literature, and provide a statement of general geographical range. Annotated faunal studies, gray literature, theses and dissertations, and websites are cited minimally. Size was estimated for many of the specimens in the photographs. The color of the typically clear to hyaline appearance of fins of amphibian larvae is changed by the color of the background on which they are photographed. In cases when it is difficult to impossible to differentiate among closely related species, we discuss multiple species in a common account. The accounts are arranged alphabetically, first by family and then by species.

Ambystomatidae (Mole and Giant Salamanders)

This family is widespread on the continent; robust body and 4 limbs and 4/5 digits normally proportioned; depressed head broadly rounded in dorsal view; digits with rounded toe tips keratinized or not, toes flattened, pointed, and flanged in *A. tigrinum* group; mandibular grooves do not bisect lower lip anteriorly; gular fold not attached to medial throat tissue; long, unbranched gill rami with numerous fimbriae; keratinized jaw sheaths sometimes present on lower jaw; eyes small and dorsal; neuromasts not obvious; lunged; balancers present except in *A. tigrinum* group; 2 genera; 20 species. *Ambystoma*: throughout the mainland; nonflowing water; dorsal fin medium and extends well onto back; toe tips not keratinized; metamorphic or pedotypic; to 70.0–100.0 mm TL as first-year larva, 130.0 mm TL as second-year larva, 330.0 mm TL as pedotype; 16 species. *Dicamptodon*: Pacific Northwest; flowing water; fleshy dorsal fin low and originates near tail-body junction; toe tips often keratinized; metamorphic, pedotypic, or paedomorphic; to 150.0 mm TL as larva, 300.0 mm TL as paedomorph; 4 species.

Ambystoma annulatum (**Ringed Salamander**)   **Fig. 7; PL 1A**

IDENTIFICATION Costal grooves 15–16; rather uniformly dark with pale lateral stripe; metamorph has silvery iridophores on dark dorsum and lacks the distinctive adult pattern; to 60.0 mm TL.

NATURAL HISTORY Ringed Salamanders deposit eggs in temporary and permanent ponds from September to April, with most clutches appearing between mid-September and late October. A clutch, consisting of 205–390 pigmented ova, each with 3 jelly layers, is laid in 2–45 or more masses that typically are attached

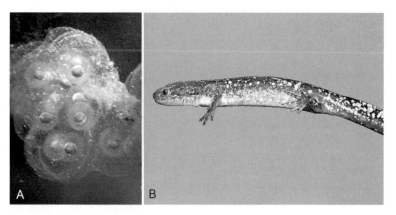

Figure 7. **Ambystoma annulatum.** (A) eggmass, (B) metamorph (46.0 mm TL; Stone Co., MO: RA). See also Plate 1A.

to vegetation or placed on the bottom. The eggs hatch in 10–21 days, and hatchlings measure 12.0–15.0 mm TL. The larvae grow to 40.0–60.0 mm TL in 180–255 days and attain 34.0–40.0 mm SVL before metamorphosis.

Reports (e.g., Strecker 1908a; see Trauth et al. 1989b) of this and other species ovipositing on land, either in secluded places or in a dry pond basin, likely result from individuals that are ready to breed before the ponds fill. These eggs with watery jelly layers have none of the adaptations for semiterrestrial oviposition as in *A. opacum.*

RANGE Ringed Salamanders occur in woodland habitats in south-central Missouri and adjacent Arkansas and Oklahoma.

CITATIONS *General*: J. D. Anderson 1965. *Development/Morphology*: Trauth et al. 2004. *Reproductive Biology*: Hutcherson et al. 1989, Noble and Marshall 1929, C. L. Peterson et al. 1991, Spotila and Beumer 1970, Strecker 1908a, Trapp 1956, Trauth et al. 1989b, 2004. *Ecobehavior*: Mathis et al. 2008, Mills and Barnhart 1999, Mills et al. 2001, Nyman et al. 1993.

Ambystoma barbouri (Streamside Salamander)   Fig. 8; PL 1B

IDENTIFICATION Costal grooves 14–15; freckled gray or brown with dorsum of tail muscle banded throughout larval ontogeny, although faint in larger larvae; metamorphs uniform dorsally to faintly mottled brown; to 50.0 mm TL.

NATURAL HISTORY Streamside Salamanders usually oviposit in low-gradient rocky streams from late December to mid-April. Pigmented ova (N = 10–1000, mean = 260, 2 jelly layers, OD 2.4–3.8 mm) are deposited as an array, usually on the bottom sides of flat rocks. The eggs hatch in 1.4–2.1 months at about 12.0 mm TL and metamorphose in 3.8–10.3 weeks (depending on temperature) at 37.0–41.0 mm TL.

RANGE *A. barbouri* occurs in north-central Kentucky and adjacent parts of Indiana, Ohio, and West Virginia. Several disjunct populations have been reported from sites in western and central Kentucky and central Tennessee.

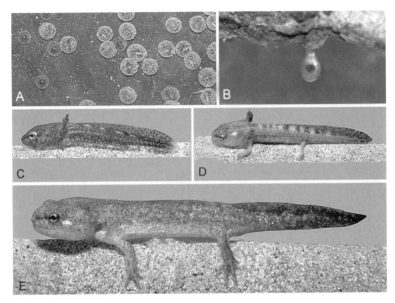

Figure 8. Ambystoma barbouri. (A) single suspended eggs on the bottom of a rock, (B) single suspended egg on a twig, (C) larva (19.0 mm TL, Perry Co., IN), (D) larva (26.0 mm TL, Fayette Co., KY), (E) metamorph (55.0 mm TL, Fayette Co., KY; all RA). See also Plate 1B.

CITATIONS *General*: Kraus 1996. *Development/Morphology*: Brandon 1961. *Reproductive Biology*: Kraus and Petranka 1980, Maurer and Sih 1996, Niemiller et al. 2009a, Regester and Miller 2000, Sih and Maurer 1992, Venesky and Parris 2009. *Ecobehavior*: Dupré and Petranka 1985, Garcia et al. 2003, Holomuzski 1989, 1991, Kats and Sih 1992, R. D. Moore et al. 1996, Petranka 1982, 1984b–c, Petranka et al. 1987, Petranka and Sih 1987, Sih and Kats 1991, Sih and Moore 1993, Sih and Petranka 1988, Storfer et al. 1999.

Ambystoma cingulatum (Frosted Flatwoods Salamander) and *A. bishopi* (Reticulated Flatwoods Salamander) Fig. 9; PL 1C

IDENTIFICATION Costal grooves 13–14; distinctive pattern includes even-edged, dark to pale brown or russet lateral stripes extending along body throughout larval life; metamorphs are mottled or splotched with silvery iridophores; pale lateral stripe of larva may persist; generally lack distinctive adult pattern; to 79.0 mm TL.

NATURAL HISTORY Flatwoods Salamanders congregate in temporary often swampy sites between October and March and oviposit clutches of 20–225 eggs (OD 2.0–2.6 mm). The eggs are laid singly or in small clumps in secluded places at the bases of grass tussocks, sometimes before the pond fills. Eggs hatch in about 2 weeks, and the larvae grow 1.7–2.5 mm/day for 2.5–4.2 months. Metamorphosis occurs over a 5–10-day period at 35.0–46.0 mm SVL. The larvae hide

Figure 9. **Ambystoma bishopi** and *A. cingulatum.* (A) larva of *A. cingulatum* (38.0 mm TL, Liberty Co., GA; DJS), (B) metamorph of *A. bishop* (52.0 mm TL, Okaloosa Co., FL; RA). See also Plate 1C.

among bottom debris during the day and may swim upward if disturbed. At night, they often stratify in the water.

RANGE   *A. bishopi* occurs in similar habitats west of the Apalachicola River into southeastern Alabama. *A. cingulatum* lives in seasonally flooded pine flatwoods habitats on the Coastal Plain from South Carolina to northern Florida and west to the Apalachicola River.

CITATIONS **General**: Martof 1968. **Development/Morphology**: Goin 1950, Mecham and Hellman 1952, Orton 1942, Telford 1954. **Reproductive Biology**: J. D. Anderson and Williamson 1976, Palis 1995, Sekerak et al. 1996. **Ecobehavior**: Bevelhimer et al. 2008, D. C. Bishop et al. 2006, Palis 1997, Palis et al. 2006, Whiles et al. 2004.

Ambystoma californiense (California Tiger Salamander) Fig. 10; PL 1D

IDENTIFICATION   Costal grooves 11–13; head notably depressed; digits flat and pointed with lateral flanges; uniformly silvery white in turbid water to bluish green with subtle mottling and reticulations in clear water; metamorph resembles adult; to 180.0 mm TL.

NATURAL HISTORY   California Tiger Salamanders usually breed between November and March in ponds in grassland and oak woodland habitats of California. Clutches with a maximum of 1340 pigmented ova are oviposited in temporary ponds that lack fish. The eggs have 3 jelly layers and are oviposited as singles or small groups attached to vegetation or scattered on the bottom. The diameter of an egg (ED) is about 3.5 mm and eggs produce hatchlings in 2–3 weeks at 10.0–13.0 mm TL. Metamorphosis occurs in 3.3–4.0 months at 58.0–76.0 mm TL. Females produce 4.7–21.9 metamorphs per clutch.

RANGE *A. californiense* occurs at scattered localities in the Central Valley and along the central coast of California south to the Santa Rita Hills.

CITATIONS **Reproductive Biology**: K. S. Baldwin and Stanford 1987, Loredo and Van Vuren 1996, Storer 1925, Trenham et al. 2000, Twitty 1941. **Ecobehavior**: Alvarez 2004, J. D. Anderson 1968c, Balfour and Stitt 2003.

Ambystoma gracile (Northwestern Salamander)   Fig. 11; PL 1E

IDENTIFICATION Costal grooves 12–13; pedotype with obvious granular glands on top of head and along crest of tail muscle; hatchling prominently striped; larva

Figure 10. **Ambystoma californiense.** (A) large larva from a turbid site (97.0 mm TL; Santa Barbara Co., CA; RA). See also Plate 1D.

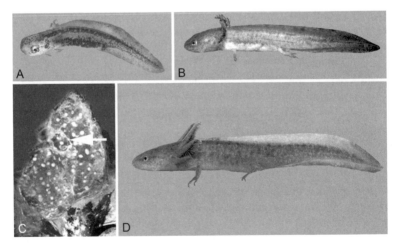

Figure 11. **Ambystoma gracile.** (A) hatchling (14.0 mm TL, Skamania Co., WA; LW), (B) small larva (68 mm TL, note missing left hand, Benton Co., OR; RA), (C) egg mass (Humboldt Co., CA; arrow = attack site of caddisfly larvae; LVD), (D) large larva (97.0 mm TL, note regenerated hind leg, Pierce Co., WA; WPL). See also Plate 1E.

brownish, sometimes greenish to yellow with blotches on tail fins; metamorph resembles adult; to 130.0 mm TL as larva, 228.0 mm TL as pedotype.

NATURAL HISTORY The breeding season of the Northwestern Salamander varies depending on elevation and latitude but oviposition occurs in both clear and turbid ponds or slow-moving streams usually between January and August. An egg mass consists of 30–270 pale brown and cream ova (OD 1.5–3.0 mm, 3 jelly layers), and sometimes is laid in smaller groups of 25–30; it consists of a voluminous, stiff (see *A. maculatum*) jelly matrix, the inner layers of which are sometimes white. A symbiotic alga (see *A. maculatum*) often invades the mass between the egg layers and commonly proliferates at stages prior to hatching. Eggs hatch in 1–2 months at 14.0–20.0 mm TL, and larvae metamorphose at about 50.0 mm TL if in first year, but some may overwinter. Larvae are commonly pedotypic at both lowland and highland sites and can reach densities of 175–1678/ha. Large individuals have granular glands on the back of the head and along the ridge of tail muscle. At high-elevation sites that lack fish, pedotypes spend considerable time resting on the bottom in open water.

RANGE *A. gracile* occurs from southeastern Alaska south to just north of San Francisco Bay, primarily west of the crest of the Sierra Nevada–Cascade ranges.

CITATIONS **General**: Snyder 1963. **Development/Morphology**: Hagen and Wilson 1999, Licht and Sever 1993. **Reproductive Biology**: Eagleson 1976, Farner and Kezer 1953, Henry and Twitty 1940, Knudsen 1960, MacCracken 2007, 2008, Marco 2001, Marco and Blaustein 1998, 2000, R. C. Snyder 1956, 1960, Watney 1941. **Ecobehavior**: J. D. Anderson 1972a, H. A. Brown 1976, Goff and Stein 1978, B. A. Henderson 1973, R. L. Hoffman and Larson 1999, J. E. Johnson et al. 2007, Licht 1975a, 1992, Neish 1971, Pearl 2003, Pearman 2002, Slater 1936, J. T. Taylor 1983a–b, 1984, T. J. Tyler et al. 1998a–b, Walls et al. 1996.

Ambystoma jeffersonianum (Jefferson Salamander) and *A. laterale* (Blue-spotted Salamander)   Fig. 12; PL 1F,G

IDENTIFICATION These two species have triploid female populations associated with them at some sites. Costal grooves 12–14; hatchling uniformly dark, sometimes greenish with a tinge of yellow on sides of head and neck; small larva with dark blotches usually in longitudinal rows along the back, often with pale, ragged-edged, lateral stripe. Older larva mottled, often with lateral stripe; metamorphs resemble adults; to 75.0 mm TL.

NATURAL HISTORY These salamanders are early spring breeders; gravid females arrive at ponds and pools in eastern deciduous forests from January to April, usually later at more northern sites. Females oviposit 150–250 pigmented ova (OD 2.0–2.5 mm, 1 jelly layer) as singles or small groups (2–40 ova/mass; a clutch comprises up to 16 masses). A mass may have a diameter of 25.0–75.0 mm and is usually attached to or surrounds a vertical piece of vegetation well below the water surface. The vitelline membrane is closely adpressed to an ovum. Eggs hatch in 2.0–6.4 weeks at 10.0–14.0 mm TL, and larvae undergo metamorphosis in 1.7–4.2 months at 40.0–55.0 mm TL.

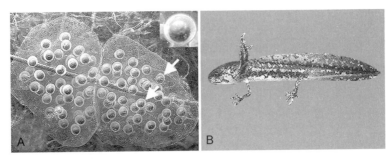

*Figure 12. **Ambystoma jeffersonianum** and A. laterale.* (A) egg mass of *A. jeffersonianum* (Hocking Co., OH; showing typical ambystomatid peripheral ova-free zone; arrows and inset show ova off-center within jelly layers because of the formation of the capsular chamber; DMD), (B) larva of *A. laterale* (45.0 mm TL, Cook Co., IL; RA). See also Plate 1F, G.

RANGE The composite ranges of *A. jeffersonianum* and *A. laterale* include most of North America south of James Bay, Ontario, east of eastern Manitoba, and south to central Kentucky. Jefferson Salamanders generally occupy the southern part of the range, and the Blue-spotted Salamanders occur at more northern sites. These species seldom if ever occur sympatrically, and diploid, triploid, and tetraploid unisexuals occur in a complicated pattern along with the diploid sexual individuals.

CITATIONS ***General***: Uzzell 1967a–d. ***Development/Morphology***: Brandon 1961, Panek 1978, Valentine and Dennis 1964. ***Reproductive Biology***: Bleakney 1957, Bolek 1998, Collins and Wilbur 1979, Parmalee et al. 2002, Piersol 1910, B. G. Smith 1911a, Stille 1954. ***Ecobehavior***: J. D. Anderson and Graham 1967, K. F. Baldwin and Calhoun 2002, Bardwell et al. 2007, Bond 1960, Branch and Altig 1983, Brodman 1995, 1996, 1999, 2004, Brodman and Jaskula 2002, Mohr 1931, J. A. Moore 1939, Nyman 1991, O'Donnell 1937, Rowe and Dunson 1995, Rubbo et al. 2006, Stauffer et al. 1983, E. L. Thompson and Gates 1982, E. L. Thompson et al. 1980, Van Buskirk and Smith 1991, Walls and Williams 2001, B. Walters 1975, K. S. Wells and Harris 2001, Wilbur 1971, 1972, 1976, 1977a.

Ambystoma mabeei (Mabee's Salamander)   Fig. 13; PL 1H

IDENTIFICATION Costal grooves 13; conspicuous pale lateral stripe at hatching, larger larvae brown to black with pale, often broken lateral stripes with ragged edges; dorsal fin heavily mottled in larger larvae; eye stripe faint; metamorph dark with flecking over the dorsum; remnant of the lateral stripe; pale area on the dorsal side of the tail; black mottling dorsally; to 63.0 mm TL.

NATURAL HISTORY Mabee's Salamanders breed in ponds and pools on the Atlantic Coastal Plain from January to March. The pigmented ova are oviposited as singles or small clumps of 2–6 (2 jelly layers) and hatch in 1.3–2 weeks at 8.0–9.0 mm TL. Larvae metamorphose after 2 months at 50.0–63.0 mm TL.

RANGE *A. mabeei* occurs around vernal ponds in bottomland forests and Carolina Bays on the Coastal Plain of Virginia and the Carolinas.

Figure 13. **Ambystoma mabeei.** (A) larva (48.0 mm TL), (B) metamorph (58.0 mm TL) from York Co., VA (RA). See also Plate 1H.

CITATIONS *General*: J. D. Hardy and Anderson 1970. *Ecobehavior*: J. D. Hardy 1969a–b, McCoy and Savitsky 2004.

Ambystoma macrodactylum (Long-toed Salamander)   Fig. 14; PL1I

IDENTIFICATION Costal grooves 12–13; quite variable in appearance throughout range, but generally rather uniformly dark to pale brown, sometimes greenish in larger individuals; metamorphs rather uniform and lack adult pattern; to 90.0 mm TL.

NATURAL HISTORY Long-toed Salamanders are found in a variety of woodland and forest habitats around ponds and slow streams from southeastern Alaska to central California. Oviposition occurs over a long period from October to July in a variety of habitats depending on latitude and elevation (sea level to near 2700 m). Egg masses typically consist of 85–415 black to brown and gray to white ova (OD 2.0–2.5 mm, ED 1.2–1.7 mm, 2 jelly layers); however, ovipositional modes vary with no obvious geographic pattern. Sometimes eggs are laid singly, in small groups of 8–10, or in small masses of 40 or more ova, and sometimes they are attached to twigs or vegetation, deposited on the open bottom, or placed on rocks. At hatching, larvae are 9.0–15.0 mm TL and metamorphose in 1.3–2 months at 48.0–90.0 mm TL.

RANGE *A. macrodactylum* occurs from southeastern Alaska, British Columbia, and southwestern Alberta south through southern Idaho and north-central California. A small disjunct population also occurs south of San Francisco Bay.

CITATIONS *General*: D. E. Ferguson 1963. *Development/Morphology*: Watson and Russell 2000. *Reproductive Biology*: J. D. Anderson 1967d, Farner and Kezer

Figure 14. Ambystoma macrodactylum. (A) silt-covered egg mass (Shoshone Co., ID; arrow = exposed jelly of one egg; RA), (B) small mass of 3 eggs attached to a stem (Thurston Co., WA; WPL), (C) egg masses (Multnomah Co., OR; CCC), (D) hatchling (26.0 mm TL, Chelan Co., WA; WPL), (E) metamorph (61.0 mm TL, Shoshone Co., ID; RA). See also Plate 1I.

1953, Kezer and Farner 1955, Knudsen 1960, Howard and Wallace 1985. *Ecobehavior*: J. D. Anderson 1968a, c, 1972a–b, Chivers et al. 1997, Pearman 2002, Slater 1936, T. J. Tyler et al. 1998a–b, Vonesh and de la Cruz 2002, Walls et al. 1993a–b, Wildy 2001, Wildy et al. 1998, 1999, Wildy and Blaustein 2001.

Ambystoma maculatum (Spotted Salamander)   Fig. 15; PL 1J

IDENTIFICATION Costal grooves 11–13; hatchling crudely striped laterally; larva uniformly black to pale brown; metamorph dark with abundant iridophores and without the distinctive adult pattern; to 75.0 mm TL.

NATURAL HISTORY Spotted Salamanders oviposit from November through May in lowland deciduous forest ponds and pools and sporadically in upland habitats. Migration to breeding ponds typically begins after the first warm rains, and breeding dates vary with latitude and temperature. Oviposition in southern populations usually begins in December and January while northern populations often don't start breeding until April. Clutches consist of 150–300 pigmented ova

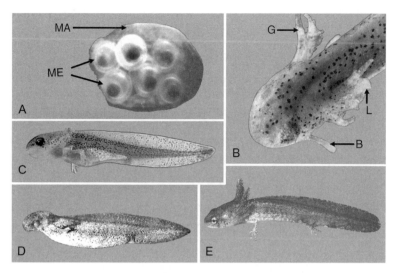

Figure 15. ***Ambystoma maculatum.*** (A) portion of an egg mass showing eggs with (ME) membranes embedded in stiff (MA) matrix (Noxubee Co., MS; RA), (B) head of a hatchling (B = balancer, G = gill, L = limb bud; MA; AMR), (C) hatchling (18.0 mm TL, stage 45, MA; AMR), (D) hatchling (11.0 mm TL, stage 36, Noxubee Co., MS; RA), (E) larva (40.0 mm TL, Will Co., IL; MR). See also Plate 1J.

with ED 5.0–6.0 mm; the 2 jelly layers are clear or milky white, and masses often harbor a symbiotic chlamydomonad alga, *Oophila ambystomatis.* Clutches may be deposited as several masses (60.0–105.0 mm diameter) that contain a few (35–90) or many (~250) eggs. Masses hatch in 4–7 weeks at 11.0–17.0 mm TL and larvae grow to 40.0–75.0 mm TL in 2–4 months and metamorphose at 35.0–75.0 mm TL. In some populations larvae may overwinter. The voluminous jelly matrix may persist for up to a month after hatching occurs.

The variability from clear to opaque white egg matrix in Spotted Salamanders is unique, and the degree of infestation of the inner jelly layers by the alga is unusual. Only *A. gracile* has similarly firm egg jelly. Larvae, commonly accompanied by those of *A. opacum* and *A. talpoideum,* hide in bottom debris during the day and at night are often observed stratified and hanging in the midwater with legs splayed as they quietly feed on passing zooplanktors.

RANGE *A. maculatum* is widely distributed through much of eastern North America from Nova Scotia south to Georgia and central Ontario south to eastern Texas.

CITATIONS *General*: J. D. Anderson 1967c. ***Development/Morphology***: Brandon 1961, S. F. Clarke 1880, Dodd 2004, Eycleshymer 1893, R. G. Harrison 1969 (reprint Duellman and Trueb 1986:130), Olsson and Löfberg 1992, Phillips 1992, Severinghaus 1930, Talentino and Landre 1991, Trauth et al. 2004. ***Reproductive Biology***: Banta and Gortner 1914, Beachy 1993a, Cliburn 1972, Collins and Wilbur 1979, Gibson and Merkle 2005, P. W. Gilbert 1942, 1944,

D. M. Hillis and Miller 1976, Keen 1975, Komoroski et al. 1998, Metts 2001, Murray 1962, Nyman 1987, Parmalee et al. 2002, Portnoy 1990, Regester et al. 2006, B. G. Smith 1907b, 1911a, L. Smith 1920, E. L. Thompson and Gates 1982, Wood and Wilkinson 1952, Woodward 1982c, A. H. Wright 1908, A. H. Wright and Allen 1909. *Ecobehavior*: Albers and Prouty 1987, A. R. Anderson and Petranka 2003, J. D. Anderson and Graham 1967, Bachmann et al. 1986, K. F. Baldwin and Calhoun 2002, Branch and Altig 1981, 1983, Branch and Taylor 1977, Brodman 1995, 1996, 1999, 2004, Brodman and Jaskula 2002, Bruce et al. 1994, Cargo 1960, Cliburn and Ward 1963, Dempster 1930, 1933, DuShane and Hutchinson 1944, Figiel and Semlitsch 1990, Formanowicz and Brodie 1982, Freda 1983, Gatz 1973, Gomez-Mestre et al. 2006, L. M. Hardy and Lucas 1991, Hay 1889a, Hoff et al. 1985, Holbrook and Petranka 2004, Hutchinson and Hewitt 1935, Hutchison 1971, Hutchison and Hammen 1958, P. H. Ireland 1973, 1989, Kaplan 1980a, Landberg and Azizi 2010, Mills and Barnhart 1999, J. A. Moore 1939, Murphy 1961, Nyman 1991, O'Donnell 1937, Petranka et al. 1998, Pinder and Friet 1994, C. H. Pope 1964, Pough 1976, Regester et al. 2008, Rowe and Dunson 1995, Ruth et al. 1993, Schneider 1968, Seale 1980, Semlitsch 1987b, Semlitsch and Walls 1993, Shoop 1974, Showalter 1940, Stauffer et al. 1983, Stenhouse et al. 1983, Tomson and Ferguson 1972, Turtle 2000, S. R. Voss 1993a, Walls 1996, 1998, Walls and Altig 1986, Walls and Jaeger 1987, 1989, Walls and Semlitsch 1991, Walls and Williams 2001, B. Walters 1975, Ward and Sexton 1981, Wassersug 1989, Weigmann and Altig 1975, Whitford and Vinegar 1966, Wilbur 1972, 1976, 1977a, Wilbur and Fauth 1990, Worthington 1968, 1969.

Ambystoma opacum (Marbled Salamander)   Fig. 16; PL 1K

IDENTIFICATION   Costal grooves 11–14; demarcation between dark dorsal and pale ventral coloration diffuse and about at the plane of dorsal margins of limb insertions; metamorph uniformly gray to brown with very fine uniform iridophore flecking, lacks adult pattern; to 76.0 mm TL.

NATURAL HISTORY   Marbled Salamanders congregate and court in dry basins where temporary pools will form. After mating, females construct nests and deposit single eggs, eventually forming clutches of 39–215 pigmented eggs that she guards until the pond is inundated by seasonal rains. Oviposition occurs between September and February, with breeding occurring earlier in northern than southern populations. Individual ovum diameter varies from 1.8–2.9 mm and eggs measure 4.0–7.0 mm; eggs sometimes are slightly adherent and their jellies tough and not watery. Eggs hatch at 10.0–20.0 mm TL and larvae grow to 76.0 mm TL in 3–7 months. Metamorphosis is in 6–9 months at 35.0–70.0 mm TL.

Marbled Salamanders lay much earlier than other sympatric species of *Ambystoma*, and their larvae prey on embryos and early larvae of sympatric species that breed later. Larvae typically stratify in the water, and their color and color intensity vary widely with water clarity; ones in acidic, swampy sites are black and ones from turbid sites are almost white.

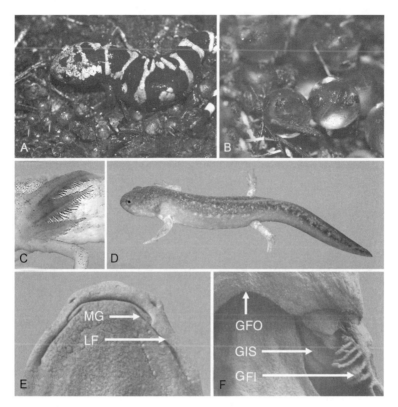

Figure 16. **Ambystoma opacum.** (A) female guarding independent eggs at a semiterrestrial site (TRK), (B) independent eggs (TRK), (C) gill structure typical of pond-type larva in nonflowing water, (D) metamorph (52.0 mm TL, Oktibbeha Co., MS; RA). SEM micrographs of the gular area of a larva: (E) MG = mandibular groove, LF = labial fold, (F) showing the GFO (gular fold), GIS (gill intrabranchial septa), and GFI (gill fimbriae) (C–F all RA). See also Plate 1K.

RANGE *A. opacum* occurs throughout most of the eastern United States from southern Illinois, southeastern Oklahoma and east Texas east to New England and Florida except for the Florida peninsula.

CITATIONS **General**: J. D. Anderson 1967b. **Development/Morphology**: Brandon 1961, M. G. Brown 1942, Dodd 2004, Trauth et al. 2004. **Reproductive Biology**: Brimley 1920, Dunn 1917a, Hassinger et al. 1970, M. E. Jackson et al. 1989, Komoroski et al. 1998, Lantz 1930, Murray 1962, Noble and Brady 1930, Noble and Richards 1932, Palis 1996, Pike 1886b, Salthe 1963, Trauth et al. 1989c, 2004, Viosca 1924. **Ecobehavior**: J. D. Anderson 1972a, J. D. Anderson and Graham 1967, Boone et al. 2002, Branch and Altig 1981, 1983, Brodman and Jaskula 2002, Chazal et al. 1996, Doody 1996, Kaplan 1980a–b, Keen 1975, Keen et al. 1984, King 1935, J. C. Mitchell et al. 1996, J. A. Moore 1939, O'Donnell 1937, Petranka 1989b, Petranka et al. 1983, Petranka and Petranka 1980, C. H. Pope 1964, Regester et al.

2006, 2008, D. E. Scott 1990, 1994, D. E. Scott et al. 2007, D. E. Scott and Fore 1995, C. K. Smith 1990, Stenhouse et al. 1983, Stewart 1956, B. E. Taylor and Scott 1997, Tomson and Ferguson 1972, Walls 1991, 1995, Walls and Altig 1986, Walls and Blaustein 1994, 1995, Walls and Roudebush 1991, Walls and Williams 2001, B. Walters 1975, Weigmann and Altig 1975, Worthington 1968, 1969.

Ambystoma talpoideum (Mole Salamander)   Fig. 17; PL 1L

IDENTIFICATION Costal grooves 10; one of two *Ambystoma* larvae (see *A. cingulatum*) with distinctive coloration—dorsum of tail banded, prominent lateral and ventral stripes; metamorphs are evenly peppered with iridophores; indications of lateral stripe may persist through metamorphosis; to 82.0 mm TL.

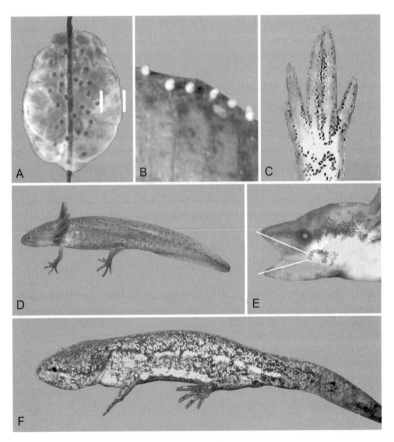

Figure 17. ***Ambystoma talpoideum.*** (A) egg mass typical of *Ambystoma* (Noxubee Co., MS; RA) with peripheral ova-free zone delimited by white lines, (B) single eggs attached to a leaf substrate (Aiken Co., SC; RDS), (C) foot of a larva (RA), (D) pedotype (117.0 mm TL, Oktibbeha Co., MS; RA), (E) larva with open mouth showing extent of labial folds and (white lines) gape of upper and lower jaws (RA), (F) metamorph (72.0 mm TL, Oktibbeha Co., MS; RA). See also Plate 1L.

NATURAL HISTORY   Mole Salamanders typically migrate during rainy periods to a variety of ponds and temporary pools without fish in or near forests. Females typically oviposit between December and March. Clutches are usually attached to twigs in the ponds and consist of 225 to more than 600 pigmented ova with 2 very watery jelly layers. Eggs are often laid in several masses of 4–45 ova; in some populations (e.g., Atlantic Coastal Plain), single eggs may be scattered on the pond bottoms (Fig. 17B). Eggs hatch in 2–7 weeks at about 10.0 mm TL, and larvae grow to 72.0–82.0 mm TL in 2–4 to 15 months. Metamorphosis occurs at 32.0–82.0 mm TL and the metamorphs resemble the adults.

In some populations, pedotypes (Fig. 17F) remain for one or more years if water persists; these individuals get much larger than larvae destined to metamorphose in their first year (Fig. 17D) and differ in coloration and size.

RANGE   *A. talpoideum* occurs from southern Illinois to southeastern Oklahoma and east Texas and east to South Carolina and northern Florida.

CITATIONS   *General*: Shoop 1964. *Development/Morphology*: Orton 1942, Semlitsch and Gibbons 1985, Trauth et al. 2004, Volpe and Shoop 1963. *Reproductive Biology*: Dodd 2004, Komoroski and Congdon 2001, Komoroski et al. 1998, Mosimann and Uzzell 1952, Patterson 1978, Raymond and Hardy 1990, T. J. Ryan and Plague 2004, T. J. Ryan and Semlitsch 2003, Semlitsch 1985, 1987d, Winne and Ryan 2001. *Ecobehavior*: J. D. Anderson and Williamson 1974, Boone et al. 2002, Branch and Altig 1981, J. P. Caldwell et al. 1980, Cliburn and Carey 1975, Fauth 1999, R. N. Harris et al. 1995, M. E. Jackson and Semlitsch 1993, Keen et al. 1984, McAllister and Trauth 1996, Mills et al. 2001, O'Donnell 1937, T. J. Ryan and Swenson 2001, D. E. Scott 1993, D. E. Scott et al. 2007, Semlitsch 1987a–c, Semlitsch and Reichling 1989, Semlitsch et al. 1988, 1990, Semlitsch and Walls 1993, Semlitsch and Wilbur 1988, 1989, Shoop 1960, B. E. Taylor et al. 1988, Trauth et al. 1993, 1995b, Walls 1995, 1996, Walls and Altig 1986, Walls and Jaeger 1987, 1989, Walls and Semlitsch 1991.

Ambystoma texanum (Small-mouthed Salamander)   Fig. 18; PL 2A

IDENTIFICATION   Costal grooves 14–15; demarcation between dark dorsal and pale ventral color discrete and ventral to the ventral margins of limb insertions; metamorph uniformly gray; to 75.0 mm TL.

NATURAL HISTORY   Small-mouthed Salamanders breed in ephemeral ponds, pools, and roadside ditches in woodland and prairie habitats and occasionally in sluggish streams from December to April. Clutches include 150–813 pigmented ova (OD about 2.0 mm, ED 6.0–6.5 mm, 2 jelly layers) laid as flimsy masses of 6–30 ova, sometimes singly or in small groups attached to twigs, leaves, and other vegetation and debris in ponds. Eggs develop for 2–8 weeks depending on temperature and hatch at about 12.0 mm TL. Larvae grow to 75.0 mm TL and metamorphose in 3–4 months at 35.0–75.0 mm TL.

RANGE   *A. texanum* occurs from Nebraska south to the Gulf of Mexico and east to Ohio and Alabama; it is absent from the Ouachita-Ozark region.

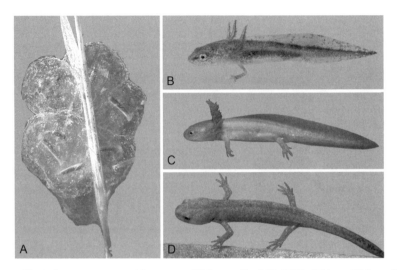

Figure 18. ***Ambystoma texanum.*** (A) egg mass (Livingston Co., MO; DLD), (B) larva (23.0 mm TL, Douglas Co., KS; RWV), (C) larva (34.0 mm TL, Union Co., IL; RA), (D) metamorph (48.0 mm TL, Union Co., IL; RA). See also Plate 2A.

CITATIONS *General*: J. D. Anderson 1967a. *Development/Morphology*: Brandon 1961, Trauth et al. 2004. *Reproductive Biology*: Burger 1950, Minton 1972, N. J. Moore and Matson 1997, Trauth et al. 2004. *Ecobehavior*: Bragg 1957c, Branch and Altig 1983, Garcia et al. 2003, Hay 1889a, Keen 1975, Maurer 1996, Maurer and Sih 1996, McWilliams and Bachmann 1989a–b, O'Donnell 1937, Parmalee et al. 2002, Petranka 1982, 1984b, 1985, Petranka and Six 1987, Punzo 1983, T. J. Ryan 2007, Seale 1980, Strecker 1909, Trauth et al. 1990, Whitaker et al. 1982, Wilbur 1972.

Ambystoma tigrinum (Eastern Tiger Salamander) and *A. mavortium* (Western Tiger Salamander)   Fig. 19; PL 2B

IDENTIFICATION Costal grooves 11–14; digits flattened with pointed tips and lateral flanges; highly variable throughout range and strongly influenced by habitat conditions—uniformly silvery to gray; sometimes lightly to boldly mottled; large coloration variations among subspecies; metamorphs from most populations resemble adults; 70.0–100.0 mm TL as first-year larva, 130.0 mm TL as second-year larva, 330.0 mm TL as pedotype.

NATURAL HISTORY Tiger Salamanders breed in a variety of permanent and temporary ponds that lack fish in many areas of the country from sea level to about 3300 m elevation. Depending on taxon, elevation, and latitude, oviposition has been reported in almost every month. *A. tigrinum* females usually deposit masses (25–165 ova/mass, perhaps 7000/clutch; OD 2.0–4.0 mm, ED 4.5–10.0 mm) with large amounts of flimsy jelly and attach them to vegetation in deeper water.

Figure 19. **Ambystoma mavortium and A. tigrinum.** (A–E) *A. mavortium*: (A) hatchling (12.0 mm TL, Evans Co., GA; DJS), (B) foot (RA) with lateral flanges (arrows) on flattened, pointed toes, (C) carnivore consuming a conspecific larva (DWP), (D) single eggs attached to vegetation (Grant Co., WA; CCC), (E) pedotype (185.0 mm TL, Yavapai Co., AZ; RA) of *A. mavortium*, (F) egg mass of *A. tigrinum* (Menard Co., IL; RA), (G) larva of *A. mavortium* (82.0 mm TL, Webster Co., MO; blanched coloration at night; RA). See also Plate 2B.

Hatching occurs at about 9.0 mm TL. As larvae, salamanders grow to about 70.0 mm TL and to 300.0 mm TL as pedotypes. Larval life varies depending on location but typically lasts 2.5–5.0 months. *A. mavortium* lays individual eggs or small groups attached to vegetation that are often arranged linearly. An egg mass may have 9–170 eggs with black to brown and gray to white ova with an OD about 2.0–3.0 mm, an ED 5.0–12.0 mm, and 3 jelly layers. Hatching occurs in 2.0–3.6 weeks at 14.0–18.0 mm TL, and metamorphosis may be relatively quick (3–4 months) or take more than 33 months. Depending on the time spent as a larva, a metamorph may be 55.0–130.0 mm TL. Larvae may also be pedotypic for various lengths of time, and others may metamorphose partially but retain a tail fin and stay in the water. Large, voracious, fast-growing larvae will eat almost anything slightly smaller than themselves and strongly influence community structure. Several life history variants involve various degrees of pedotypy (e.g., ponds in eastern United States and llanos of west Texas) and carnivorous morphotypes; carnivores have enlarged heads and sometime modified teeth.

RANGE The Eastern Tiger Salamander occurs from extreme southeastern Manitoba south to eastern Texas and southeastern New York south to northern Florida and southeastern Louisiana, excluding the Mississippi Embayment and the Appalachian Mountains. The Western Tiger Salamander occurs in isolated populations in southern British Columbia, eastern Washington and Idaho, and the Great Plains of Alberta, Saskatchewan, and Manitoba south to northern and southeastern Arizona and into Mexico and east to eastern South Dakota, eastern Kansas, Oklahoma, and southern Texas. Because of their value as fish bait, larvae of these taxa have been introduced into many areas in the western United States.

CITATIONS *General*: Gehlbach 1967. *Development/Morphology*: Banta 1912, Brandon 1961, Fernandez and Collins 1988, Hoy 1871, Olsson and Löfberg 1992, Pierce et al. 1983, Powers 1903, 1907, Sexton and Bizer 1978, Sheen and Whiteman 1998, W. W. Tanner et al. 1971, Trauth et al. 2004, Whiteman et al. 1998. *Reproductive Biology*: J. D. Anderson et al. 1971a–b, Arndt 1989, Burger 1950, Collins and Wilbur 1979, Hamilton 1948, Hassinger et al. 1970, Kaplan 1979, 1980b, Leonard and Darda 1995, Lindberg 1995, Micken 1968, Parmalee et al. 2002, Regester et al. 2006, B. G. Smith 1911a, Tucker 1999, Webb and Roueche 1971. *Ecobehavior*: J. D. Anderson 1972a, J. D. Anderson and Graham 1967, Bizer 1978, Brandon and Bremer 1967, Brodman 2004, Brodman and Jaskula 2002, T. E. Brophy 1980, Brunkow and Collins 1998, Collins 1981, Collins and Cheek 1983, Collins et al. 1993, Collins and Holomuzski 1984, Dalrymple 1970, De Neff and Sever 1977, Denoël et al. 2006, Dodson and Dodson 1971, Duncan 1999, Ghioca and Smith 2008, Glass 1951, Hay 1889a, E. A. Hoffman and Pfennig 1999, Holomuzski 1986a, Hutchinson and Hewitt 1935, Hutchison 1971, Kaplan 1979, 1980a–b, Keen at al. 1984, Lannoo and Bachmann 1984a–b, K. L. Larson et al. 1999, W. Larson 1968, Lee and Franz 1974, Leff and Bachmann 1986, 1988, Loeb et al. 1994, Micken 1971, J. A. Moore 1939, Mould and Sever 1984, Nietfeldt et al. 1980, Norris 1989, O'Donnell 1937, Parris et al. 2005, Pfennig and Collins 1993, Pfennig et al. 1994, 1998, 1999, Regester et al. 2008, Reilly et al. 1992, F. L. Rose and Armentrout 1976, Seale 1980, Semlitsch 1983b, Sever et al. 1987, Shrode 1972, K. M. Smith and Ghioca-Robrecht 2008, Sredl and Collins 1991, 1992, Storfer and White 2004, D. H. Taylor 1972, Tomson and Ferguson 1972, Trauth et al. 1990, J. D. Tyler and Buscher 1980, Ultsch et al. 2004, Ward and Sexton 1981, Wassersug and Seibert 1975, Whiteman and Brown 1996, Whiteman et al. 1995, 1996, 1998, 2003, 2012, Wilbur 1972, 1977a, Woodward and Johnson 1985, Ziemba et al. 2000, Zerba and Collins 1992.

Dicamptodon aterrimus (**Idaho Giant Salamander**), *D. copei* (**Cope's Giant Salamander**), *D. ensatus* (**California Giant Salamander**), **and** *D. tenebrosus* (**Coastal Giant Salamander**)   **Fig. 20; PL 2C–E**

IDENTIFICATION Costal grooves 11–13; coloration quite variable with size and region, but ranges from pale to dark brown with abundant iridophore frosting and marbling; posterior part of tail often with contrasty marks; belly of *D. copei* darker than sympatric *D. tenebrosus*; metamorphs, and sometimes large larvae, usually

Figure 20. **Dicamptodon spp.** (A) egg array and attendant parent (Lincoln Co., OR; EDB), (B) head of a pedotype (283.0 mm TL, Lane Co., OR; RA), (C) egg array (Humboldt Co., CA; LVD), A, B, C of *D. tenebrosus*, (D) larva of *D. aterrimus* (170.0 mm TL, Latah Co., ID; CRP), (E) larva of *D. tenebrosus* (74.0 mm TL, Deschutes Co., OR; RA). See also Plate 2C–E.

show indication of reticulate pattern of adults that starts on the head and develops posteriorly; to 150.0 mm TL as larva, greater than 300.0 mm TL as pedotype.

NATURAL HISTORY Female Giant Salamanders live in mountain streams and deposit and guard eggs under large rocks or within rubble of mountain streams. Based mostly on *D. tenebrosus*, oviposition occurs from May to March when 70–200 (23–28 in *D. copei*, 135–200 in *D. aterrimus*), large (OD 6.0–6.9 mm, ED in long dimension to 33 mm), nonpigmented eggs with 5 jelly layers (2 visible by unaided eye) with a short pedicel involving outer 2 jelly layers, are laid as a suspended to pendent array. Eggs hatch in about 9.2 months at 16.0–40.0 mm TL; larvae may subsist on residual yolk for up to 9 months and usually metamorphose in 2–3 years at 85.0–150.0 mm TL. Eggs described by Storer (1925) were incorrectly identified. Most individuals in most populations metamorphose, but pedotypy is common in some areas with individuals growing to 350.0 mm TL. Individuals of *D. copei* seldom metamorphose, and pedotypes mature at about 200.0 mm TL.

RANGE *D. aterrimus* is known from much of northern Idaho and adjacent Montana. *D. copei* occupies streams in western Washington and adjacent Oregon.

D. ensatus is restricted to the Coast Range from northwestern California to just south of San Francisco Bay. *D. tenebrosus* occurs from southwestern British Columbia to northern California west of the Cascade Mountains.

CITATIONS **General**: J. D. Anderson 1969. **Development/Morphology**: Kessel and Kessel 1943a–b, 1944, Nussbaum 1970, Schuierer 1958, Wake and Shubin 1998. **Reproductive Biology**: Dethlefsen 1948, T. Evans et al. 2005, Henry and Twitty 1940, L. L. C. Jones et al. 1990, Nussbaum 1969a, Steele et al. 2003a. **Ecobehavior**: Antonelli et al. 1972, S. C. Bishop 1943, Feral et al. 2005, H. M. Ferguson 2000, H. S. Fitch 1936, Franz 1970c, C. R. Johnson and Schreck 1969, Loafman and Jones 1996, Metter 1963, Nussbaum and Clothier 1973, M. S. Parker 1991, 1994, Rundio and Olson 2001, K. R. Russell et al. 2002, Sagar et al. 2007, Salthe 1963, Schuierer 1958, Storer 1925 (eggs are of *Ambystoma gracile*, see Henry and Twitty 1940), Welsh and Lind 2002.

Amphiumidae (Amphiumas)

Members of this family occur east of the Rocky Mountains; nonflowing water, burrower with distinctly elongate, cylindrical body with four disproportionately tiny limbs with 1–3 digits without keratinized tips; snout depressed, somewhat pointed in dorsal view; mandibular grooves do not bisect lower lip anteriorly; gular fold absent; gills with branched rami and few fimbriae present for short period after hatching; gill slit 1; keratinized jaw sheaths absent; eyes small and dorsal; neuromasts not obvious on small individuals; lungs present; balancers absent; dorsal fin on tail low and faint in hatchling, soon atrophies; paedomorphic; about 60.0 mm TL at hatching, to 1162 mm TL as paedomorph; genus, 1 genus, 3 species. *Amphiuma*: as for family.

***Amphiuma means* (Two-toed Amphiuma), *A. pholeter* (One-toed Amphiuma), and *A. tridactylum* (Three-toed Amphiuma)   Fig. 21; PL 2F**

IDENTIFICATION Costal grooves 57–64; uniformly dark black, brown, or slightly maroon dorsally and lighter below; 40.0–70.0 mm TL as larva, to 1162.0 mm TL as paedomorph. *A. means*: digits 2/2; to 1162.0 mm TL as paedomorph. *A.*

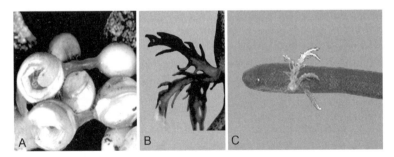

*Figure 21. **Amphiuma means** and **Amphiuma tridactylum**.* (A) egg rosary of *A. means* (Lake Co., FL; BM), (B) gills in silhouette, (C) head of a hatchling *A. tridactylum* (62.0 mm TL, Oktibbeha Co., MS; RA). See also Plate 2F.

pholeter: digits 1/1; head and eyes notably small; to 330.0 mm TL as paedomorph.
A. tridactylum: digits 3/3; to 65.0 mm TL as larva, 1045.0 mm TL as paedomorph.

NATURAL HISTORY Species of *Amphiuma* live in lakes, ponds, slow streams, and wetlands. Females oviposit from August to June, laying 50–200 nonpigmented ova (*A. means*: OD about 8.0 mm, ED 9.0–11.0 mm; *A. tridactylum*: OD 3.4–4.5 mm, ED 9.0 mm) as a rosary in semiterrestrial nest guarded by female. Ova are spaced 5.0–10.0 mm apart and hatch in 5 months at 40.0–65.0 mm TL. Larvae resorb gills in about 2 weeks and can survive for an average of 125 days on yolk reserves remaining at hatching.

These large nocturnal predators cruise along the bottom and among debris of ponds and lakes at night. They crush crayfish and fish by slowly flexing their heads downward almost 90° once the item is in their throat. Individuals vary in temperament, but aggressive ones willingly give a damaging bite. Amphiumas occur in many sorts of water bodies and are likely more common than typically known. They move overland on rainy nights.

RANGE *A. means* occurs on the Coastal Plain from southern Virginia to western Louisiana including peninsular Florida. *A. pholeter* is found in steephead creeks in extreme southern Alabama and adjacent areas of the Florida panhandle. *A. tridactylum* ranges from eastern Texas to south-central Alabama and north throughout the Mississippi Embayment to southern Illinois.

CITATIONS **General**: Means 1996, Salthe 1973a–b. **Development/Morphology**: C. L. Baker 1945, Hay 1888, Neill 1964, Ryder 1889, Trauth et al. 2004, Ultsch and Arceneaux 1988. **Reproductive Biology**: Fontenot 1999, Fritts 1966, J. A. Weber 1944. **Ecobehavior**: C. L. Baker 1945, L. C. Baker 1937a–b, Bancroft et al. 1983, Cagle 1948, Chaney 1951, Enge 1998, Gunzburger 2003, Hamilton 1950, Knepton 1954, F. L. Rose 1966a–b, 1967, Saumure and Doody 1998, Snodgrass et al. 1999, Sorensen 2004, Trauth et al. 1990.

Cryptobranchidae (Hellbenders)

These large salamanders occur east of the Rocky Mountains in flowing water; robust body of typical proportions but strongly depressed, especially large rounded head; digits 4/5, fleshy with rounded, nonkeratinized tips; mandibular grooves do not bisect lower lip anteriorly; gular fold present and not attached to medial throat tissue in larva, absent in paedomorph; gills with nonbranched rami and abundant fimbriae lost at metamorphosis at about 180.0 mm TL; gill slits 4 in larva, 1 in paedomorph; keratinized jaw sheaths absent; small eyes dorsal; neuromasts not obvious; lunged; balancers absent; dorsal fin low, originates near tail-body junction in paedomorph; fleshy folds along body and trailing edges of limbs in paedomorphs; costal grooves 14–15, indistinct in paedomorphs. Larva: uniformly gray to brown with increasing contrasting darker marks with age. Paedomorph: pale to dark brown or reddish, especially western populations have large dark marks dorsally; venter dirty white to brownish; 30.0–45.0 mm TL at hatching, 100.0–180.0 mm TL when loose gills at about 1.5–2.0 years, to 740.0 mm TL as paedomorph; paedomorphic; 1 genus, 1 species. *Cryptobranchus*: as for family.

Cryptobranchus alleganiensis (Hellbender)   **Fig. 22; PL 2G**

IDENTIFICATION As for family.

NATURAL HISTORY Hellbenders live in clear, cool, permanent streams, where males prepare nests beneath large rocks and wait for passing females. Oviposition occurs from August to November and more than one female may oviposit in a given male's nest. Clutch size varies from about 136 to 334 and represents about 75% of the enlarged ova in a female. Multiple females may use the same nest and as many as 1946 eggs have been found in one of these joint nests in New York. Freshly laid eggs have yellowish ova (OD 6.0 mm, ED 18.0–20.0 mm) and are laid as a rosary. Partial rosaries of eggs, apparently washed out of nests by episodic rains or the guarding activities of the attendant males, are often found in streams during the breeding season. Eggs hatch in about 45 days in Missouri and 68–75 days in more eastern populations; at hatching larvae are 30.0–45.0 mm TL and have incompletely formed limbs and digits. Gills are lost after 1.5–2.0 years at 120.0–180.0 mm TL.

This large salamander feeds on relatively large prey, and crayfishes are a common food item. The fleshy folds along the body and limbs presumably serve as respiratory surfaces, and in warm water these salamanders arch their back upward and rock back and forth to enhance gas exchange. Some individuals will bite, but their jaws are not nearly as strong as those of *Amphiuma*.

RANGE The eastern population occurs in the Ohio River drainage south and east to the Appalachian Mountains. A western population occurs in south-central Missouri and adjacent Arkansas.

CITATIONS *General*: Dundee 1971. ***Development/Morphology***: Dodd 2004, Fauth et al. 1996, Grobman 1943, McGregor 1897, Nickerson and Mays 1973, B. G. Smith 1912a–b; Trauth et al. 2004, Valentine 1989. ***Reproductive Biology***:

Figure 22. ***Cryptobranchus alleganiensis.*** (A) embryo (19.0 mm TL, Tazewell Co., VA; RWV), (B) single egg showing (arrow) remnant connectors as part of a rosary (Tazewell Co., VA; RWV), (C) larva (163.0 mm TL, KY; MR). See also Plate 2G.

Jensen et al. 2004, C. L. Peterson et al. 1989a, B. G. Smith 1907a, Topping and Ingersol 1981. *Ecobehavior*: Alexander 1927, Coatney 1982, Harlan and Wilkinson 1981, R. E. Hillis and Bellis 1971, W. J. Humphries 2007, W. J. Humphries and Pauley 2000, 2005, Mays and Nickerson 1971, Netting 1929, Nickerson et al. 2003, Nickerson and Mays 1973, Noeske and Nickerson 1979, C. L. Peterson et al. 1989a–b, C. L. Peterson and Wilkinson 1996, Pitt and Nickerson 2006; Reese 1904, 1906, Salthe 1963, Taber et al. 1975, C. H. Townsend 1882, Trauth et al. 1992, Ultsch and Duke 1990, Wheeler et al. 2003.

Plethodontidae (Lungless Salamanders)

Members of this family that have free-living larvae occur east of the Rocky Mountains; flowing water; nasolabial groove, a familial diagnostic trait, not present in larvae; streamlined body with 4 normally proportioned limbs; head rounded to angular in dorsal view, varies in shape in lateral view; digits usually 4/5, 4/4 in 3 taxa, with nonkeratinized, rounded or pointed tips; mandibular grooves do not bisect lower lip anteriorly; gular fold present and not attached to adjacent throat at midline, usually fits rather tightly around throat; keratinized jaw sheaths absent; lungless; balancers absent; dorsal fin low with rounded tip, originates at tail-body junction; 7 genera, 55 species. *Desmognathus*: mandibular grooves bisect lower lip anteriorly; digits 4/5, tips pointed, often keratinized; gill rami very short and nonbranched, few fimbriae filiform; metamorphic; 19 species; undescribed species known.

In the following taxa, the mandibular grooves do not bisect lower lip anteriorly, the gill rami are long with numerous, nonbranched fimbriae. *Eurycea*: digits usually 4/5, 4/4 in 2 cases; neuromasts not obvious except as foci of dorsolateral spots; east and west of Mississippi River; metamorphic or paedomorphic; 27 species; undescribed species known; *Hemidactylium*: digits 4/4; neuromasts not obvious; east and west of Mississippi River; metamorphic; 1 species; *Gyrinophilus*: digits 4/5; neuromasts usually obvious on head; east of Mississippi River; metamorphic or paedomorphic; 4 species; *Pseudotriton*: digits 4/5; neuromasts usually obvious on head; east of Mississippi River; metamorphic; 2 species; *Stereochilus*: digits 4/5; neuromasts quite obvious on head; east of Mississippi River; metamorphic; 1 species; *Urspelerpes*: digits 4/5; neuromasts not obvious; east of Mississippi River; metamorphic; 1 species.

Desmognathus auriculatus (Southern Dusky Salamander) Fig. 23; PL 2I

IDENTIFICATION Costal grooves 14–15; usually uniformly dark with 7–9 dorsolateral pale spots; gills more bushy (22–40 fimbriae) than most *Desmognathus*; 21.0–32.0 mm TL.

NATURAL HISTORY The Southern Dusky Salamander lives in seepage sites often associated with swampy areas and slow-moving streams in bottomland hardwood forest on the Coastal Plain of the eastern United States. Females deposit 7–41 nonpigmented ova (ED 1.7–2.9 mm) in nests in *Sphagnum* moss, in cypress logs, and beneath

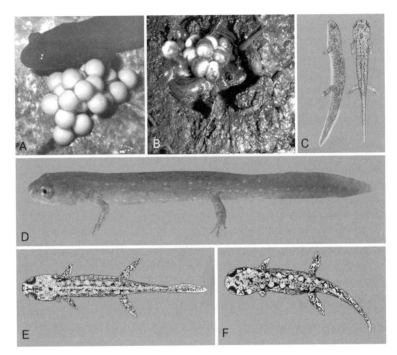

Figure 23. **Desmognathus auriculatus and relatives.** Egg clumps of *D. fuscus* from (A) Pennsylvania (MGP) and (B) West Virginia (KRP), (C) dorsal and lateral views of a hatchling *D. auriculatus* (18.0 mm TL, Alachua Co., FL; modified from Goin 1951), (D) larva (27.0 mm TL; RWV) of *D. auriculatus* (see also Plate 2I), (E) hatchlings of *D. fuscus* (18.1 mm TL, Franklin Co., MA), (F) hatchlings of *D. santeetlah* (10.9 mm TL, Monroe Co., TN; modified from Tilley 1981). See also Plates 2I, 3E.

bark or logs from July through September. Eggs hatch in about 1 month at 9.0–18.5 mm TL, and metamorphosis occurs in 5.0–6.7 months at 22.0–32.0 mm TL.

RANGE *D. auriculatus* occurs on the Coastal Plain from southern Virginia through the northern half of peninsular Florida to eastern Texas.

CITATIONS *Development/Morphology*: S. C. Bishop 1943, M. L. Cook and Brown 1974, Dodd 1998, Goin 1951, Means 1974, Neill 1951a, Valentine 1963. *Reproductive Biology*: Means 1974, Neill and Rose 1949, Trauth et al. 1990. *Ecobehavior*: M. L. Cook and Brown 1974, Dodd 1998, T. H. Eaton 1953, Goin 1951, Neill 1951a, Rubenstein 1971, Valentine 1963.

Desmognathus brimleyorum (Ouachita Dusky Salamander)   PL 2J

IDENTIFICATION Costal grooves 13–14; brown streaks along sides of body and tail; 11–13 pale spots with irregular edges in dorsolateral rows on body and basal part of tail; throat and venter not pigmented; to 38.0 mm TL.

NATURAL HISTORY Ouachita Dusky Salamanders live under large rocks in or at the edge of streams. Females oviposit between 20 and 36 nonpigmented ova (OD

3.2–4.5 mm) between March and September. Metamorphosis occurs at 27.0–31.0 mm SVL. Brooding females guard eggs in buried chambers constructed in mud and may move the eggs if conditions get too dry.

RANGE *D. brimleyorum* lives in rocky streams between 100 and 800 m in the Ouachita Mountains of west-central Arkansas and adjacent Oklahoma.

CITATIONS ***General***: Means 1999. ***Development/Morphology***: S. C. Bishop 1943, Means 1974, Rubenstein 1971, Trauth et al. 2004, Valentine 1963. ***Reproductive Biology***: Strecker 1908a, Trauth 1988, Trauth et al. 1990. ***Ecobehavior***: S. C. Bishop 1943, Karlin et al. 1993, Shipman et al. 1999.

Desmognathus conanti (Spotted Dusky Salamander), *D. fuscus* (Northern Dusky Salamander), and *D. planiceps* (Flat-headed Salamander)   PL 2L

IDENTIFICATION Costal grooves 14; 9–17 gill fimbriae on each side; dorsum brownish to black; 5–8 dorsolateral spots, diameters greater than spaces between them, usually extend onto tail; 16.0–35.0 mm TL.

NATURAL HISTORY The two species of Dusky Salamanders occur in sheltered sites along streams and rivulets in forested habitats of eastern North America. Oviposition of 3–40 (mean = 14) nonpigmented, yellowish ova occurs from May to September. Eggs hatch in 45–60 days at 12.0–20.0 mm TL, and larvae metamorphose after about 6.7–12.0 months at 16.0–35.0 mm TL depending on locality.

RANGE *D. conanti* occurs south from southern Illinois to the Gulf Coast and east to southwestern South Carolina with disjunct populations west of the Mississippi River in northern Louisiana and adjacent Arkansas and in northeastern Arkansas. *D. fuscus* occurs at sites from sea level to moderate elevations in Maine and adjacent Canada south to Virginia and the Carolinas.

CITATIONS ***Development/Morphology***: Davic 2005, Hilton 1909, Montague 1977, 1987, Rossman 1958, Rubenstein 1971, Trauth et al. 2004, Valentine 1963. ***Reproductive Biology***: Dennis 1962, Dodd 2004, Hom 1987, R. L. Jones 1986, Juterbock 1986, 1990, Noble and Evans 1932, Organ 1961b, Verrill 1863, I. W. Wilder 1913, 1917, Wood 1948, Wood et al. 1955, Wood and Fitzmaurice 1948. ***Ecobehavior***: Brode 1961, Burton 1976, R. F. Hall 1977, R. L. Jones 1986, Montague 1979, Montague and Poinski 1978, Orr and Maple 1978, J. P. Price et al. 2012, Spight 1967, Valentine 1963, H. H. Wilder 1899, 1904.

Desmognathus marmoratus (Shovel-nosed Salamander)   Fig. 24; PL 3A

IDENTIFICATION Costal grooves 13–14; gills often in bright contrast to dark ground color; more gracile than other large *Desmognathus* larvae; color highly variable from totally black to many mottled and spotted patterns; two dorsolateral rows of pale spots in some populations; border between dorsal pigmented and ventral nonpigmented areas distinct; about 70.0 mm TL.

NATURAL HISTORY Shovel-nosed adults live in cold, moderate to fast-flowing, gravelly bottomed streams and females attach eggs on the undersides of rocks in late spring to early summer. The nonpigmented ova are in clusters and attended

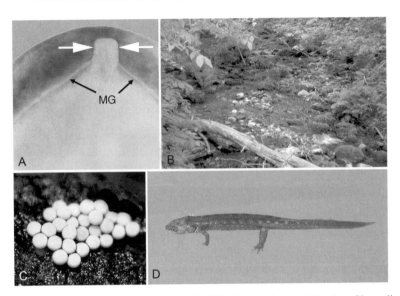

*Figure 24. **Desmognathus marmoratus.*** (A) labial folds showing (arrows) bisection of lower lip by (MG) mandibular grooves typical of all *Desmognathus* (compare Figs. 24A and 42A), (B) Appalachian stream habitat typical of plethodontid larvae, (C) egg clump (Macon Co., NC; ELJ), (D) larva (56.0 mm TL; RWV). See also Plate 3A.

by the female. Eggs hatch in 10–12 weeks at about 11.0 mm TL. Larvae metamorphose at 26.0–38.0 mm TL after 10–20 months.

RANGE *D. marmoratus* can be found in mountain streams between 300 and 1600 m elevation in the central Appalachian Mountains from southwestern Virginia to northeastern Georgia.

CITATIONS *General*: Martof 1963. *Reproductive Biology*: Dodd 2004, Martof 1962, C. H. Pope 1924. *Ecobehavior*: T. H. Eaton 1941, 1956.

Desmognathus monticola (Seal Salamander)   PL 3B

IDENTIFICATION Costal grooves 13–14; gill fimbriae/side 16–17; dorsum medium brown, 4–5 dorsolateral round spots between limb insertions; to about 50.0 mm TL.

NATURAL HISTORY Seal Salamanders can be found in hidden sites in small streams and seeps. Females oviposit from July to September and deposit 13–40 nonpigmented ova laid as clusters. Eggs hatch in 3–12 weeks at 11.0–20.0 mm TL, and larvae metamorphose at 35.0–50.0 mm TL after 10–11 months.

RANGE *D. monticola* inhabits cool streams between 1200 and 1500 m elevation in hardwood forests of the Appalachian Mountain ranges from Pennsylvania southeast to southwestern Alabama.

CITATIONS *Development/Morphology*: Dodd 2004, Rubenstein 1971. *Reproductive Biology*: Brady 1924, Bruce 1989, 1996, Bruce and Hairston 1990, Organ 1961b, C. H. Pope 1924. *Ecobehavior*: Orr and Maple 1978.

Desmognathus ochrophaeus group: *D. abditus* (Cumberland Dusky Salamander), *D. apalachicolae* (Apalachicola Dusky Salamander), *D. carolinensis* (Carolina Mountain Dusky Salamander), *D. imitator* (Imitator Salamander), *D. ochrophaeus* (Allegheny Mountain Dusky Salamander), *D. ocoee* (Ocoee Salamander), and *D. orestes* (Blue Ridge Dusky Salamander)   Fig. 25; PL 2H, 2K, 3C

IDENTIFICATION Costal grooves 14–15; toe tips not keratinized; gracile body form, rounded snout; extremely variable from unicolored to dorsolateral series of 4–6 pale spots, sometimes with a longitudinal stripe lateral to the spots; to about 20.0 mm TL.

NATURAL HISTORY Dusky Salamanders and their relatives live in streams and seeps in forested habitat at relatively high elevations. Nonpigmented eggs (6–37) are deposited in hollowed depressions beneath the ground, in mud crevices, under logs in the mud, and in similar places. Oviposition occurs from March to October and perhaps into the spring. Females remain with the cluster until they hatch in 6–8 weeks; yolk reserves at hatching may last for 10–21 weeks. Metamorphosis is in 6–10 months or less at about 20.0 mm TL.

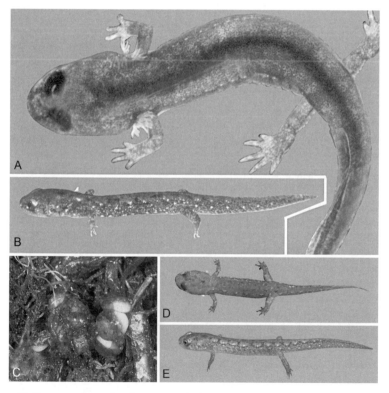

Figure 25. **Desmognathus ochrophaeus** **and relatives.** (A) larva of *D. ochrophaeus* (24.0 mm TL, Washington Co., TN; RAP), (B) larva of *D. ocoee* (16.0 mm TL, Macon Co., NC; RA), (C) cluster of eggs of *D. ocoee* (HM), (D) larva of *D. apalachicolae* (26.0 mm TL, Liberty Co., FL; DBM), (E) larva of *D. carolinensis* (21.0 mm TL, Yancy Co., NC; RA). See also Plates 2H, 2K, 3C.

The larvae of most of these are found in seepy, rocky, or talus areas at higher elevations, but *D. apalachicolae* is a Coastal Plain form found in first- and second-order streams in ravine headwaters.

RANGE *D. abditus* occurs on the Cumberland Plateau of Tennessee from Wartburg, Morgan Co. to Tracy City, Grundy Co. *D. apalachicolae* is restricted to drainages of the Choctawhatchee, Chattahoochee, and Apalachicola rivers in southeastern Alabama adjacent to Georgia and Florida. *D. carolinensis* is distributed on wet rock faces, seeps, and similar habitats between Linville Falls and McKinley Gap on Blue Ridge Divide and Iron Mountain Gap on the North Carolina–Tennessee border to Pigeon River valley (= Blue Ridge, Black, Bald, and Unaka mountains). *D. imitator* is restricted to seeps and streams in the central Smoky Mountains above 900 m. *D. ochrophaeus* is found in the Adirondack Mountains from southern Quebec south to southwestern Virginia and southeastern Kentucky. *D. ocoee* occurs in disjunct populations on the Appalachian Plateau of northeastern Alabama and the southwestern Blue Ridge Physiographic Province south of the Pigeon River. *D. orestes* is known from southwestern Virginia south on the Blue Ridge Divide to near Linville Falls in North Carolina and west to Avery and Carter cos. of northeastern Tennessee.

CITATIONS *General*: Means 1993, Tilley 1973b, 1985, Valentine 1964a. *Development/Morphology*: S. C. Bishop 1924, 1941a, S. C. Bishop and Chrisp 1933, Dodd 2004, T. H. Eaton 1954, 1956, Martof and Rose 1963, Means 1974, Means and Karlin 1989, Neill 1950, Nicholls 1949, Organ 1961b, Rubenstein 1971, Tilley 1969, 1972. *Reproductive Biology*: Beachy 1993a, 1995a–b, Bernardo 2000, Bernardo and Agosta 2003, Bruce 1989, Camp 2000, Koenings et al. 2000, Pfingsten 1965, Sever and Houck 1985, Tilley 1968, 1972, 1973a–b, 1980, Tilley and Tinkle 1968, Wood and Wood 1955. *Ecobehavior*: S. C. Bishop 1941a, Bruce 1996, Dunn 1917b, T. H. Eaton 1954, 1956, Fitzpatrick 1973, R. F. Hall 1977, Hess and Harris 2000, Mushinsky 1976, Orr and Maple 1978, C. H. Pope 1924, Tilley 1968, 1974, 1980.

Desmognathus quadramaculatus (Black-bellied Salamander) and *D. folkertsi* (Dwarf Black-bellied Salamander)   Fig. 26; PL 3D

IDENTIFICATION Costal grooves 14–15; small larva: dark gray to black with white spots associated with lateral line pores; 6–8 dorsolateral spots between limbs. Large larva: lighter dorsum usually with dorsolateral spots sometimes with reddish dorsal tail stripe. *D. quadramaculatus*: to 70.0 mm TL; *D. folkertsi*: to 38.0 mm TL.

NATURAL HISTORY Black-bellied Salamanders are found under stones in streams and rivulets. In May–June, females deposit 22–65 large, nonpigmented eggs in flattened clusters on the undersides of rocks. Eggs hatch in 1–4 months at 8.0–16.0 mm SVL; larvae metamorphose in 8–48 months at 35.0–54.0 mm SVL.

RANGE *D. folkertsi* is known in Lumpkin, Towns, and Union cos., Georgia in drainages of the Hiwassee and Chattahoochee rivers. *D. quadramaculatus* lives in rapidly flowing streams between 400 and 1700 m elevation in the central Appalachian Mountains from West Virginia southwest to northeastern Georgia.

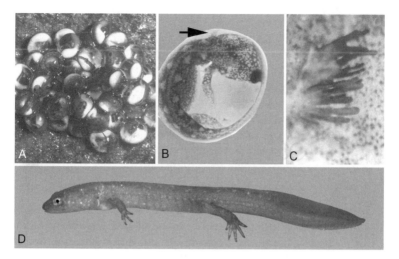

*Figure 26. **Desmognathus quadramaculatus**.* (A) melded clump of eggs (Clay Co., NC; RA), (B) embryo (9.0 mm TL, Macon Co., NC; arrow = tiny gills; RCB), (C) larval gills with very short rami and few, filiform fimbriae, (D) larva (64.0 mm TL, Clay Co., NC; RA). See also Plate 3D.

CITATIONS *General*: Valentine 1974. ***Development/Morphology***: R. M. Austin and Camp 1992, Dodd 2004, Rubenstein 1971. ***Reproductive Biology***: Beachy and Bruce 2003, S. C. Bishop and Chrisp 1933, Bruce 1988b, Bruce and Beachy 2003, Camp and Marshall 2006, Camp et al. 2010a–b; Montague 1987, Organ 1961b, Salthe 1963, C. K. Smith et al. 1996a, Tilley 1973a, c. ***Ecobehavior***: Beachy 1993b, 1997, S. C. Bishop 1941b, Bruce 1985b, 1996, T. H. Eaton 1941, 1956, Fitzpatrick 1973, Organ 1961b, Orr and Maple 1978, C. H. Pope 1924, Tilley 1968, 1973a–b.

Desmognathus santeetlah (Santeetlah Dusky Salamander)   PL 3E

IDENTIFICATION   Costal grooves 16; russet to brownish, 4–5 prominent dorsolateral spots, sometimes extend well onto tail, white line from eye prominent; to 29.0 mm TL.

NATURAL HISTORY   Dusky Salamanders live along seeps and stream headwater rivulets at high elevations in the Great Smoky and adjacent mountains. In May through July females oviposit 17–20 nonpigmented eggs in a cluster. Eggs hatch at 8.0–12.0 mm SVL and larvae grow to 12.0–20.0 mm TL in 45–60 days. Larvae metamorphose at 9.0–20.0 mm SVL after less than a year.

RANGE   *D. santeetlah* occurs at high elevations in a small area in the Great Smoky, Unicoi, and Great Balsam mountains on the border between Tennessee and North Carolina.

CITATIONS *General*: Tilley 2000. ***Development/Morphology***: Dodd 2004, Tilley 1981. ***Reproductive Biology***: Beachy 1993a, R. L. Jones 1986.

Desmognathus welteri (Black Mountain Salamander)   PL 3F

IDENTIFICATION Costal grooves 14; greenish to brownish, nearly uniform with dorsolateral rows of faint spots; to 35.0 mm TL.

NATURAL HISTORY The Black Mountain Salamander is found along steep gradient streams and rivulets flowing over rocky gravel courses. Eggs are laid among leaf packs and beneath submerged objects from March to December. Clutches consist of 18–57 nonpigmented ova (ED 4.4–4.6 mm); hatchlings measure 11–13 mm SVL.

RANGE *D. welteri* populations are known from the Cumberland Mountains and Cumberland Plateau in western Kentucky and adjacent Virginia and Tennessee.

CITATIONS *Reproductive Biology*: Petranka 1998, C. K. Smith et al. 1996b. *Ecobehavior*: Barbour and Hays 1957.

Eurycea aquatica (Brown-backed Salamander), *E. bislineata* (Northern Two-lined Salamander), *E. cirrigera* (Southern Two-lined Salamander), *E. junaluska* (Junaluska Salamander), and *E. wilderae* (Blue Ridge Two-lined Salamander)   Fig. 27; PL 3G,H,J,K

IDENTIFICATION Epigean sites; costal grooves 13–20; digits 4/5; small individuals pale brown, often with dorsolateral pale spots; older larvae sometimes with darker pigment patterns in posterior third of tail. *E. aquatica* and *E. junaluska* are darker than others, and *E. wilderae* often has obvious black marks on the posterior third of the tail and iridophores on the venter. 40.0–77.0 mm TL.

NATURAL HISTORY Two-lined Salamander and relatives are common along streams and rivulets in the Eastern United States. Dates of oviposition vary by latitude, with southern populations laying between December and March and more northern populations, from April to June. Clutch size also varies geographically, from 18 in Massachusetts to 53 in Mississippi, 15 and 30 in New York, and 39 in Ohio. Nonpigmented ova (OD 2.3–2.8 mm, ED 3.5–4.5 mm, 3 jelly layers) are laid as an array on the undersides of rocks, or when rocks are not available (e.g., some populations on the Coastal Plain) on root fibers, on undersides of leaves, or under logs. The pedicel consists of 2 jelly layers. Nests are in shallow water (3.9–4.6 cm) with sandy or gravelly substrates without silt. Eggs hatch at 11.0–14.0 mm TL in 1–6 months, and larvae grow to 18.0–32.0 mm TL in 1–3 years.

RANGE *E. aquatica* is found from north-central Alabama near Birmingham northeast to northwestern Georgia. *E. bislineata* occurs from Nova Scotia and Quebec south and east to Virginia and Ohio. *E. cirrigera* is known from eastern Illinois and central Virginia (generally south of the range of *E. bislineata*) southwest to Mississippi River exclusive of peninsular Florida, and the range of *E. wilderae* is in the higher elevations of the Appalachian Mountains. *E. junaluska* is restricted to a small area in Graham Co., North Carolina and Sevier and Monroe cos., Tennessee. *E. wilderae* occurs in the Appalachian Mountains from southwestern Virginia to northern Georgia.

CITATIONS *General*: Mittleman 1966, F. L. Rose 1971, Sever 1999a–c. *Reproductive Biology*: Bruce 1982b, 1988a, Graham et al. 2010, Guy et al. 2004,

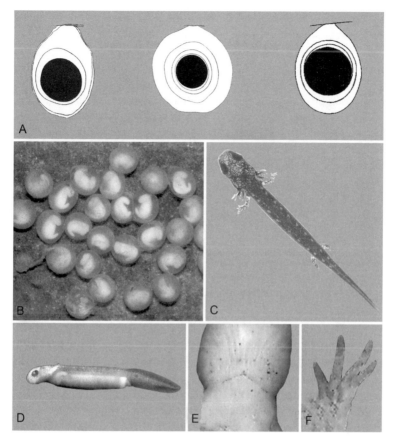

Figure 27. **Eurycea bislineata and relatives.** (A) schematic drawings of individual suspended eggs of (left) *Eurycea bislineata* with two membranes involved in the pedicel, (middle) *Gyrinophilus porphyriticus,* and (right) *Stereochilus marginatus,* each with one membrane in the pedicel (all modified from Noble and Richards 1932), (B) orderly array of suspended, pendent eggs of *E. wilderae* on the bottom of a rock (Watauga Co., NC; DMD), (C) hatchling of *E. bislineata* (12.0 mm TL, MA; AMR), (D) embryo of *E. bislineata* (9.0 mm TL, MA; AMR), (E) gular fold (Winston Co., MS; RA) of *E. cirrigera,* (F) foot with medium toes with pointed tips of *E. cirrigera* (RA). See also Plates 3H, 3J–K, 4K.

Jakubanis et al. 2008, Noble and Richards 1930, T. J. Ryan 1998, Salthe 1963, Sever 1983, Sever et al. 1987. ***Ecobehavior***: Barr and Babbitt 2002, Beachy 1993b, 1997, T. R. Brophy and Pauley 2001, Bruce 1982b, Burton 1976, Burton and Likens 1975, R. S. Caldwell and Houtcooper 1973, Duellman 1951, Duellman and Wood 1954, Hudson 1955, J. E. Johnson and Golberg 1975, McDowell 1995, J. C. Mitchell and Brown 2005a, Niemiller and Miller 2007, Petranka 1984a, F. L. Rose and Bush 1963, T. J. Ryan 1997, 1998, S. Smith and Grossman 2003, Stoneburner 1978, S. R. Voss 1993b, H. H. Wilder 1899, I. W. Wilder 1924a–b, 1925, Wiltenmuth 1997, Wood 1949, 1953b.

Eurycea chisholmensis (Salado Salamander), *E. naufragia* (Georgetown Salamander), and *E. tonkawae* (Jollyville Plateau Salamander)   Fig. 28; PL 3I

IDENTIFICATION Epigean sites; costal grooves 16; digits 4/5; *E. chisholmensis*: densely dark without well-defined iridophores; no dark eye ring; upper lip lacks dark pigment. *E. naufragia*: dorsum gray in fine reticulate pattern; palms with some pigment; dark canthal line; yellow dorsal fin; light areas around dorsolateral row of iridophores rosette-shaped. *E. tonkawae*: dorsum dark greenish brown, tail pale to yellow, dark canthal and middorsal body line, dorsolateral and ventrolateral lines of iridophores surrounded by lighter areas; eyes well developed. *E. chisholmensis*: mean total length 57.5 mm; *E. naufragia*: 50.1 mm TL; *E. tonkawae*: 38.0–54.0 mm TL.

NATURAL HISTORY Not much known about these species but presumably similar to other epigean Texas *Eurycea*. Juveniles of *E. tonkawae* are present March–August.

RANGE *E. chisholmensis* is known from springs at Salado, Bell Co., Texas. *E. naufragia* is from the springs and caves near Georgetown, Williamson Co., Texas, and *E. tonkawae* is from the Jollyville Plateau, Travis and Williamson cos., Texas.

CITATIONS *Development/Morphology*: Chippindale et al. 2000. *Ecobehavior*: Bowles et al. 2006.

Eurycea latitans (Cascade Caverns Salamander), *E. tridentifera* (Comal Blind Salamander), and *E. troglodytes* (Valdina Farms Salamander) Fig. 29

IDENTIFICATION Hypogean sites; digits 4/5; *E. latitans*: costal grooves 14–15; short stout legs, prominent forehead, flat snout, and small eyes beneath skin; pale and somewhat translucent with tan to brown netlike pattern of melanophores, some white specks. *E. tridentifera*: costal grooves 11–12; white to yellowish with

Figure 28. **Eurycea naufragia** **and relatives.** Paedomorphs: (A) *E. naufragia* (51.0 mm TL, Williamson Co., TX; DMH), (B) *E. tonkawae* (54.0 mm TL, Williamson Co., TX; RWV). See also Plate 3I.

Figure 29. **Eurycea latitans and relatives.** (A) single suspended egg of *E. latitans* (Bexar Co., TX; modified from Barden and Kezer 1944), (B) paedomorph of *E. troglodytes* (preserved, 73.0 mm TL; RA) and heads of (upper) *E. troglodytes* and (lower) *E. tridentifera* (Medina Co., TX; JPB).

translucent skin, reduced eyes under skin. *E. troglodytes*: costal grooves 13–17; pale gray or cream with translucent skin, pale yellow stripe sometimes present on sides and top of tail. *E. latitans*: to 105.0 mm TL; *E. tridentifera*: 38.0–73.0 mm TL; *E. troglodytes*: to 78.0 mm TL.

NATURAL HISTORY Not well known. Clutch size of 7–18 nonpigmented ova in *E. tridentifera*.

RANGE *E. latitans* occurs in Cascade Caverns, Kendall Co., Texas, *E. tridentifera* is from caves in Bexar and Comal cos., Texas, and *E. troglodytes* is known from the Valdina Farms Sinkhole, Medina Co., Texas.

CITATIONS *E. latitans*: ***General***: B. C. Brown 1967a. ***Development/Morphology***: J. K. Baker 1957, 1961, Burger et al. 1950, R. W. Mitchell and Reddell 1965, H. M. Smith and Potter 1946. ***Ecobehavior***: Sweet 1984; *E. tridentifera: **General***: Sweet 1977a. ***Development/Morphology***: J. K. Baker 1957, R. W. Mitchell and Reddell 1965. ***Ecobehavior***: Sweet 1984; *E. troglodytes: **General***: J. K. Baker 1961. ***Development/Morphology***: J. K. Baker 1957, R. W. Mitchell and Reddell 1965. ***Ecobehavior***: Sweet 1984.

Eurycea longicauda (Long-tailed Salamander) and *E. guttolineata* (Three-lined Salamander)   Fig. 30; PL 3L

IDENTIFICATION Epigean sites; costal grooves 13–14; digits 4/5; dorsum tan to brown uniform or with dorsolateral pale spots; venter immaculate; older ones often start to show adult chevron pattern posteriorly on the tail and may have a weakly developed middorsal stripe; venter immaculate; to 60.0 mm TL.

NATURAL HISTORY The Long-tailed and Three-lined Salamanders are often found under logs and other cover objects along the margins of shaded seepages, streams, and rivulets in floodplain forests of eastern North America and the Ozark Uplift. Females oviposit apparently from October to May, but few egg clutches have been collected, and those suggest deposition in dark subsurface streams, seeps, and caves. Between 8 and 106 nonpigmented ova (OD 2.5–3.0 mm,

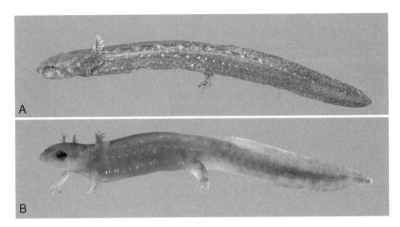

Figure 30. **Eurycea guttolineata and E. longicauda.** (A) larva of *E. longicauda* (38.0 mm TL, Lee Co., GA; DJS), (B) larva of *E. guttolineata* (38.0 mm TL, Forrest Co., MS; JRL). See also Plate 3L.

ED 8.0 mm, 2 jelly layers) are laid as an array on the undersides of rocks, rarely on vegetation in nonflowing water. Eggs hatch in 1–3 months at 10.0–19.0 mm TL and grow to 50.0–60.0 mm TL in 3.4–15.0 months. Larvae metamorphose at 22.0–27.0 mm SVL, and salamanders become mature in 1–2 years postmetamorphosis.

RANGE *E. guttolineata* is distributed south of the range of *E. longicauda* from northeastern Virginia south to the Gulf of Mexico and west to the Mississippi River. *E. longicauda* occurs from southern New York southwest through western Virginia and Tennessee to northern Alabama, and across the Mississippi River in the Ozark Region of Missouri, Arkansas, northeastern Oklahoma, and extreme southeastern Kansas.

CITATIONS ***Development/Morphology***: Dodd 2004, Sinclair 1951, Trauth et al. 2004. ***Reproductive Biology***: J. D. Anderson and Martino 1966, Bruce 1970, 1982a, Franz 1965, 1967, Freeman and Bruce 2001, P. H. Ireland 1974, J. L. Marshall 1999, McDowell 1989, 1992, McDowell and Shepherd 2003, Mohr 1943. ***Ecobehavior***: Bruce 1982a, Franz and Harris 1965, S. L. Freeman and Bruce 2001, Rudolph 1978.

Eurycea lucifuga (Cave Salamander)   Fig. 31; PL 4A

IDENTIFICATION Epigean sites; costal grooves 14–15; digits 4/5; ventral fin extends to vent; dorsum tan to brown and uniformly marked or with dorsolateral light spots; throat well pigmented and extends medially at first gill; soles of feet pigmented; dorsal pigmentation extends ventrally onto belly below a line connecting ventral margins of limb insertions; to 58.0 mm TL.

NATURAL HISTORY Cave Salamanders occur along streams and rivulets at cave entrances. They are also found in forest habitats with nearby rock walls or cliff faces. Females deposit 49–120 nonpigmented ova (OD 2.4–3.2 mm, ED

Figure 31. **Eurycea lucifuga.** (A) larva (44.0 mm TL; RWV). See also Plate 4A.

3.5–5.0 mm, 2 jelly layers) as an array from late September to early April; the eggs are sometimes unattached. Hatching occurs at 8.0–17.5 mm TL; the larvae grow to 70.0 mm TL. Larvae metamorphose after 6–18 months, but usually in less than a year, at 31.0–37.0 mm SVL and 50.0–70.0 mm TL.

RANGE *E. lucifuga* is known from limestone area habitats in the Ozark Region of northeastern Oklahoma east to Indiana and northwestern Virginia and south to central Alabama.

CITATIONS *General*: Hutchison 1966. ***Development/Morphology***: Dodd 2004, McAtee 1906, Trauth et al. 2004. ***Reproductive Biology***: Banta and McAtee 1906, Barden and Kezer 1944, Carlyle et al. 1998, N. B. Green 1968, N. B. Green et al. 1967, McDowell 2008, C. W. Myers 1958, Ringia and Lips 2007, Salthe 1963, Trauth et al. 1990. ***Ecobehavior***: Hutchison 1956, 1958, Rudolph 1978, Sinclair 1950.

Eurycea multiplicata (Many-ribbed Salamander)   PL 4B

IDENTIFICATION Epigean sites; costal grooves 19–20; digits 4/5; 7–10 costal grooves between adpressed limbs; uniformly yellowish brown to tan dorsally with uniform sprinkling of melanophores; to 55.0 mm TL, often smaller.

NATURAL HISTORY Many-ribbed Salamanders occur frequently beneath rocks and other cover along streams or under rocks in the stream channel proper; these streams may be permanent or seasonally ephemeral. Relatively small clutches of eggs (3–21) are deposited from September to June. The eggs have nonpigmented ova (OD 1.9–2.6 mm, ED 4.5–6.1 mm, 2 jelly layers) and hatch in 4–6 weeks. Metamorphosis occurs in 5–8 months.

RANGE *E. multiplicata* is known from between 100 and 760 m elevation in the Ozark and Ouachita mountains of eastern Arkansas and Oklahoma and adjoining rocky lowlands in south-central and southwestern Missouri.

CITATIONS *General*: Dundee 1965a. ***Development/Morphology***: G. A. Moore and Hughes 1941, Trauth et al. 2004. ***Reproductive Biology***: P. H. Ireland 1976, Spotila and Ireland 1970, Trauth et al. 1990. ***Ecobehavior***: S. C. Bishop 1943, 1944, Dundee 1947, P. H. Ireland 1976, Loomis and Webb 1951, McAllister and Fitzpatrick 1985, Rudolph 1978, Whitham and Mathis 2000.

Eurycea nana (San Marcos Salamander)   Fig. 32; PL 4C

IDENTIFICATION Epigean sites; costal grooves 16–17, with 6–7 costal grooves between adpressed limbs; digits 4/5; iris with dark ring; dorsum brown with small yellow spots or flecks in two rows down the sides, venter yellow and translucent; to 56.0 mm TL.

NATURAL HISTORY San Marcos Salamanders live in large mats of blue-green alga that cover much of the substrate and among plants and beneath rocks and other bottom debris along the north shore of Spring Lake that forms the headwaters of the San Marcos River. Although eggs have never been found, gravid females have been collected in every month, and 3–4% of mature females examined are gravid at any given time. Circumstantial evidence suggests that females oviposit throughout the year, and that maximum clutch size is 2–73 (mean = 34.7) eggs/clutch; one pair deposited 27–59 eggs over 3 clutches. The ova were nonpigmented (OD 1.5–2.0 mm) and laid as an array or as single eggs. Hatching took place after 3–4 weeks at about 8.0 mm TL; a mean population density of 115/m^2 has been noted.

RANGE This species is found only in Spring Lake, San Marcos, Hays Co., Texas and in the San Marcos River about 150 m below the dam.

CITATIONS *General*: B. C. Brown 1967b. *Development/Morphology*: J. K. Baker 1957, 1961, S. C. Bishop 1941b, Chippindale et al. 1993, Schwetmen 1967. *Reproductive Biology*: Najvar et al. 2007. *Ecobehavior*: Berkhouse and Fries 1995, Epp and Gabor 2008, Fries 2002, Thaker et al. 2006, Tupa and Davis 1976.

Eurycea neotenes (Texas Salamander) and *E. pterophila* (Fern Bank Salamander)   PL 4D

IDENTIFICATION Epigean sites; costal grooves 15–17; digits 4/5; 5–7 costal grooves between adpressed limbs; ventral fin originates near the middle of the tail; pale brown to yellow, two rows of pale flecks on sides of body, dark bar from eye to naris, venter cream and translucent; to 103.0 mm TL.

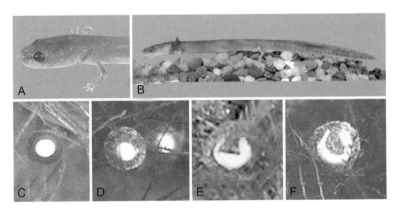

Figure 32. ***Eurycea nana.*** (A) head (42.0 mm TL; RWV), (B) paedomorph (48.0 mm TL; RA), (C–F) eggs and embryos (JNF) from Hays Co., TX. See also Plate 4C.

NATURAL HISTORY Texas Salamanders and their larvae occur in spring runs and small creeks around cave entrances. Females oviposit from April to July and deposit 6–19 nonpigmented, yellowish ova (OD 1.5–2.0 mm, ED 3.5–4.5 mm, 3 jelly layers) as singles or in an array. Eggs hatch in 2.0–3.5 weeks at about 7.0 mm TL and larvae mature in 9 months. The Texas Salamander occurs in epigean sites but certainly moves through underground water.

RANGE *E. neotenes* is found in springs in Bexar and Williamson cos., Texas, and *E. pterophila* occurs in and near Fern Bank Springs, Hays Co., Texas.

CITATIONS ***General***: B. C. Brown 1967c. ***Development/Morphology***: J. K. Baker 1957, 1961, S. C. Bishop 1943, S. C. Bishop and Wright 1937, B. C. Brown 1942, Burger et al. 1950, H. M. Smith and Potter 1946, Sweet 1977b. ***Reproductive Biology***: J. K. Baker 1957, Barden and Kezer 1944, Bruce 1976, Burger et al. 1950, R. W. Mitchell and Reddell 1965. ***Ecobehavior***: McAllister and Fitzpatrick 1989, Roberts et al. 1995, Sweet 1982, 1984.

Eurycea quadridigitata (Dwarf Salamander) and *E. chamberlaini* (Chamberlain's Dwarf Salamander)   Fig. 33; PL 4E

IDENTIFICATION Epigean sites; costal grooves 15–18; digits 4/4; dorsal fin extends well onto body at least early in ontogeny; 6–10, 5–9, and 4–6 fimbriae on gills; somewhat resembles the adult form at early stage, lateral pattern formed by melanophore stippling, 12–20, usually 14–17 pale spots between limb insertions more common in *E. chamberlaini*, sometimes with a middorsal stripe, gills silvery, lower lip and belly not pigmented; to 40.0 mm TL.

NATURAL HISTORY Dwarf Salamanders live around or in rivulets, seeps, bogs, swamps, and temporary pools. Female Dwarf Salamanders oviposit 12–62 ova in late summer and fall, while those of Chamberlain's Dwarf Salamander lay 35–64 from November to February. Ova are nonpigmented (OD about 2.0 mm, ED 3.0–4.0 mm), have 2 jelly layers, and are laid as an array of 3–6 or as singles attached to leaves and other debris. In some populations, females deposit eggs in dry depressions in traditional pond sites where they develop normally and hatch in advanced stages shortly after the pond fills. Eggs hatch in 1.0–1.3 months at

Figure 33. **Eurycea chamberlaini and E. quadridigitata.** (A) *E. quadridigitata* hatchling (8.3 mm TL; Alachua Co., FL; modified from Goin 1951), (B) *E. quadridigitata* larva (33.0 mm TL, Berkeley Co., SC; RA). See also Plate 4E.

11.0–12.0 mm TL and larvae grow to 48.0 mm TL. Metamorphs appear in 2–3 months at about 24.0 mm TL.

RANGE *E. chamberlaini* occurs on the Coastal Plain of North and South Carolina, Georgia, and Alabama and is sympatric with *E. quadridigitata* in some areas of the Carolinas and Georgia. *E. quadridigitata* lives on the Coastal Plain from North Carolina to Texas and most of peninsular Florida.

CITATIONS **General**: Mittleman 1967. **Development/Morphology**: Bailey 1937, T. H. Eaton 1956, Goin 1951, J. R. Harrison 1973, J. R. Harrison and Guttman 2003, C. H. Pope 1924, Semlitsch 1980, Trauth et al. 2004. **Reproductive Biology**: J. C. Mitchell and Gibbons 2012, Salthe 1963. **Ecobehavior**: Brimley 1923, Franklin 2000, Semlitsch and McMillan 1980, B. E. Taylor et al. 1988, Trauth 1983.

Eurycea rathbuni (Texas Blind Salamander), *E. robusta* (Blanco Blind Salamander), and *E. waterlooensis* (Austin Blind Salamander)   Fig. 34; PL 4F,J

IDENTIFICATION Hypogean sites; costal grooves 11–12; digits 4/5; depressed snout, elongate appendages, reduced eyes (more prominent in juveniles), and sparse pigmentation (scattered melanophores in juveniles); more or less uniform silvery luster and sparse melanophores; eyes small to nearly absent. *E. rathbuni*: to 134.0 mm TL; *E. robusta*: only holotype known, 100.8 mm TL; *E. waterlooensis*: 68.6 mm TL.

NATURAL HISTORY Little is known about the reproductive biology of these species, although we suspect that females oviposit throughout the year in cave streams at water temperatures of about 20°C. Clutch size is up to 39 based on oviductal egg counts of a gravid female. Small juveniles are present throughout the year. Courtship and feeding have been observed in captive situations.

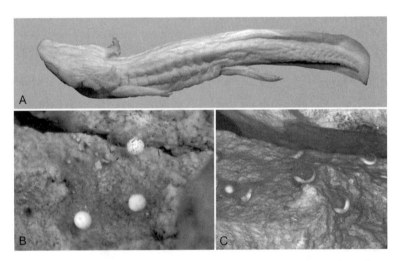

Figure 34. **Eurycea rathbuni** and *E. robusta*. (A) lateral view of a paedomorph of *E. robusta* (preserved, 101.0 mm TL, Hays Co., TX; RA), (B–C) embryos of *E. rathbuni* (Hays Co., TX; JF). See also Plate 4F.

RANGE *E. rathbuni* is known from several localities in the Purgatory Creek system and along the San Marcos Fault in San Marcos, Hays Co., Texas. *E. robusta* is known only from the aquifer beneath the Blanco River, Hays Co., Texas, and *E. waterlooensis* is known only from three of the water outlets at Barton Springs, Austin, Travis Co., Texas.

CITATIONS *E. rathbuni*: ***Development/Morphology***: Emerson 1905, R. W. Mitchell and Reddell 1965, Stejneger 1892, 1896. ***Reproductive Biology:*** *E. rathbuni*: Bechler 1988. ***Ecobehavior***: Norman 1900, Uhlenhuth 1919, 1921; *E. robusta*: Potter and Sweet 1981; *E. waterlooensis*: D. M. Hillis et al. 2001.

Eurycea sosorum (Barton Springs Salamander)   Fig. 35; PL 4G

IDENTIFICATION Epigean sites; costal grooves 13–16; digits 4/5; dorsum gray, brown to purplish gray; venter cream and translucent, tail fin low, orange eye stripe; to 62.0 mm TL.

NATURAL HISTORY Presumably similar to epigean *Eurycea* of the Edwards Plateau in south-central Texas. Hatchlings have been found in November, March, and April, and females with well-developed eggs are found in September through January.

RANGE *E. sosorum* is known only from a group of spring outlets in the Barton Creek drainage in Austin, Travis Co., Texas.

CITATIONS ***Development/Morphology***: Chippindale et al. 1993, Sweet 1984.

Eurycea spelaea (Grotto Salamander)   Fig. 36; PL 4H

IDENTIFICATION Hypogean sites; costal grooves 16–19; digits 4/5; neuromasts not obvious; larvae in caves uniformly pink to grayish with coarse distribution of small melanophores, venter pale; individuals in surface streams more densely pigmented; to 120.0 mm TL.

NATURAL HISTORY The Grotto Salamander lives in hidden sites in caves, perhaps just above the water level on wet surfaces. Females oviposit from August to

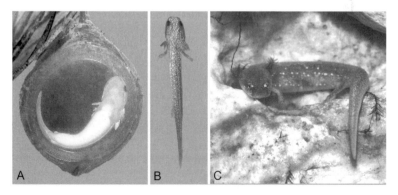

Figure 35. **Eurycea sosorum.** (A) egg attached to vegetation (LO), (B) hatchling (12.0 mm TL; DAC), (C) paedomorph (all from Travis Co., TX; LO). See also Plate 4G.

Figure 36. **Eurycea spelaea.** (A) oviposited eggs; inset: a single egg (Cambden Co., MO; modified from Barden and Kezer 1944) and larvae, (B) larva (72.0 mm TL, Phelps Co., MO; TRJ), (C) larva (66.0 mm TL, Shannon Co., MO; RA). See also Plate 4H.

February and clutches consist of 13–19 nonpigmented ova (OD 2.0–2.2 mm, ED 3.1–9.0 mm, 3 jelly layers). Metamorphosis takes 2–3 years and occurs at about 70.0 mm TL.

Larvae occur in caves or surface streams that issue from caves. This is the only hypogean salamander that metamorphoses. If metamorphosis occurs in the dark, the eyes degenerate and the eyelids fuse; in the light, the eyes remain functional.

RANGE *E. spelaea* is known from the caves and grottos of the Ozark region of southern Missouri and southeastern Kansas, northwestern Oklahoma, and northern Arkansas.

CITATIONS *General*: Brandon 1965a, 1970. *Development/Morphology*: Besharse and Brandon 1976, S. C. Bishop 1944, C. C. Smith 1968, Trauth et al. 2004. *Reproductive Biology*: Barden and Kezer 1944, C. C. Smith 1960, Trauth et al. 1990. *Ecobehavior*: Brandon 1971b, Dodd 1980, L. J. Hendricks and Kezer 1958, Rudolph 1978.

Eurycea tynerensis (Oklahoma Salamander)   PL 4I

IDENTIFICATION Epigean sites; costal grooves 19–21; digits 4/5; 7–11 costal grooves between adpressed limbs; dorsum uniformly gray to slightly streaked or with slight mottling of black and cream; venter not pigmented except few melanophores on chin; dorsum of tail with brownish stripe; to 80.0 mm TL.

NATURAL HISTORY The Oklahoma Salamander is found in gravel beds of streams and spring runs. Females oviposit in summer months and deposit up to 11 nonpigmented ova (OD about 1.6 mm). Larvae reach sexual maturity at 26.0 mm SVL, and the prevalence of metamorphosing and paedomorphic individuals varies among populations.

RANGE *E. tynerensis* has been collected in adjacent areas of northeastern Oklahoma, southwestern Missouri, and northeastern Arkansas.

CITATIONS *General*: Dundee 1965b. *Development/Morphology*: G. A. Moore and Hughes 1939, Trauth et al. 2004, Tumlison et al. 1990c. *Reproductive Biology*: Trauth et al. 1990. *Ecobehavior*: Cline and Tumlison 2001, Cline et al. 1989, Mathis and Unger 2012, McKnight and Nelson 2007, Rudolph 1978, Tumlison and Cline 1997, 2002, 2003, Tumlison et al. 1990a–c.

Eurycea wallacei (Georgia Blind Salamander)   Fig. 37; PL 4L

IDENTIFICATION Hypogean sites; costal grooves 11–13; digits 4/5; snout depressed and square in dorsal view; eyes reduced to pigmented specks; limbs spindly; neuromasts not obvious; dorsal fin restricted to tail; uniformly silvery white with a fine speckling of scattered melanophores; to 76.0 mm TL.

NATURAL HISTORY Georgia Blind Salamanders have been collected from only a deep well and several cave streams. Little is known about their reproductive biology. Presumably females oviposit 12–13 ova between May and November. Individuals measuring 25.0–50.0 mm TL are juveniles, and poorly known hypogean paedomorphs are found in artesian wells and caves.

RANGE *E. wallacei* has been collected in parts of the Florida-Georgia cave region. In Florida, sites are in Bay, Holmes, Jackson, and Washington cos. in drainages of the Chipola and Choctawhatchee rivers and Ecofina Creek. In Georgia, it is known from caves in Decatur and Dougherty cos. in the Flint River drainage.

CITATIONS *General*: Brandon 1967c. *Development/Morphology*: Carr 1939, Mansell 1971, Pylka and Warren 1958, Valentine 1964b. *Ecobehavior*: Carr 1939, Lee 1969, Mansell 1971, Peck 1973.

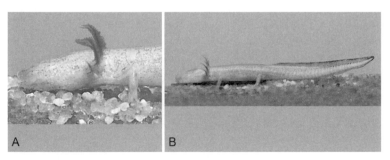

Figure 37. **Eurycea wallacei.** (A) head and (B) entire body of a paedomorph (68.0 mm TL, Jackson Co., FL; RA). See also Plate 4L.

Gyrinophilus gulolineatus (Berry Cave Salamander), *G. palleucus* (Tennessee Cave Salamander), and *G. subterraneus* (West Virginia Spring Salamander)   Fig. 38; PL 5A

IDENTIFICATION Hypogean sites; costal grooves 17–19; anterior corner of eye to snout tip 4–5 times eye diameter; usually appear uniformly pale pink to maroon; some populations have contrasty but diffuse dark marks dorsally; *G. gulolineatus* has a black, medial gular stripe; *G. subterraneus* is almost uniformly but faintly mottled; to 230.0 mm TL as paedomorph.

NATURAL HISTORY Not well known although presumably similar to epigean *G. porphyriticus*, although the larval period may be protracted. Small larvae have been found in December and February, suggesting oviposition likely occurs in the autumn.

RANGE *G. gulolineatus* is known from Knox, McMinn, and Roane cos., Tennessee, and *G. palleucus* has been collected in cave systems in south-central Tennessee, adjacent Alabama, and extreme northwestern Georgia. *G. subterraneus* occurs in a single cave system in Greenbrier Co., West Virginia.

CITATIONS ***General***: Brandon 1967a. ***Development/Morphology***: Besharse and Holsinger 1977, Brandon 1965b, 1971a, Bruce 1979, Dent and Kirby-Smith 1963, Lazell and Brandon 1962, McCrady 1954, Yeatman and Miller 1985. ***Ecobehavior***: Brandon 1967d, B. T. Miller and Niemiller 2005.

Figure 38. **Gyrinophilus palleucus and relatives.** (A) head of a paedomorph of *G. gulolineatus* (208.0 mm TL, Roane Co., TN; RA), (B) paedomorph of *G. palleucus* (189.0 mm TL, TN; RWV), (C) paedomorph of *G. subterraneus* (172.0 mm TL, Greenbrier Co., WV; RWV). See also Plate 5A.

Gyrinophilus porphyriticus (Spring Salamander)   Fig. 39; PL 5B

IDENTIFICATION Epigean sites; costal grooves 17–18; anterior corner of eye to snout tip less than 3.3 times eye diameter; snout truncate at least in larger specimens; variable but grossly tan to brownish or reddish maroon with variable small dark marks scattered over the dorsum; to 100+ mm TL.

NATURAL HISTORY Spring Salamander larvae live in hidden sites in streams and rivulets, and between February and August, females oviposit an array of nonpigmented eggs (15–106; larger numbers occur in northern populations) on the undersides of rocks. Eggs (OD 3.5–4.0 mm, ED 7.5–9.0 mm) have 3 jelly layers, the outer of which is thick and adhesive; they are attached singly with a short attachment stalk, missing in some cases). They hatch at 18.0–23.0 mm TL; larvae grow to 100+ mm TL in 3–6 years and metamorphose in the spring or summer at 55.0–70.0 or up to 160.0 mm TL. Metamorphic salamanders are influential predators in their environment and often feed on other salamanders. Larval sizes and ages at metamorphosis vary throughout the range.

RANGE *G. porphyriticus* is distributed throughout the Appalachian Mountains and Plateau from Maine to northeastern Mississippi.

CITATIONS ***General***: Brandon 1967b. ***Development/Morphology***: Besharse and Holsinger 1977, Birchfield and Bruce 2000, Dodd 2004, T. H. Eaton 1956, Organ 1961a, Valentine and Dennis 1964. ***Reproductive Biology***: Bruce 1972b, 1978a, 1980, H. T. Green 1925, Niemiller et al. 2009b, Noble and Richards 1932, Organ 1961a, Salthe 1963. ***Ecobehavior***: Bruce 2003, Burton 1976, Burton and Likens 1975, Culver 1973, Lowe 2005.

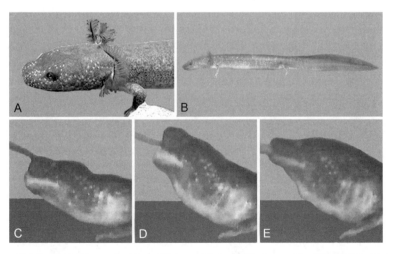

Figure 39. ***Gyrinophilus porphyriticus.*** (A) larva (62.0 mm TL, Scioto Co., OH; DMD), (B) larva (71.0 mm TL, Clay Co., NC; RA), (C–E) video frames of a larva eating an earthworm (each frame 10 ms after the preceding with one frame omitted between each two shown; SMD). See also Plate 5B.

Hemidactylium scutatum (Four-toed Salamander)   Fig. 40; PL 5C

IDENTIFICATION Costal grooves 13–14; 4 toes; dorsal fin extends onto body during at least early ontogeny; eyes large; neuromasts not obvious; brownish to gray, may have pale middorsal stripe, lateral part of body with dashed pattern, eye line dark; to 28.0 mm TL.

NATURAL HISTORY Four-toed Salamanders live in forests surrounding boggy, swampy areas and ponds. Adults move to breeding sites in late winter or early spring, and nesting begins in February and March in southern populations and in April and May in more northern populations. Females often nest in clumps of *Sphagnum*, within rotten logs, and under surface objects at the edges of ponds and swamps. Several females may oviposit at same site between February and May and as many as 1110 eggs from 30–35 females have been reported. Females deposit 20–145 nonpigmented ova with 2 jelly layers (OD 2.5–3.0 mm) singly but eggs often adhere to each other or to vegetation. They hatch in 3.3–8.6 weeks at 12.0–14.0 mm TL. Metamorphosis is in 21–120 days at 11.0–15.0 mm SVL and 17.0–25.0 mm TL.

RANGE *H. scutatum* occurs throughout eastern North America east of western Arkansas and adjacent Oklahoma in many disjunct populations.

CITATIONS *General*: Neill 1963a. *Development/Morphology*: Berger-Bishop and Harris 1996, S. C. Bishop 1920, Chalmers and Loftin 2006, Dodd 2004, S. J. Price and Jaskula 2005, Trauth et al. 2004. *Reproductive Biology*: Blanchard 1922, 1923, 1936, Collins and Wilbur 1979, Desroches and Pouliot 2005, Elliott et al. 2002, P. W. Gilbert 1941, Goodwin and Wood 1953, R. N. Harris et al. 1995, Herman and Enge 2007, Salthe 1963, Trauth et al. 1990, Thurow 1997a, Vaglia et al. 1997, Wood 1953a, 1955. *Ecobehavior*: R. N. Harris and Gill 1980, Hess and Harris 2000, Parmalee et al. 2002, K. S. Wells and Harris 2001.

Pseudotriton montanus (Mud Salamander)   Fig. 41; PL 5D

IDENTIFICATION Costal grooves 16–18; variable from uniformly tan, brownish to russet, sometimes with distinct punctate marks; venter white; to 60.0 mm TL.

Figure 40. **Hemidactylium scutatum.** (A) adult female with a group of single eggs; inset: small larva (28.0 mm TL; Wolfe Co., KY; both DMD). See also Plate 5C.

Figure 41. **Pseudotriton montanus.** (A) egg array (modified from Goin 1947), (B) Larva (58.0 mm TL, OH; DMD), (C) Larva (73.0 mm TL; RWV). See also Plate 5D.

NATURAL HISTORY Mud Salamanders occur in wet muddy habitats in hard-wood forests near swamps and small streams. The few clutches known were found in a small cavity in a seep, attached to dead leaves in a spring-fed ditch, and hanging from rootlets in an undercut bank. Presumably females oviposit from November to February and deposit an array of 27–192 nonpigmented ova (ED about 6.0 mm, each with a stalk). They hatch at 12.0–13.0 mm TL in 1–4 months, and larvae metamorphose in 9.5–30.0 months at 35.0–44.0 mm SVL, 65.0–98.0 mm TL.

RANGE *P. montanus* occurs from southeastern Pennsylvania and New Jersey south through the mid-Atlantic states to the northern third of peninsular Florida and west to southeastern Louisiana. The species is absent from the higher elevations in the Appalachian Mountains but occurs in the Interior Lowlands from southwestern West Virginia and southern Ohio to western Kentucky and eastern Tennessee.

CITATIONS *General*: Martof 1975a. *Development/Morphology*: Birchfield and Bruce 2000, Dodd 2004, Wood 1946. *Reproductive Biology*: Brimley 1923, Bruce 1974, 1975, 1978b, Fowler 1946a, Goin 1947. *Ecobehavior*: Bruce 1975, 2003.

Pseudotriton ruber (Red Salamander)   Fig. 42; PL 5E

IDENTIFICATION Costal grooves 15–17; variable from uniformly tan, brownish to russet, sometimes with indistinct punctate to mottled marks; venter dull white; to 86.0 mm TL.

NATURAL HISTORY Red Salamanders occur near swampy sites and small streams; larvae and juveniles are often found in thick accumulations of leaf litter in streams. Breeding is variable depending on geography. Clutches have been found in southern populations during summer months and in northern populations in the fall and spring. Females often deposit 50–100 nonpigmented ova (OD

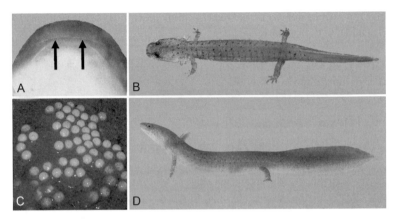

Figure 42. **Pseudotriton ruber.** (A) ventral view of snout showing (arrow) medial extent of mandibular grooves typical (i.e., mandibular grooves do not bisect lower lip) of all salamander larvae except *Desmognathus*, (B) larva (60.0 mm TL, Franklin Co., TN; RA), (C) array of eggs (modified from B. T. Miller and Niemiller 2005), (D) larva (98.0 mm TL, Watauga Co., NC; RWV). See also Plate 5E.

about 4.0 mm, ED about 6.0 mm) as an array attached by stalks to the undersides of rocks. They hatch in 1.4–3.0 months at 11.0–25.0 mm TL and grow to about 100.0 mm TL in 1.5–3.5 years. Metamorphosis is at 34.0–46.0 mm SVL, to 86.0–110.0 mm TL.

RANGE *P. ruber* occurs from northeastern Ohio to southeastern New York southeast to southeastern Louisiana. It is absent from the Atlantic Coastal Plain and peninsular Florida.

CITATIONS *General*: Martof 1975b. *Development/Morphology*: Birchfield and Bruce 2000, S. C. Bishop 1928, Bruce 1974, Dodd 2004, T. H. Eaton 1956. *Reproductive Biology*: S. C. Bishop 1925, Bruce 1972a–b, 1978c, Gordon 1966, B. T. Miller and Niemiller 2005, B. T. Miller et al. 2008. *Ecobehavior*: Bruce 2003, Cecala et al. 2007, Dunn 1915, Semlitsch 1983a.

Stereochilus marginatus (Many-lined Salamander)   Fig. 43; PL 5F

IDENTIFICATION Costal grooves 16–18; 7–9 costal grooves between adpressed limbs; wedge-shaped head somewhat flattened and pointed; neuromasts obvious on snout; distinctive pale ventrolateral stripe, sides of body with dashed pattern, and dorsum purplish; larvae resemble adults; to 90.0 mm TL.

NATURAL HISTORY *S. marginatus* lives in swampy, boggy permanently aquatic wetland sites. Courtship takes place in the fall and females oviposit in winter and early to spring. Clutches are made up of 16–121 nonpigmented, yellowish ova (OD 2.0–2.5 mm, ED 3.0–3.5 mm, 2 jelly layers). Eggs are laid as singles or in a loose array attached to stones or vegetation, sometimes above water. The stalk is short and involves only the outer layer, which is thick, soft, and adhesive. Larvae metamorphose in 1.5–3.0 years and mature in 3–4 years.

RANGE *S. marginatus* is known from the Coastal Plain from southwestern Virginia to northeastern Florida.

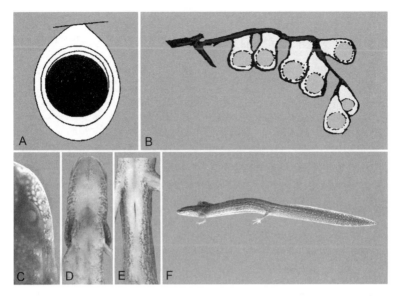

Figure 43. ***Stereochilus marginatus.*** (A) schematic of an individual suspended, pendent egg with one membrane involved in the pedicel (modified from Noble and Richards 1932; see also Fig. 27A), (B) schematic of single eggs attached to a twig (modified from Wood and Rageot 1963), (C–E) larva: (C) side of head, (D) gular fold, (E) vent (RA), (F) larva (62.0 mm TL, Bladen Co., NC; RWV). See also Plate 5F.

CITATIONS *General*: Rabb 1966. ***Development/Morphology***: Birchfield and Bruce 2000. ***Reproductive Biology***: Bruce 1971, Noble and Richards 1932; Salthe 1963, Wood and Rageot 1963. ***Ecobehavior***: Foard and Auth 1990, Rabb 1956.

Urspelerpes brucei (Patch-nosed Salamander)   Fig. 44; PL 5G

IDENTIFICATION Costal grooves 15; sparsely pigmented with most of upper surface of snout not pigmented; less than 52.0 mm TL.

NATURAL HISTORY This recently described salamander occurs in first-order streams. The three known females had 6–14 nonpigmented ova about 1.5–2.0 mm diameter. Males collected in April and May had well-developed nasal cirri and enlarged mental glands.

RANGE *U. brucei* is known from one small area in the Appalachian foothills at the foot of the Blue Ridge Escarpment in Stephens Co., northeastern Georgia, and in Oconee Co., northwestern South Carolina.

CITATIONS Camp et al. 2009, J. C. Mitchell and Gibbons 2012

Proteidae (Mudpuppy and Waterdogs)

These large salamanders occur east of the Rocky Mountains; sites with flowing water but found in permanent nonflowing water; 4 limbs normally proportioned; head depressed in lateral view, often angular in dorsal view; digits 4/4 with rounded, nonkeratinized tips; mandibular grooves do not bisect lower lip anteriorly; neuromasts not obvious; gular fold present and attached medially to throat

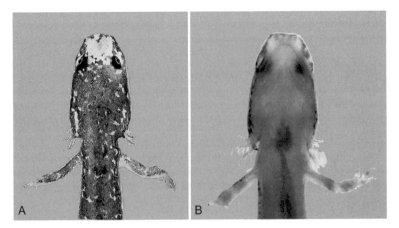

*Figure 44. **Urspelerpes brucei.** (A) dorsal and (B) ventral views of the anterior part of the body (20 mm SVL; Stephens Co., GA; JB). See also Plate 5G.

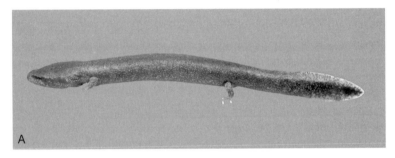

*Figure 45. **Necturus alabamensis.** (A) paedomorph (183.0 mm TL, Winston Co., AL; BM). See also Plate 5H.

tissue; nonbranched gill rami long, fimbriae numerous; keratinized jaw sheaths absent; eyes small and dorsal; neuromasts not obvious; lunged; balancers absent; low dorsal fin originates near tail-body junction; paedomorphic; 1 genus, 5 species; undescribed species known. The number of species in this family and their ranges are poorly resolved. *Necturus*: as for family.

Necturus alabamensis (Black Warrior River Waterdog)   Fig. 45; PL 5H

IDENTIFICATION Costal grooves 16. Larva: striped. Paedomorph: large dorsal spots on reddish brown dorsum; to 220.0 mm TL as paedomorph.

NATURAL HISTORY Gravid females of *N. alabamensis* have been collected from December to February in streams and rivers, but essentially nothing is known about their breeding biology. Larvae are 28.0–50.0 mm TL as larva and 150.0–248.0 mm TL as paedomorph.

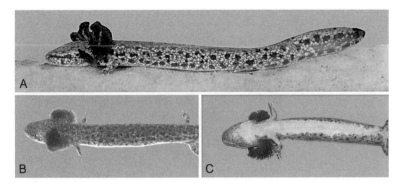

Figure 46. **Necturus beyeri.** (A) paedomorph (172.0 mm TL; RWV), (B) dorsal and (C) ventral views of a paedomorph (183.0 mm TL, Noxubee Co., MS; RA). See also Plate 5I.

RANGE *N. alabamensis* is restricted to the upper drainage of Black Warrior River, Alabama.

CITATIONS *Development/Morphology*: Bart et al. 1997, Hecht 1958, Viosca 1937. *Reproductive Biology*: Bart et al. 1997. *Ecobehavior*: Durflinger Moreno et al. 2006, Neill 1963b.

Necturus beyeri (Gulf Coast Waterdog)   Fig. 46; 5I

IDENTIFICATION Costal grooves 16. Larva and paedomorph: dorsum brown, gray, or maroon, dorsum spotted, venter spotted or not; to about 80.0 mm TL as larva, 160.0–223.0 mm TL as paedomorph.

NATURAL HISTORY Gulf Coast Waterdogs oviposit from December to April in streams and rivers on the Gulf Coast. Between 40 and 70 nonpigmented ova (OD about 6.0 mm) are deposited as an array beneath submerged objects. Eggs hatch in at least 2 months at 21.0–26.0 mm TL. Larvae commonly congregate in leaf mats in streams and sometimes occur in small ditches.

RANGE West of Mississippi River, *N. beyeri* is known from eastern Texas and adjacent west-central Louisiana. East of Mississippi River, it occurs on the Coastal Plain of southeastern Louisiana east to at least the Mobile River drainage. *N. maculosus* may be sympatric with *N. beyeri* since the opening of the Tennessee and Tombigbee Waterway in 1975.

CITATIONS *Development/Morphology*: Hecht 1958. *Ecobehavior*: Bart and Holzenthal 1985, Brenes and Ford 2006, Shoop 1965, Shoop and Gunning 1967.

Necturus lewisi (Neuse River Waterdog)   Fig. 47; PL 5J

IDENTIFICATION Costal grooves 14. Larva and paedomorph: wide, tan, dorsal stripe, at least in larva, sides darker with pale flecks; dorsum and venter spotted; 140.0–276.0 mm TL as paedomorph.

NATURAL HISTORY Females of *N. lewisi* oviposit in the summer in streams and small rivers. Up to 35 nonpigmented ova (ED 8.0–9.0 mm) are deposited in

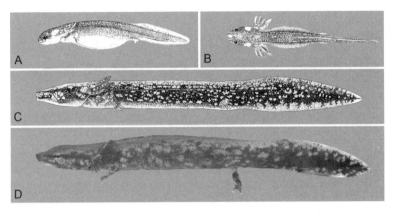

Figure 47. Necturus lewisi. (A) lateral and (B) dorsal views of a hatchling (20.7 mm TL), (C) posthatching larva (65.0 mm TL; from Wake Co., NC; modified from Ashton and Braswell 1979), (D) paedomorph (172.0 mm TL; RWV). See also Plate 5J.

an array on the bottoms of submerged surfaces. Eggs have 2 jelly layers and hatch at 21.0–24.0 mm TL. The adult pattern is attained at about 45.0 mm SVL, and salamanders mature in 6 years.

RANGE *N. lewisi* is restricted to the Neuse and Tar rivers, North Carolina.

CITATIONS **General**: Ashton 1990. **Development/Morphology**: Ashton and Braswell 1979, S. C. Bishop 1926, Brimley 1924, Hecht 1958. **Reproductive Biology**: Ashton and Braswell 1979, Fedak 1971. **Ecobehavior**: Ashton 1985, Braswell and Ashton 1985.

Necturus maculosus (Mudpuppy)   Fig. 48; PL 5K

IDENTIFICATION Costal grooves 15. Larva: dark dorsally with pale to reddish dorsolateral stripes that fade by year 4. Paedomorph: quite variable throughout its range; *N. m. louisianensis* usually with larger dark spots near unmarked central part of belly; *N. m. maculosus* with various spotted to uniform coloration dorsally and usually a spotted venter; to 78.0 mm TL as larva, 220.0–228.0 mm TL as paedomorph.

NATURAL HISTORY Females oviposit from May to June in many different habitats, including streams, rivers, lakes, canals and ditches. Nests are located below large rocks or logs and usually guarded by the female. They consist of 15–174 nonpigmented ova (OD 4.3–6.5 mm, ED about 13 mm, 3 jelly layers) that hatch at 20.0–25.0 mm TL in 1.3–2.3 months. Larvae grow to 50.0–60.0 mm TL in first year, and are sexually mature in about 5–6 years at 130.0–200.0 mm TL. These commonly nocturnal salamanders spend the other times under submerged objects and among debris; they often occur in large bodies of water and may move to deeper water during the winter.

RANGE *N. maculosus* occurs from southern Manitoba east to central New York and southwest to a disjunct taxon in Louisiana, Arkansas, southern Missouri,

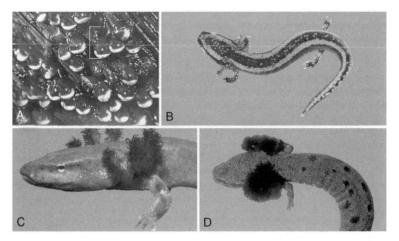

Figure 48. ***Necturus maculosus.*** (A) egg array (Muskegon Co., MI; suspended eggs lying against tilted wooden substrate; extent of center egg noted by white bracket; RHH), (B) larva (48.0 mm TL, Muskegon Co., MI; SLB), (C) paedomorph (190.0 mm TL, Tazewell Co., VA; WPL), (D) paedomorph (211.0 mm TL; RA). See also Plate 5K.

southeastern Kansas, and eastern Oklahoma. Populations are exotic in Maine (Hunter et al. 1999) and Connecticut (Klemens 1993).

CITATIONS ***Development/Morphology***: Cahn and Shumway 1926, Dodd 2004, Eycleshymer and Wilson 1910, Hecht 1958, McAllister et al. 1981, Trauth et al. 2004. ***Reproductive Biology***: S. C. Bishop 1932, Cagle 1954, Collins and Wilbur 1979, K. L. Fitch 1959, J. P. Harris 1959a–b, 1961, Parmalee et al. 2002, Salthe 1963, Shoop 1965, B. G. Smith 1911b, Trauth et al. 1990. ***Ecobehavior***: S. C. Bishop 1924, 1926, Briggler and Moser 2008, Brimley 1924, Cochran 1995, Cochran and Lyons 1985, Eycleshymer 1906, Lockhart 2000, Shoop and Gunning 1967, Sinclair 1950, Ultsch et al. 2004, Ultsch and Duke 1990, R. E. Weber et al. 1985.

Necturus punctatus (Dwarf Waterdog)   Fig. 49; PL 5L

IDENTIFICATION Costal grooves 14–16; body notably slender and elongate. Larva: uniformly gray to maroon. Paedomorph: uniformly gray dorsum or spotted in some populations; unmarked center of belly; 100.0–200.0 mm TL as paedomorph.

NATURAL HISTORY Dwarf Waterdogs oviposit from April to September in slow streams and nonflowing water. The eggs have black ova (OD 1.0–1.3 mm, ED 3.2 to more than 3.9 mm, 1 jelly layer), are laid as an array, and hatch in 5 months at about 70.0 mm TL. Sexual maturity is reached at greater than 75.0 mm SVL.

RANGE *N. punctatus* occurs on the Atlantic Coastal Plain from southeastern Virginia to southern Georgia.

Figure 49. ***Necturus punctatus.*** Paedomorphs: (A) 98.0 mm TL; RWV (B) 184.0 mm TL; RA. See also Plate 5L.

CITATIONS *General*: Dundee 1998. ***Development/Morphology***: Brimley 1924, Hecht 1958. ***Reproductive Biology***: Fedak 1971, Meffe and Sheldon 1987. ***Ecobehavior***: Brimley 1924.

Rhyacotritonidae (Torrent Salamanders)

This small family of salamanders occurs only in the Pacific Northwest; flowing water; body and 4 limbs normally proportioned; snout broadly rounded in dorsal view; digits 4/5 with rounded, nonkeratinized tips; 14–15 costal grooves; mandibular grooves do not bisect lower lip anteriorly; gular fold present and not attached to throat; gill rami very short, fimbriae few; keratinized jaw sheaths absent; eyes large, protuberant, and set lateral; neuromasts not obvious; lungs small; balancers absent; dorsal fin low, originates on tail posterior to tail-body junction, ventral fin originates on tail muscle; metamorphic; to 60.0 mm TL, often smaller; larva and adult same coloration, yellowish to golden dorsum with various amounts of small black marks; metamorphic; 1 genus, 4 species. *Rhyacotriton*: as for family.

Rhyacotriton cascadae (Cascades Torrent Salamander), *R. kezeri* (Columbia Torrent Salamander), *R. olympicus* (Olympic Torrent Salamander), and *R. variegatus* (Southern Torrent Salamander)   Fig. 50; PL 6A,B

IDENTIFICATION As for family; species not diagnosed.

NATURAL HISTORY Not much is known about the life history of these stream-dwelling species. Oviposition likely occurs over a long season, with probable peaks in the spring and early summer or late fall. Two clutches of 9 eggs each of *R. kezeri* were collected in December near a spring mouth in cracks in sandstone; eggs of the other three species have not been found. Female *R. variegatus* may contain up to 16 large, nonpigmented ova (OD 2.9–5.6 mm, 6 jelly layers, some reports say 3). Eggs with nonsticky jelly are laid singly as a group and communal nesting likely, although parental care probably absent, eggs not attended; the inner jelly is clear, the outer is pitted and accumulates debris. Larvae occur in sheltered

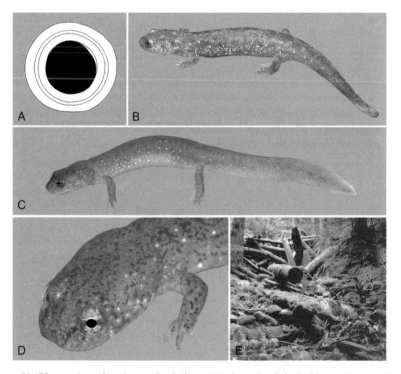

Figure 50. **Rhyacotriton olympicus and relatives.** (A) schematic of single *Rhyacotriton* egg (modified from Noble and Richards 1932), (B) larva of *R. kezeri* (45.0 mm TL, Pacific Co., OR; WPL), (C) larva of *R. olympicus* (48.0 mm TL, Jefferson Co., WA; WPL), (D) head of a larva of *R. cascadae* (53.0 mm TL, Skamania Co., WA; WPL), (E) mountain stream habitat (Benton Co., OR; RA) where *Ascaphus* and *Dicamptodon* are sympatric. See also Plate 6A, B.

spots among gravel or under stones (8.0–11.0 °C) in upper reaches of streams and springs. They hatch at 11.0–26.0 mm TL in 7.0–9.7 months, although they probably do not feed for a considerably longer period. Larvae grow to 70.0 mm TL and metamorphose in about 3.0–3.5 years at 30.0–68.0 mm TL. Hatchlings are pale yellow with 2 rows of circular marks along back. They have few predators and population densities may reach 12.9/m². They mature in 1.0–1.5 years after metamorphosis.

RANGE *R. cascadae* lives in the Cascade Mountains from just north of Mt. St. Helens, Washington to Lane Co., Oregon. *R. kezeri* is known from the Coast ranges of Oregon and Washington south of the Chehalis River to Yamhill Co., Oregon. *R. olympicus* is restricted to the Olympic Peninsula of Washington, and *R. variegatus* is found in the Coast Range from south of Yamhill Co., Oregon south to Mendocino Co., California.

CITATIONS *General*: J. D. Anderson 1968b. *Development/Morphology*: Gaige 1917, G. S. Myers 1943, Worthington and Wake 1971. *Reproductive Biology*: Karraker 1999, Karraker et al. 2005, MacCracken 2004, Noble and Richards 1932, Nussbaum 1969b, Nussbaum and Tait 1977, Salthe 1963, Tait and Diller 2006, C. E. Thompson et al. 2011. *Ecobehavior*: Hicks et al. 2008, G. S. Myers 1943, Nijhuis and Kaplan 1998, Rundio and Olson 2001, K. R. Russell et al. 2002, Stebbins 1955, Steele et al. 2003b, Welsh and Lind 1996.

Salamandridae (Newts)

These salamanders occur throughout much of North America; usually nonflowing water, rarely in slow streams; slender body and 4 limbs normally proportioned; digits 4/5, narrow and pointed, sometimes with keratinized tips; mandibular grooves do not bisect lower lip anteriorly; gular fold present and not attached medially to throat, often fits rather loosely against throat; nonbranched gill rami long, fimbriae numerous; keratinized jaw sheaths absent; medium-sized eyes set laterally; neuromasts not obvious; lunged; balancers present; dorsal fin low, originating well onto back; striped pattern as hatchlings; 2 genera, 7 species. *Notophthalmus*: east of Rocky Mountains; nonflowing water; eye line present; metamorphic or pedotypic; 3 species. *Taricha*: widespread in western North America; nonflowing or flowing water; eye line absent; metamorphic; 4 species.

Notophthalmus meridionalis (Black-spotted Newt)   Fig. 51; PL 6C

IDENTIFICATION Small larva: dorsum dark, including fins, dark eye line with white below; large larva: ventrolateral flanks speckled, dorsum nearly uniform gray to brown; to 35.0 mm TL.

NATURAL HISTORY The Black-spotted Newt oviposits in March and April in permanent and seasonally ephemeral ponds and pools. Up to 300 eggs with oval outer jelly layer are attached singly to the leaves of submerged plants. Data from *N. viridescens* are presumably applicable although a well-defined eft stage may not exist.

RANGE *N. meridionalis* occurs from southeastern Texas to northeastern Mexico.

CITATIONS *General*: Mecham 1968. *Reproductive Biology*: Strecker 1922.

*Figure 51. **Notophthalmus meridionalis.*** (A) single egg, (B) hatchlings: (top) 9.0 mm TL, (bottom) 11.0 mm TL, (C) larva (21.0 mm TL), (D) larva (31.0 mm TL; all DLM; culture from Cameron Co., TX). See also Plate 6C.

*Figure 52. **Notophthalmus perstriatus.*** (A) larva (50.0 mm TL, Putnam Co., FL; RWV), (B) pedo-type (76.0 mm TL, Alachua Co., FL; RA). See also Plate 6D.

Notophthalmus perstriatus (Striped Newt)   Fig. 52; PL 6D

IDENTIFICATION Larva: dorsum generally dark but with lighter blotches on the sides; dark above and pale below, distinct line through eye that continues to top of hind leg insertion. Pedotype: similar to metamorph but more uniform; stripes faint to absent. Eft: similar to a drab adult; may be absent in some populations; 25–37 mm TL as larva, pedotypes as large as metamorphosed adults and with sexually dimorphic traits.

NATURAL HISTORY   Striped Newts oviposit from February to June in temporary sinkhole ponds in Sandhills and cypress and bay ponds in Pine Flatwoods.
Pigmented ova are attached singly to leaves. They hatch between April and July
and larvae grow to 25.0–37.0 mm TL, undergoing metamorphosis in about
6 months. Pedotypes attain the size of metamorphosed adults.

RANGE   *N. perstriatus* occurs from south of the Savannah River in southeastern Georgia to the northern third of peninsular Florida and west to Baker County
in Georgia and the Apalachicola National Forest in northern Florida.

CITATIONS   *General*: Mecham 1967a. ***Development/Morphology***: S. A. Johnson and Franz 1999, Mecham and Hellman 1952. ***Reproductive Biology***: S. A.
Johnson 2002, Moler 1992. ***Ecobehavior***: Christman and Franz 1973, Dodd 1993.

Notophthalmus viridescens (Eastern Newt)   Fig. 53; PL 6E

IDENTIFICATION   Larva: light to medium tan or brown, dark line through eye;
black flecks on body, flecks of iridophore on tail fins. Pedotype: resembles metamorphosed adult. Eft: brilliant red with red dorsolateral dots circled in black;
occurs at higher elevations and latitudes; drab brown dorsum in lowland areas; to
42.0 mm TL as larva, 110.0 mm TL as pedotype and eft; efts largest at higher
elevations, smaller at lowland sites.

NATURAL HISTORY   Eastern Newts oviposit January–June or later depending
on elevation and latitude. They use all sorts of permanent or semipermanent
aquatic habitats, including ponds, lakes, swamps, ditches, and slow streams.
Clutches consist of 80–450 pigmented ova (OD 1.5 mm, ED 6.5–8.0 mm, 3 jelly

Figure 53. ***Notophthalmus viridescens.*** (A) larva (62.0 mm TL, Alachua Co., FL; DMD), (B) pedotype (68.0 mm TL; RWV), (C) larva (31.0 mm TL, Bryan Co., GA; DJS). See also Plate 6E.

layers) laid as singles and attached to and often wrapped by leaves on submerged plants. Eggs may be laid over a long period, and hatching occurs in 1–2 months at 6.0–9.0 mm TL. Larvae grow to 43.0 mm TL and metamorphose in 2–3 months at about 35.0 mm TL. When present, efts may live in the forest for 2–8 years before returning for the so-called second metamorphosis into reproductive adults. Adults that occasionally leave the water, especially if temporary sites dry out, attain an eft-like morphology (i.e., drier skin and no fins).

The occurrence of efts and their coloration, size, and toxicity vary geographically (i.e., brightest red, largest, and highest toxicity in Appalachian Mountains). Usually metamorphic, but pedotypes may grow to the size of metamorphosed adults and have similar coloration.

RANGE   *N. viridescens* is widespread in eastern North America from southern Canada through the eastern states to east Texas.

CITATIONS *General*: Mecham 1967b. *Development/Morphology*: Chalmers and Loftin 2006, Dodd 2004, Fowler 1946b, Goin 1951, P. S. Hampton 2006, R. N. Harris 1989, Jordan 1893, McCallum et al. 2003, Trauth et al. 2004. *Reproductive Biology*: Brandon and Bremer 1966, Chadwick 1944, Collins and Wilbur 1979, Gage 1891, Healy 1973, 1974, A. A. Humphries 1966, McLaughlin and Humphries 1978, Noble 1926, 1929, Parmalee et al. 2002, Pike 1886a, C. H. Pope 1919b, 1921, 1928. *Ecobehavior*: T. E. Brophy 1980, Burton 1976, Chadwick 1950, Fauth and Resetarits 1991, Formanowicz and Brodie 1982, Hamilton 1940, R. N. Harris 1987, R. N. Harris et al. 1988, Reilly and Lauder 1988, K. G. Smith 2005, Takahashi and Parris 2008, B. E. Taylor et al. 1988, B. Walters 1975, P. J. Walters and Greenwald 1977, Wilbur 1987, Worthington 1968, 1969.

Taricha granulosa (Rough-skinned Newt)   Fig. 54; PL 6F

IDENTIFICATION Striped hatchling pattern soon replaced by dark mottling or crudely marked with dashes and blotches with lateral series of pale spots; dorsum brownish; head shape angular; 50.0–75.0 mm TL.

NATURAL HISTORY Rough-skinned Newts oviposit between November and July in lakes, ponds, and slow streams. The tan (less than half of hemisphere) and cream ova (OD 1.8–2.0 mm, ED 2.5–4.0 mm, 2–3 jelly layers) are usually laid as singles, but sometimes as small clumps of 5–22; their jelly is stiff and rubbery. Eggs hatch in 2.8–3.7 weeks at 7.0–18.0 mm TL, and larvae grow to 50.0–70.0 mm TL, metamorphosing at 23.0–25.0 mm TL. Larvae sometimes overwinter at high elevations.

RANGE Native populations of *T. granulosa* occur from southeastern Alaska to just south of San Francisco Bay west of the Cascade ranges at elevations from sea level to about 2800 m. Isolated populations that presumably have been introduced are known from Latah Co., Idaho and Saunders Co., Montana.

CITATIONS *Development/Morphology*: Twitty 1942, 1966. *Reproductive Biology*: Farner and Kezer 1953. *Ecobehavior*: Marangio 1978, M. G. Oliver and McCurdy 1974, J. T. Taylor 1984.

Figure 54. **Taricha granulosa.** (A) single eggs attached to vegetation (Sonoma Co., CA; RA), (B–C) ova: (B) late cleavage, (C) early tailbud stages (Thurston Co., WA; WPL), (D) larva (68.0 mm TL, Klickitat Co., WA; WPL). See also Plate 6F.

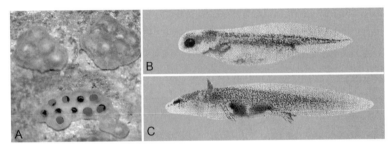

Figure 55. **Taricha rivularis.** (A) flattened, melded egg clumps attached to the bottom of stones in streams (Sonoma Co., CA; RA), (B) embryo, (C) small larva (modified from Twitty 1936). See also Plate 6G.

Taricha rivularis **(Red-bellied Newt)**   **Fig. 55; PL 6G**

IDENTIFICATION Uniformly dark to medium brown to black after disappearance of striped hatchling pattern; to about 45.0 mm TL.

NATURAL HISTORY Red-bellied Newts oviposit in streams that flow through the redwood forests of northern California from February to May. Clutches consist of 5–16 dark gray to chocolate brown ova (OD 2.7 mm) pigmented to well below the equator with greenish gray vegetal pole, and arranged as a flattened clump (about 25.0 mm diameter, usually 1 egg thick) on undersides of stones. The jelly is stiff and rubbery, and several females may oviposit on same rock (up to 70 masses have been found on the underside of one rock). Hatching occurs in about 2.9 weeks (balancers absent or small) at 10.3–11.0 mm TL, and metamorphosis occurs at 20.0–25.0 mm TL. Juveniles mature in about 5 years.

Figure 56. **Taricha sierra** and **T. torosa.** (A–D) *T. torosa*: (A) globular, melded egg clump attached to vegetation and rocks (Sonoma Co., CA; RA), (B) small larva (21.0 mm TL, Tulare Co., CA; RA), (C) foot, (D) gular fold (RA), (E) larger larva (40.0 mm TL, Kern Co., CA; RWH) of *T. sierra*. See also Plate 6H.

RANGE   *T. rivularis* is found from Humboldt Co. south to Sonoma Co., California.

CITATIONS   *General*: Twitty 1964. *Development/Morphology*: Twitty 1942, 1966. *Reproductive Biology*: Salthe 1963. *Ecobehavior*: Licht and Brown 1967.

Taricha torosa (California Newt) and *T. sierra* (Sierra Newt)   Fig. 56; PL 6H

IDENTIFICATION   From hatching throughout larval ontogeny, dorsum uniformly yellowish with dorsolateral black stripes on body and tail muscle; 50.0–60.0 mm TL.

NATURAL HISTORY   Female California Newts oviposit in coastal habitats from December through May. In some areas separate groups seemingly oviposit in ponds versus streams. The pale brown, gray to olive and cream ova pigmented to the equator (OD 2.0–2.8 mm, ED 4.5–5.3 mm, 2 jelly layers) are deposited in small groups of 7–47, attached to the roots and undersides of rocks in streams and vegetation in ponds. Their jelly is stiff and rubbery and the masses globular. Eggs hatch at 7.6–14.4 mm TL in 1.0–1.5 months. Larvae grow to 50.0–62.0 mm TL and metamorphose in 1.7–2.0 months; some may overwinter. Populations of the Sierra Newt breed in streams in the Sierra Nevada, while those of the California Newt in the Coast Range breed in streams or ponds.

RANGE   *T. sierra* is from localities in the Sierra Nevada Range of California up to 2000 m elevation, and *T. torosa* occurs in the Coastal Ranges from Mendocino to San Diego cos.

CITATIONS   *Development/Morphology*: Storer 1925, Twitty 1942, 1966. *Reproductive Biology*: Daniel 1937a–b, Ritter 1897, M. R. Miller and Robbins

1954. *Ecobehavior*: Brame 1956, 1968, Carroll et al. 2005, S. M. R. Jennings and Cook 1998, Marangio 1978, C. H. Marshall et al. 1990, Storer 1925.

Sirenidae (Dwarf Sirens and Sirens)

These unusual salamanders occur east of the Rocky Mountains; burrower in non-flowing water; disproportionately elongate, cylindrical body with small, front limbs with 3–4 digits, hind limbs absent throughout life; head rounded in dorsal view; mandibular grooves do not bisect lower lip anteriorly; gular fold absent; long gill rami branched, fimbriae numerous, rami atrophy during aestivation; keratinized jaw sheaths present; eyes small and dorsal; neuromasts not obvious in small individuals; lunged; balancers absent; unmarked, transparent dorsal fin originates well forward on body of larvae, opaque, fleshy fin of paedomorph originates at tail-body junction; paedomorphic; 2 genera, 4 species. *Pseudobranchus*: digits 3/0; gill slits 1; 2 species. *Siren*: digits 4/0; gill slits 3; 2 species; undescribed species known.

Pseudobranchus axanthus (Southern Dwarf Siren) and *P. striatus* (Northern Dwarf Siren)   Fig. 57; PL 6I

IDENTIFICATION Costal grooves 34–36 from axilla to posterior angle of vent; head narrow and pointed, body particularly slim; fins clear and originating on body in larvae, opaque and originate near tail-body junction in paedomorphs; toe tips not keratinized. Larva: dorsum dark with lateral and middorsal pale stripes; gill rami with pale pigment; fins clear. Paedomorph: unicolored to muted stripes nearly entire length of the body; about 10.0–20.0 mm TL as larva, 155.0–250.0 mm TL as paedomorph.

NATURAL HISTORY Dwarf Sirens live in ponds, lakes, and sloughs and have been reported to oviposit from February to November (*axanthus*) or November to March (*striatus*), but egg laying is likely to occur any month. Clutches may reach 60 lightly pigmented ova (OD 2.5–2.7 mm, ED 5.5–6.0 mm, 4 jelly layers); because

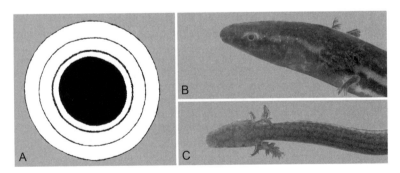

*Figure 57. **Pseudobranchus** spp.* (A) schematic drawing of single egg (modified from Noble and Richards 1932), (B–C) views of the head of a paedomorphic *P. striatus*: (B) lateral (195.0 mm TL, FL; WPL) and (C) dorsal (187.0 mm TL, FL; RA). See also Plate 6I.

eggs are attached singly or widely separated as small groups (up to 10) on aquatic plants, especially the rootlets of Water Hyacinth (*Eichhornia crassipes*), actual clutch size is unknown. Outer jelly is thick, opaque, and adhesive. Eggs hatch at 13.0–16.0 mm TL in 3–4 weeks, and larvae grow to adult size in a year but perhaps do not breed until the fourth year. The sex ratio apparently is essentially 1:1, but males have seemed uncommon in some studies because of the difficulty of distinguishing between the testes and ovaries.

RANGE   *P. striatus* is know from the Coastal Plain from southeastern South Carolina south to Volusia County and Hernando County, Florida, and through the western two-thirds of the Florida panhandle and into southeastern Georgia. Populations of *P. axanthus* occur from scattered localities in peninsular Florida.

CITATIONS *General*: Martof 1972. *Development/Morphology*: Goin 1942, 1947, Goin and Crenshaw 1949, Neill 1951b, Netting and Goin 1942, Schwartz 1952. *Reproductive Biology*: Goin 1947, Moler and Kezer 1993, Noble 1930, Pfaff and Vause 2002. *Ecobehavior*: J. R. Freeman 1967, Ultsch 1971, Wakeman and Ultsch 1975.

Siren intermedia (Lesser Siren)   Fig. 58; PL 6J,K

IDENTIFICATION Costal grooves 31–37 from axilla to posterior angle of vent; because of larval condition, usually impossible to sex externally except that large males often have enlarged head musculature that presents a bullish appearance. Larva: pronounced red band across tip of snout, along side of head, across the head behind the eyes, and sometimes middorsal stripe. Paedomorph: may keep pale snout band, but other marks disappear as the adult pattern appears; geographically variable, but dorsum greenish to gray with variable amounts of iridophores; about 10.0 mm TL as hatchling, to 690.0 mm TL as paedomorph.

NATURAL HISTORY Lesser Sirens live in an assortment of semipermanent and permanent habitats, including shallow lakes, ponds, ditches, sloughs, marshes, and swamps where aquatic vegetation and organic sediment are abundant. Oviposition

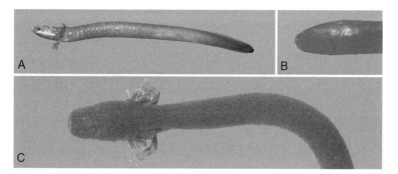

Figure 58. **Siren intermedia.** (A) juvenile (80.0 mm TL, pale coloration with black tail as photographed at night, Noxubee Co., MS; RA), (B) head of an individual encased in an aestivation cocoon (AA), (C) anterior part of a paedomorph (240.0 mm TL; Union Co., IL; RA). See also Plate 6J–K.

occurs from January to April, when 60–1500 lightly pigmented ova (OD 2.5–3.0 mm, ED 6.0–6.5 mm, 3 jelly layers, crystalline inclusions in outer layer) are laid as single eggs in large aggregates (200–500 eggs, masses measure up to 65 mm) at the bases of aquatic macrophytes, in fibrous mats of rootlets, and in depressions in bottom debris of ponds; usually the nests are guarded by a parent. Southern populations are likely to breed earlier than northern ones. Eggs hatch in 1.5–2.5 months at about 11.0 mm TL, and the larvae are brightly marked with red.

Courtship and fertilization modes are unknown, but adults often have many bite marks during the presumed breeding period. RA observed fast head bobs by the uppermost individual of two animals resting on top of each other with the posterior three-quarters of their bodies in a burrow.

A voracious, nocturnal, mostly benthic forager that often moves upward in dense stands of aquatic vegetation. Individuals click and yelp when in a strange place or when attacked by other sirens. These salamanders occur in many types of nonflowing water and are probably more abundant than normally recognized; they are commonly associated with vegetation and soft soil along the banks and readily gulp air when oxygen levels are low. They aestivate in burrows when the water dries and form a cocoon of multiple layers of shed skin. The highly modified, partial metamorphosis occurs slowly over a large size range.

RANGE *S. intermedia* is encountered on the Coastal Plain from Virginia to the southern tip of Texas and adjacent Mexico and north in the Mississippi Embayment to southwestern Michigan. It is absent from the southern third of peninsular Florida.

CITATIONS **General**: Martof 1973a, 1974a–b. **Development/Morphology**: Altig 1973, S. C. Bishop 1943, Cope 1885, Goin 1942, 1957, Middleton 1971, Neill 1949, Sugg et al. 1988, Trauth et al. 2004, H. H. Wilder 1891. **Reproductive Biology**: Collette and Gehlbach 1961, Godley 1983, McDowell 1997, Noble and Marshall 1932, Sever et al. 1996, Trauth et al. 1990. **Ecobehavior**: Altig 1967, 1973, Asquith and Altig 1987, Brodman 2008, W. B. Davis and Knapp 1953, Dunn 1924, Fauth 1999, Fauth and Resetarits 1999, Frese and Britzke 2001, Frese et al. 2003, Gehlbach et al. 1970, 1973, Gehlbach and Kennedy 1978, Raymond 1991, Reno et al. 1972, Scroggin and Davis 1956, Snodgrass et al. 1999, A. M. Sullivan et al. 2000, Weigmann and Altig 1975.

Siren lacertina (Greater Siren)   Fig. 59; PL 6L

IDENTIFICATION Costal grooves 36–40 from axilla to posterior angle of vent. Larva: pronounced yellow, and red to white or silver marks across tip of snout, stripes on head and laterally on body, middorsal stripe sometimes present. Paedomorph: snout band usually persists, but other marks disappear as the adult pattern appears, geographically variable but dorsum greenish to gray with variable numbers of iridophores; about 10.0 mm TL as hatchling, 50.0–60.0 mm TL as larva, to 980.0 mm TL as paedomorph.

NATURAL HISTORY Greater Sirens occur in semipermanent to permanent bodies of water such as lakes, ponds, streams, and canals that are often larger and deeper than those frequented by Lesser Sirens. Females oviposit from February

Figure 59. **Siren lacertina.** (A) schematic diagram of single egg (modified from Noble and Richards 1932), (B) larva (30.0 mm TL, Alachua Co., FL; BM), (C) juvenile (82.0 mm TL, FL; RWV). See also Plate 6L.

through April and may deposit 500–1400 brown and white ova (OD about 4.0 mm, 3 jelly layers, crystalline inclusions in outer layer) laid as singles or in small groups, perhaps attached to vegetation. Fertilization has never been observed and spermatophores have never been reported. Accordingly, we assume fertilization is external and occurs at the time of deposition. Eggs hatch in about 2 months at 13.0–16.0 mm TL.

RANGE *S. lacertina* is found on the Coastal Plain from northern Virginia and southern Maryland south through peninsular Florida to southwestern Alabama.

CITATIONS **General**: Martof 1973b, 1974a–b. **Development/Morphology**: Goin 1947, 1961, Neill 1949, H. H. Wilder 1891. **Reproductive Biology**: Hanlin and Mount 1978, Ultsch 1973. **Ecobehavior**: Aresco 2001, Frick 1999, Hanlin 1978, Hanlin and Mount 1978, Martof 1969, Pryor et al. 2006, Snodgrass et al. 1999, Sorensen 2004, Ultsch et al. 2004, Viosca 1925.

ORDER ANURA

Frogs and Toads

By stage 25, when the tadpole morphotype has been attained, several morphological features have proven to be useful for identifications (McDiarmid and Altig 1999). We incorporate these traits into a series of nondichotomous keys that, as do those presented earlier for salamanders, also rely on geography. We encourage readers to send us feedback on these keys so we can make progress on proper identifications of this difficult group of organisms.

The spiracle (Figs. 60A, 91A, 105D) differs in position and number: dual and lateral in pipids and rhinophrynids, sinistral (on left side) in bufonids, hylids, and ranids, sinistral but low on the side in scaphiopodids, or single and midventral (on chest in *Ascaphus*, near vent tube in microhylids). The presence and configuration of the centripetal wall (i.e., medial or dorsal depending on location of the spiracle) are also informative traits.

The dorsal fin originates well anterior on the body or near the dorsal tail-body junction (most common), and the tip may be broadly rounded to pointed. The flagellum (e.g., *Hyla femoralis* and particularly *Xenopus*) is an extension of the tail tip that sometimes moves independently of the remainder of the tail. The maximum height of the dorsal fin is greater than the ventral fin except in pipids and some scaphiopodids. Whether the greatest height of the dorsal fin is equal to (i.e., fin low) or higher than (i.e., fin high) the plane of the dorsal surface of the body is often used as a subjective measure. A measure of the maximum tail height (or individual fins) and its location posterior to the body terminus provides a rough impression of tail shape; even with considerable intraspecific variation (see Fig. 92B, E), fin configurations provide useful information.

The most consistently accurate measurement of body length (Fig. 60A) among species and stages is taken from the tip of the snout to the intersection of the axis of the tail myotomes with the body wall (i.e., body terminus). Tail length is measured from the body terminus to the tip of the tail. Measurements of greatest body height and its location from the tip of the snout and body width and its location provide an impression of body shape. *Compressed* describes the shape of any structure that is higher than wide (i.e., ellipse oriented vertically in cross-section), and *depressed* describes a structure that is wider than high (ellipse oriented horizontally in cross-section).

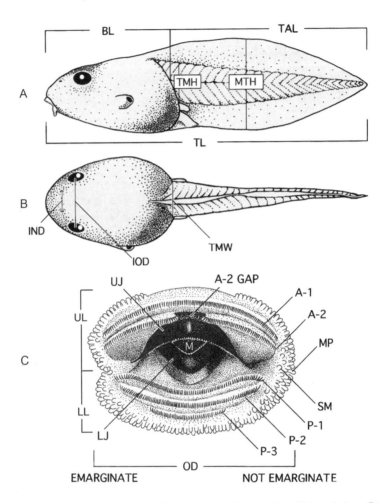

Figure 60. **Standard measurements and morphology of a tadpole.** (A) lateral view, (B) dorsal view, (C) typical oral apparatus. Notations: A-1 and A-2 = first and second anterior labial tooth rows; A-2 gap = medial gap in labial tooth row A-2; BL = body length; IND = internarial distance; IOD = interorbital distance; LJ = lower jaw sheath; LL = lower labium; M = mouth; MP = marginal papillae; MTH = maximum tail height; OD = oral disc diameter; P-1, P-2, and P-3 = first, second, and third lower labial tooth rows; SM = submarginal papillae; TAL = tail length; TL = total length; TMH = tail muscle height; TMW = tail muscle width; UJ = upper jaw sheath (KS); UL = upper labium.

Two major states describe the position and orientation of the vent tube: *dextral* (Fig. 105C), aperture opens to the right of the plane of the ventral fin, and *medial* (Fig. 63A), aperture opens parallel with the plane of the ventral fin. There are many other detailed variations that can be used (Johnston and Altig 1986). Eyes are dorsal (i.e., no part of the eye included in a dorsal silhouette) or lateral (i.e., margin of eyes included in dorsal silhouette; Fig. 6E–I).

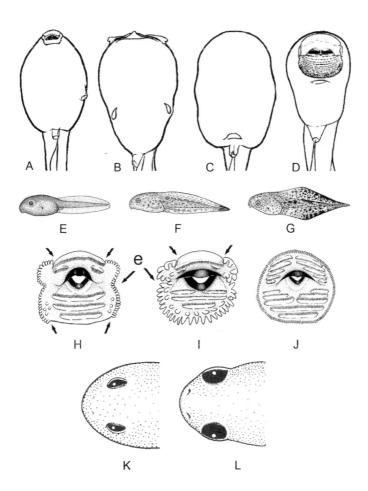

Figure 61. **Tadpole morphology.** Spiracles: (A) sinistral (single on left side; most common), (B) dual and lateral (e.g., pipids and rhinophrynids), (C) midventral near vent (e.g., microhylids), (D) midventral on chest (e.g., *Ascaphus*; all modified from Altig and McDiarmid 1999; all KS). Examples of tadpole fins: (E) low with a rounded tip (e.g., *Anaxyrus*; see also Plate 7A), (F) low with a pointed tip (e.g., *Acris* sp.; see also Plate 8E), (G) tall with a pointed tip (e.g., *Hyla* sp.; all PCU; see also Plate 8I). Configurations of marginal papillae (arrows): (H) wide dorsal and ventral gaps (e = emarginate) as in bufonids, (I) wide dorsal gap and emarginate as in ranids, (J) complete around oral disc and nonemarginate as in *Ascaphus* and scaphiopodids (all PCU). Examples of tadpole eyes positioned (K) dorsally (e.g., *Ascaphus*, bufonids, leptodactylids, ranids, and scaphiopodids) versus (L) laterally (e.g., hylids and microhylids; all KS).

The oral apparatus (Fig. 60C) of most tadpoles consists of two poorly defined labia (anterior, upper; posterior, lower). For the standardization of terms, the oral apparatus is always described in face view with the anterior labium uppermost. The lateral and posterior margins of the disc are free from the body and have various configurations of marginal papillae: complete around oral disc (e.g., *Ascaphus* and scaphiopodids), a dorsal gap that is narrow in scaphiopodids or wide in hylids and ranids, or wide dorsal and ventral gaps in bufonids. Size, shape, and density of marginal papillae (number/unit distance) are less commonly reported unless major differences are involved. Marginal papillae of hylids are small, short, and densely packed, and within *Hyla*, they are often accompanied by many adjacent submarginal papillae. Marginal papillae of ranids are large, short, and less densely arranged than in hylids. Lateral indentations in the margin of the disc (= emarginate; Figs. 60C left side, 117A) occur in bufonids and ranids. Staining mouthparts with crystal violet (i.e., dip wet tips of forceps into the powder, swirl the amount that sticks to the forceps in a small dish of water, momentarily swirl the mouthparts of a tadpole in that mixture; overstaining is easy, but repeat if a darker stain is needed; the stain will eventually leach out of the specimens; discard this solution after each usage) greatly enhances the ability to see translucent structures.

Various numbers of transverse rows of keratinized, pigmented labial teeth occur on each labium. These teeth, which have numerous shapes and patterns of cusps (e.g., Figs. 75B, 81A, 108A, 124E; Altig and Pace 1974, Gosner 1959, Vera Candioti and Altig 2010), occur on tooth ridges elevated above the local surfaces of the labia. Each erupted tooth usually has 1–3 replacement teeth interdigitated below it (Fig. 83J), and the total stack is termed a tooth series. Small muscles (Fig. 81B) change the orientation of the tooth rows during feeding. The labial tooth row formula (= LTRF) indicates the number of upper (numerator) and lower (denominator) tooth rows and which ones have medial gaps. Tooth rows are numbered from the margin of the oral disc toward the mouth on the upper labium and from the mouth toward the posterior edge of the labium on the lower labium. An LTRF of 5(5)/6(1) indicates that there are 5 upper and 6 lower rows, and the fifth upper (nearest the mouth) and first lower (nearest the mouth) have medial gaps. Tooth densities (teeth/mm) vary in ways that can be used as a subjective comparative feature; extremes of "coarse" (larger, widely spaced) and "fine" (smaller, closely spaced) are general descriptive terms. Keratinized and darkly pigmented jaw sheaths with serrated edges give the cartilages that serve as jaws in a tadpole (upper = suprarostral; lower = infrarostrals + Meckel's cartilages) a wear-resistant cutting edge. The jaw sheaths vary in shape and size among taxa but are generally similar in all taxa except *Ascaphus*; in this case, the upper sheath is large and platelike and lies almost parallel with the substrate, and the lower sheath is small and U-shaped.

The oral structures of microhylid, pipid, and rhinophrynid tadpoles deviate considerably from the above morphologies. Microhylids lack keratinized structures but have a pair of semicircular, fleshy oral flaps pendent over the mouth. The two flaps are separated by an inverted, U-shaped notch, the edges of the flaps may be papillate or not, and the medial margins may be convergent or divergent. Pipid tadpoles lack all derivatives of the oral disc, and the mouth is a simple, transverse slit with a single, long, motile barbel at each corner. The oral structures of rhinophrynids are similar to pipids except several pairs of short, nonmotile papillae occur around the slitlike mouth.

The nondichotomous key applies to free-living, feeding tadpoles and, because of the lack of more comprehensive data, emphasizes individuals in stages in the middle of larval ontogeny (stages 30–36). Tadpoles that lack an oral disc and keratinized mouthparts, and those with LTRFs of 2/2, greater than 2/3, and 2/3 are covered in sequence; geographic boundaries are used specifically for tadpoles with an LTRF of 2/3. Users of the key should check the species accounts for additional information. Also, note that tadpoles that lack parts or all of the keratinized oral structures because of an infection with chytrid fungus are not covered in the key.

KEY TO TADPOLES

ORAL DISC AND KERATINIZED MOUTHPARTS ABSENT

I. Spiracles dual and lateral without medial walls; snout depressed, body globular

   A. African exotic in southern California and southern Arizona; single long, motile barbel at each corner of slitlike mouth ....
....*Xenopus laevis* (**African Clawed Frog**)
Fig. 100A

   B. Southern Texas; several short, nonmotile barbels around slitlike mouth ....
.... *Rhinophrynus dorsalis* (**Burrowing Toad**)
Fig 122B, D–E

II. Single midventral spiracle on abdomen; fleshy hemispherical oral flaps pendent over V-shaped mouth, body depressed, rounded in dorsal view

   A. Margins of oral flaps divergent and smooth

     1. East of a line extending from south-central Texas coast northeast to central Missouri and east over most of the area south of the latitude of southern Illinois; dorsum mostly dark to black, paler at night; belly dark with pale blotches and mottling; lateral pale tail stripe bright and distinct basally; sympatric: *Gastrophryne olivacea* and *Hypopachus variolosus* west of Mississippi River, no confamilial taxa east of Mississippi River ....
.... *Gastrophryne carolinensis* (**Eastern Narrow-mouthed Toad**)
Fig. 98C–D, Plate 10G

     2. Eastern three-quarters of Texas north to southern Iowa and east to central Missouri, also south-central Arizona; dorsum olive-gray to brown to brassy; venter white or coppery, any contrasting marks small; lateral tail

stripe faint to absent; sympatric: *Gastrophryne carolinensis, Hypopachus variolosus* ..... *Gastrophryne olivacea* (**Western Narrow-mouthed Toad**)

Plate 10H

B. Margins of oral flaps convergent and papillate

Southern Texas and much of Mexico; dorsum uniform to mottled brownish, usually with pale middorsal stripe; venter faintly marked; sympatric: *Gastrophryne carolinensis, G. olivacea* .... *Hypopachus variolosus* (**Sheep Frog**)

Fig. 99A–D

TYPICAL ORAL APPARATUS PRESENT, LTRF 2/2

I. Arizona east to central Texas; ephemeral nonflowing water; dorsum of tail muscle usually banded; vent medial; eyes dorsal; marginal papillae with wide dorsal and ventral gaps, papillae terminate near lateral ends of LTR A-1; nostrils notably large and not exactly round; A-2 gap narrow

[Southeastern Arizona north to southeastern Colorado and east to central Texas .... *Anaxyrus debilis* (**Green Toad**)]

Plate 7E

[South-central Arizona south into Mexico ....
.... *Anaxyrus retiformis* (**Sonoran Green Toad**)]

II. East of Rocky Mountains to Atlantic Coast; ephemeral or permanent nonflowing water or slow streams; vent dextral; eyes dorsal (*Acris*) or lateral (*Pseudacris*); marginal papillae with only wide dorsal gap, papillae extend well medial of lateral ends of LTR A-1; nostrils medium and round

A. Permanent nonflowing water and slow streams; all individuals at all sites and stages with LTRF 2/2 (i.e., LTR P-3 never present); eyes dorsal; labial teeth coarsely spaced; A-2 gap wide; fins low; dorsum of tail muscle usually banded, tail tip usually black

1. Eastern Texas east to western Florida and Virginia and north to southern New York, absent from most of Atlantic Coastal Plain; A-2 gap about 120% of one side of LTR A-2; spiracular tube short; chin not boldly marked; sympatric: *Acris gryllus* ....
.... *Acris crepitans* (**Northern Cricket Frogs**)

Fig. 74A–B, Plate 8E

[West Texas north to southeastern Nebraska and east to central Ohio outside range of *A. crepitans*; sympatric: *Acris crepitans, A. gryllus* ....
.... *Acris blanchardi* (**Blanchard's Cricket Frog**)]

2. Southeastern Louisiana north to southeastern Tennessee and east to southeastern Virginia including all of Florida Peninsula; A-2 gap greater than 100% of one side of LTR A-2; LTR P-1 without medial gap; spiracular tube long; chin often boldly marked; sympatric: *Acris crepitans* ....
.... *Acris gryllus* (**Southern Cricket Frog**)

Fig. 75C, Plate 8F

B. Most common at ephemeral habitats; some individuals at some sites and stages with LTRF 2/2 if LTR P-3 absent; eyes lateral; labial teeth finely spaced; A-2 gap narrow; fins medium to tall; dorsum of tail rarely banded, tail tip never black

    1. Eastern Texas north to southeastern Manitoba and east to Nova Scotia and south to the northern third of Florida Peninsula; A-2 gap about 30% of one side of LTR A-2; LTR P-1 without medial gap; marginal papillae uniserial midventrally, sometimes with a medial gap in marginal papillae the width of LTR P-3; fins medium to tall, dorsum usually uniform, tail fin marks vary with locality and size from nearly absent to fine speckles to large blotches, dorsum of tail usually banded in younger stages, tail musculature never bicolored; throat speckled; sympatric: *Pseudacris "triseriata"* ....
    .... ***Pseudacris crucifer* (Spring Peeper)**
                          Fig. 89B–C, Plate 9I

    2. Several taxa widespread eastward from Rocky Mountains to Atlantic Coast; A-2 gap 20–30% of one side of LTR A-2; LTR P-1 without medial gap; LTR P-3 length 35–40% of P-2 length; midventral marginal papillae uniserial; sympatric: *Pseudacris crucifer* [**note**: LTRF 2/2 might occur in any of the Trilling Chorus Frogs] ....
    .... ***Pseudacris "triseriata"* (Trilling Chorus Frogs)**
                          Fig. 94D–F, Plate 10C

TYPICAL ORAL APPARATUS PRESENT, LTRF > 2/3

  I. Most of British Columbia coast south in Cascade and Coast ranges to just north of San Francisco Bay range; LTRF 2–3/9–12; upper jaw sheath massive and flat, lower sheath small and U-shaped, some tooth rows bi- or multiserial; LTRs A-1 and A-2 about equal in length, only LTR P-1 with a medial gap; oral disc emarginate, marginal papillae complete; eyes dorsal; vent medial; spiracle midventral on chest; body depressed; nostrils round with tubular extensions; highly variable coloration, usually an eyespot at tail tip; sympatric: none of similar morphology
    .... ***Ascaphus truei* (Coastal Tailed Frog)**
                          Fig. 96C, E, Plate 10F

    [Northern Rocky Mountains and Blue, Seven Devil, and Wallowa mountains in Idaho and Oregon .... ***Ascaphus montanus* (Rocky Mountain Tailed Frog)**]

  II. Pacific to Atlantic coasts; LTRF 2/4–6/7; LTRs A-1 and A-2 about same length; jaw sheaths typical; all tooth rows uniserial; oral disc emarginate; marginal papillae with dorsal gap; eyes dorsal; vent dextral; spiracle sinistral ....
    .... **Key Ranid > 2/3 (True Frogs) (p. 150)**

  III. Pacific to Atlantic coasts; LTR A-1 much shorter than A-2; marginal papillae complete; oral disc not emarginate; marginal papillae complete; vent medial; eyes dorsal .... **Key Scaphiopodid (Spadefoots) (p. 151)**

  IV. Atlantic coastal areas of Florida from Key West north to at least Jacksonville, and Gulf Coast north to at least Tampa Bay; LTRF 2/4; LTRs A-1 and A-2 about

same length; marginal papillae with dorsal gap, marginal papillae extend well medial of ends of LTR A-1; oral disc not emarginate; vent dextral; eyes lateral; ephemeral nonflowing water; sympatric: none of similar oral morphology ....
.... *Osteopilus septentrionalis* (**Cuban Treefrog**)

Fig. 85B–C

Key Ranid > 2/3

I. Southwestern British Columbia south to Elk Creek, Mendocino Co., California; exotic near Duckwater, Nevada; nonflowing water; LTRF 3/4; A-2 gap about 150% of one side of LTR A-2; LTR P-1 with medial gap; dorsal fin with prominent arch to nearly parallel with tail muscle; tail length about 1.8 times body length; dorsum reddish to maroon, sometimes with gold flecking; sympatric: *Rana cascadae* .... *Rana aurora* (**Northern Red-legged Frog**)

Fig. 117A–B, Plate 12B

[Elk Creek, Mendocino Co., California south into Baja California; often with LTRF 2/3; sympatric: none of similar morphology ....
.... *Rana draytonii* (**California Red-legged Frog**)]

Fig. 117D

II. Olympic and Cascade mountains from near Canadian border south to Mt. Lassen, California; lake margins and associated pools; LTRF 3/4; A-2 gap about 70% of one side of LTR A-2; LTR P-1 without medial gap; dorsal fin with prominent arch; tail length about 1.8 times body length; dorsum uniformly brownish; sympatric: *Rana aurora* .... *Rana cascadae* (**Cascades Frog**)

Fig. 119B, Plate 12D

III. Transverse ranges in southern California and central Sierra Nevada, 1370–3660 m; slow streams and associated nonflowing water; LTRF 2/4–4/4; A-2 gap about 150% of one side of LTR A-2 length; LTR P-1 with medial gap; dorsal fin nearly parallel with tail muscle; tail length 1.8–2.4 times body length; dorsum dark brown, sometimes mottled; sympatric: none of similar LTRF ....
.... *Rana muscosa* (**Mountain Yellow-legged Frog**)

Fig. 120B

[Sierra Nevada of California and extreme western Nevada; sympatric: none of similar LTRF and habitat ....
.... *Rana sierra* (**Sierra Nevada Yellow-legged Frog**)]

Plate 12F

IV. Most of Alaska to southern British Columbia, south and east across Canada to Arkansas and east-central Alabama and northeast to northern Quebec, disjunct populations in Wyoming and Colorado; usually ephemeral, nonflowing water, often wooded; LTRF 2–4/4; A-2 gap about 50% of one side of LTR A-2; LTR P-1 without medial gap; dorsal fin low; tail length 1.7 times body length; dorsum uniformly dark, often with profusion of iridophores, pale lip line usually present, fins without contrasting marks; sympatric: none with similar LTRF ....
....*Lithobates sylvaticus* (**Wood Frog**)

Fig. 114B, Plate 11K

V. LTRF 4–5/3

South-central Arizona; rocky streams; dorsum and fins boldly marked; sympatric: no congeners with similar LTRF ....
.... *Lithobates tarahumarae* (**Tarahumara Frog**)
Fig. 115A, C–D

VI. LTRF 6/6–7

West-central Oregon south in Sierra Nevada-Cascade and Coast ranges to southern California outside the Central Valley; streams; A-2 gap about 25% of one side of LTR A-2; LTR P-1 with medial gap; sympatric: none of similar LTRF ....
.... *Rana boylii* (**Foothill Yellow-legged Frog**)
Fig. 118A, Plate 12C

Key Scaphiopodid

I. 40 mm TL or less; keratinized knob absent on roof of anterior part of buccal cavity; metatarsal spade of advanced tadpoles sickle-shaped; carnivore morphotypes absent .... *Scaphiopus*

A. Southeastern California across southern Arizona and New Mexico east across Rocky Mountains to southeastern Oklahoma and south into Mexico; ephemeral, often xeric, nonflowing water; LTRF typically 4/4; spiracle nearer eye than vent; tail tip obtusely pointed; dorsum freckled brassy with irregular black marks on posterior third of tail muscle; sympatric: *Scaphiopus hurterii*
.... *Scaphiopus couchii* (**Couch's Spadefoot**)
Plate 12G

B. Northeastern third of Arkansas and adjacent Missouri and across much of the eastern states; usually ephemeral, nonflowing water; LTRF usually 6/6, majority of rows with medial gaps; spiracle nearer vent than eye and low on side; tail tip broadly rounded; brown to bronze with no contrasting marks on tail muscle; sympatric: no congeners ....
.... *Scaphiopus holbrookii* (**Eastern Spadefoot**)
Fig. 124F–G, Plate 12H

C. Eastern third of Texas and Oklahoma and adjacent Arkansas and Louisiana; ephemeral nonflowing water; LTRF 4/4–5/5; tail tip broadly rounded; gray to bronzy brown, light band at tail base if develop in clear water; similar to *S. holbrookii*; sympatric: *Scaphiopus couchii* ....
.... *Scaphiopus hurterii* (**Hurter's Spadefoot**)

II. To 70 mm TL, often smaller; keratinized knob present on roof of anterior part of buccal cavity; LTRF usually 6/6; metatarsal spade of advanced tadpoles wedge-shaped; normal and carnivore (i.e., greatly hypertrophied jaw musculature; jaw sheaths robust and cuspate) morphotypes present at some sites in some years; identification in sympatry tenuous .... *Spea*

A. East of Rocky Mountains from southern Alberta, Saskatchewan, and Manitoba south to southeastern Arizona and east to eastern Missouri; ephemeral, often xeric, nonflowing water; sympatric: *Spea multiplicata* ....
.... *Spea bombifrons* (**Plains Spadefoot**)
Fig. 125B–C, Plate 12I

B. Central and southern California; ephemeral, often xeric, nonflowing water; to 70 mm TL, often smaller; LTR A-1 much shorter than A-2; marginal papillae complete;sympatric: no congeners .... ***Spea hammondii* (Western Spadefoot)**
Fig. 126B, Plate 12J

C. South-central British Columbia south to southern Nevada and Wyoming and northwestern Colorado; ephemeral, often xeric, nonflowing water; sympatric: *Spea bombifrons* .... ***Spea intermontana* (Great Basin Spadefoot)**
Fig. 127A, D, Plate 12K

D. Southeastern Utah south into Mexico and east to northwestern Oklahoma panhandle and western third of Texas; ephemeral, often xeric, nonflowing water; sympatric: *Spea bombifrons* .... ***Spea multiplicata* (Mexican Spadefoot)**
Fig. 128A, C, Plate 12L

TYPICAL ORAL APPARATUS PRESENT, LTRF 2/3

I. Hawaii
  A. Eyes dorsal; vent medial
    1. Neotropical exotic on Maui and Oahu; in phytotelmons; only dorsal gap in marginal papillae; oral disc not emarginate; body uniformly dark; sympatric: none of similar morphology ....
.... ***Dendrobates auratus* (Green and Black Dart-poison Frog)**
Fig. 73A, C

    2. Neotropical exotic on all major islands; nonflowing water and slow streams; wide dorsal and ventral gaps in marginal papillae; oral disc emarginate; sympatric: none of similar morphology ....
.... ***Rhinella marina* (Cane Toad)**
Fig. 72A

  B. Eyes dorsal; vent dextral; oral disc emarginate
    1. Oriental exotic on all major islands; dorsum faintly marbled to uniform brown; all sizes similar in coloration; sympatric: *Lithobates catesbeianus* .... ***Glandirana rugosa* (Japanese Wrinkled Frog)**
Fig. 101A–B

    2. Mainland exotic on all major islands; greenish to brownish dorsum with distinct black dots on dorsum and dorsal fin; individuals smaller than 25 mm TL with dark ground color and dappling of gold iridophores; sympatric: *Glandirana rugosa* .... ***Lithobates catesbeianus* (American Bullfrog)**
Fig. 105B, F, Plate 10L

II. Pacific Coast east to include Sierra Nevada–Cascade ranges
  A. Oral disc emarginate; jaw sheaths medium; marginal papillae with wide dorsal and ventral gaps; eyes dorsal; vent medial; spiracle sinistral; body depressed; nostrils large and not round ..... **Key Pacific Toad (Toads) (p. 154)**

B. Oral disc not emarginate; jaw sheaths medium to wide; marginal papillae with dorsal gap; eyes lateral; vent dextral; spiracle sinistral; body compressed; nostrils round .... **Key Pacific Treefrog (p. 155)**

C. Oral disc emarginate; jaw sheaths medium to wide; marginal papillae with wide dorsal gap; eyes dorsal; vent dextral; spiracle sinistral; body depressed; nostrils variable .... **True Frogs**

1. Jaw sheaths wide, front surface of lower sheath convex; demarcation between keratinized (black) and nonkeratinized (white) parts of jaws abrupt and discrete; nostrils large, easily seen .... *Lithobates pipiens* **group**

   Scattered populations in California; nonflowing water; A-2 gap about 200% of one side of LTR A-2; LTR P-1 with or without medial gap; LTR P-3 length 80–90% of P-2 length; ventral marginal papillae medium; general coloration relatively uniform throughout or with subtle mottling and small contrasty tail marks; neuromasts obscure; belly musculature barely visible through skin of preserved tadpole; sympatric: *Rana draytonii* .... *Lithobates pipiens* **(Northern Leopard Frog)**
   Fig. 113D, G, Plate 11H

2. Jaw sheaths narrow to medium, front surface of lower sheath not strongly convex; demarcation between keratinized (black) and nonkeratinized (white) parts of jaws diffuse; nostrils small, difficult to see .... *Lithobates catesbeianus* **group and relatives**

   a. Scattered exotic populations throughout much of the area west of Sierra Nevada–Cascade ranges; larger, permanent nonflowing water; greenish to brownish dorsum with distinct black dots on body and dorsal fin; individuals smaller than 25 mm TL with dark ground color and dappling of gold iridophores; sympatric: *Lithobates clamitans, Rana luteiventris* .... *Lithobates catesbeianus* **(American Bullfrog)**
   Fig. 105B, F, Plate 10L

   b. Exotic in northeastern Washington, northwestern Montana; permanent, often swampy, nonflowing water; dorsum gray to brown and many small markings without discrete borders, distal third of tail often with contrasting marks; A-2 gap over 300% of one side of LTR A-2; LTR P-1 with medial gap; LTR P-3 length about 50% of P-2 length; sympatric: *Lithobates catesbeianus, Rana luteiventris* .... *Lithobates clamitans* **(Green Frog)**
   Fig. 107B, D–E, Plate 11B

   c. Central British Columbia south to central Nevada and east to northeastern Colorado, disjunct populations in the southern part of range; dorsum uniform or finely marbled; sympatric: *Lithobates catesbeianus, L. clamitans* .... *Rana luteiventris* **(Columbia Spotted Frog)**
   Fig. 121C, F–G

III. East of Sierra Nevada–Cascade ranges to include Rocky Mountains

A. Oral disc emarginate; jaw sheaths medium; marginal papillae with wide dorsal and ventral gaps; eyes dorsal; vent medial; spiracle sinistral; body depressed; nostrils large and not round .... **Key Middle Toad (Toads) (p. 155)**

B. Oral disc not emarginate; jaw sheaths medium to wide; marginal papillae with dorsal gap; eyes lateral; vent dextral; spiracle sinistral; body compressed; nostrils round .... **Key Middle Treefrog (Treefrogs and Chorus Frogs) (p. 157)**

C. Oral disc emarginate; jaw sheaths medium to wide; marginal papillae with wide dorsal gap; eyes dorsal; vent dextral; spiracle sinistral; body depressed; nostrils variable .... **Key Middle Frog (True Frogs) (p. 158)**

IV. East of Rocky Mountains to Mississippi River

A. Southern Texas; ephemeral nonflowing water; vent medial; eyes dorsal; oral disc not emarginate; marginal papillae, with only wide dorsal gap, extend slightly medial to ends of LTR A-1; uniformly dark or paler and slightly mottled; A-2 gap narrow; sympatric: none of similar morphology but can be mistaken for a bufonid or small ranid .... *Leptodactylus fragilis* **(Mexican White-lipped Frog)** Fig. 97B–C

B. Oral disc emarginate; jaw sheaths medium; marginal papillae with wide dorsal and ventral gaps; eyes dorsal; vent medial; spiracle sinistral; body depressed; nostrils large and not round .... **Key Central Toad (Toads) (p. 160)**

C. Oral disc not emarginate; jaw sheaths medium to wide; marginal papillae with dorsal gap; eyes lateral; vent dextral; spiracle sinistral; body compressed; nostrils round .... **Key Central Treefrog (Treefrogs and Chorus Frogs) (p. 162)**

D. Oral disc emarginate; jaw sheaths medium to wide; marginal papillae with wide dorsal gap; eyes dorsal; vent dextral; spiracle sinistral; body depressed; nostrils variable .... **Key Central Frog (True Frogs) (p. 164)**

V. East of Mississippi River to Atlantic Coast

A. Oral disc emarginate; jaw sheaths medium; marginal papillae with wide dorsal and ventral gaps; eyes dorsal; vent medial; spiracle sinistral; body depressed; nostrils large and not round .... **Key Eastern Toad (Toads) (p. 166)**

B. Oral disc not emarginate; jaw sheaths medium to wide; marginal papillae with dorsal gap; eyes lateral; vent dextral; spiracle sinistral; body compressed; nostrils round .... **Key Eastern Treefrog (Treefrogs and Chorus Frogs) (p. 167)**

C. Oral disc emarginate; jaw sheaths medium to wide; marginal papillae with wide dorsal gap; eyes dorsal; vent dextral; spiracle sinistral; body depressed; nostrils variable .... **Key Eastern Frog (True Frogs) (p. 171)**

Key Pacific Toad

I. Southeastern Alaska to southern California and into Baja California; ephemeral and permanent nonflowing water and slow streams; A-2 gap about 45% of one side of LTR A-2; LTR P-1 without medial gap; LTR P-3 length about 90% of P-2 length; small individuals throughout range and large individuals in most parts of range are

black dorsally, large specimens may have coppery iridophores posteroventrally and laterally, profuse dorsal iridophores occur in some southern populations; dorsum and sides of tail muscle uniformly dark; fins clear to moderately and uniformly opaque; snout slopes gradually in lateral view, slightly pointed in dorsal view; sympatric: *Anaxyrus californicus*, *A. canorus* .... ***Anaxyrus boreas* (Western Toad)**
Fig. 63D, Plate 7B

II. Central Sierra Nevada from Eldorado to Fresno cos., California above 1460 m; alpine lakes and marshy meadows; A-2 gap 75–90% of one side of LTR A-2; LTR P-1 with medial gap; LTR P-3 length about 45–50% of P-2 length; individuals of all stages sooty black, fins semi-opaque, dorsum of tail muscle uniformly dark; snout rounded in dorsal and lateral views; sympatric: *Anaxyrus boreas* .... ***Anaxyrus canorus* (Yosemite Toad)**
Plate 7C

III. Monterey County south in coastal California to northwestern Baja California; backwaters of rocky streams; LTR P-3 length about 60% of P-2 length; small individuals appear black, but at most stages brassy to golden, nonmelanic pigment covers most of body; dorsum of tail muscle banded, blotched in larger specimens; fins largely unmarked; snout curves sharply downward in lateral view; sympatric: *Anaxyrus boreas* .... ***Anaxyrus californicus* (Arroyo Toad)**
Plate 7I

Key Pacific Treefrog

I. Southwestern coastal California south into Mexico; slow flowing streams and associated nonflowing water; jaw sheaths medium; dorsum of tail muscle often banded; sympatric: *Pseudacris hypochondriaca* ....
.... ***Pseudacris cadaverina* (California Chorus Frog)**
Plate 9G

II. Pacific Northwest from northern California, adjacent Oregon, Washington, and British Columbia to southern Alaska; nonflowing water, often ephemeral habitats; jaw sheaths wide and robust; A-2 gap about 40% of one side of LTR A-2; LTR P-1 with medial gap; LTR P-3 about 60% of P-2 length; dorsum of tail muscle never banded; sympatric: none in group...
.... ***Pseudacris regilla* (Northern Pacific Treefrog)**
Fig. 92B

[Southern California, Nevada, and adjacent Arizona south into Baja California; sympatric: *Pseudacris cadaverina* ....
.... *Pseudacris hypochondriaca* (**Baja California Treefrog**)]
Fig. 92E, Plate 9J

[Central California, eastern Oregon and Idaho and western Montana; sympatric: none in group... *Pseudacris sierra* (**Sierran Treefrog**)]

Key Middle Toad

I. Small relictual ranges in east-central California and adjacent Nevada; sympatric: no congeners

[Spring runs and sloughs; Deep Springs Valley east of Big Pine, Inyo Co., California .... *Anaxyrus exsul* (**Black Toad**)]

Fig. 63E

[Spring runs and river marshes; Amargosa River drainage near Beatty, Nye Co., Nevada .... *Anaxyrus nelsoni* (**Amargosa Toad**)]

Fig. 63I

II. Larger ranges with sympatric congeners; length of LTR P-3 less than 40% of P-2 length
Southeastern California north to southeastern Nevada and east across Rocky Mountains; ephemeral nonflowing water; A-2 gap 15–30% of one side of LTR A-2; LTR P-1 with medial gap; dorsal fin noticeably arched; at least large individuals mottled with iridophores; pale throat patch; long sloping snout in lateral view; sympatric: none with short P-3 .... *Anaxyrus cognatus* (**Great Plains Toad**)

Plate 7D

III. Larger ranges with sympatric congeners; length of LTR P-3 medium, 40–70% of P-2 length

A. Eastern British Columbia east and south to central Alberta and south to central Colorado; ephemeral and permanent nonflowing water and slow streams; A-2 gap about 45% of one side of LTR A-2; LTR P-1 without medial gap; small individuals throughout range and large individuals in most parts of range appear black, large specimens may have coppery iridophores posteroventrally and laterally, in some southern areas, profuse dorsal iridophores occur; dorsum and sides of tail muscle uniformly dark; fins clear to moderately but uniformly opaque; pale throat patch small to absent; snout slopes gradually in lateral view; sympatric: *Anaxyrus punctatus*, *A. woodhousii* .... *Anaxyrus boreas* (**Western Toad**)

Fig. 63D, Plate 7B

B. Disjunct populations from southern Nevada and adjacent Utah southeast to southwestern New Mexico; usually in slow streams; larger specimens with many iridophores that impart a brassy coloration, especially laterally; tail muscle blotched to uniform, slightly banded in smaller individuals; ephemeral and permanent habitats of flowing water; sympatric: *Anaxyrus punctatus*, *A. woodhousii* .... *Anaxyrus microscaphus* (**Arizona Toad**)

Plate 7H

C. Southeastern California, southern Nevada, and adjacent Arizona, Utah, southwestern Idaho, and eastern Oregon and Washington, and east across Rocky Mountains; ephemeral nonflowing water; A-2 gap 40–50% of one side of LTR A-2; LTR P-1 with medial gap; dorsal fin not notably arched; small specimens appear black but scattered golden iridophores visible at slight magnification; larger specimens with iridophores on body and dorsum of tail muscle; ephemeral nonflowing water; sympatric: *Anaxyrus boreas*, *A. microscaphus*, *Incilius alvaria* .... *Anaxyrus woodhousii* (**Woodhouse's Toad**)

Fig. 70B–C, Plate 8B

D. Southern half of Arizona and adjacent California and New Mexico; ephemeral nonflowing water; A-2 gap about 55% of one side of LTR A-2; LTR P-1 without medial gap; snout with uniform, medium curvature in lateral view; large specimens appear uniformly brassy or brownish; tail muscle with small, irregular, low-contrast dark marks but never bicolored; sympatric: *Anaxyrus woodhousii* .... ***Incilius alvaria* (Sonoran Desert Toad)**
Fig. 71A–B, Plate 8C

IV. Larger ranges with sympatric congeners; lengths of LTRs P-3 and P-2 about equal

Southeastern California east to central Oklahoma and south into Mexico; ephemeral nonflowing water and slow streams; A-2 gap about 13% of one side of LTR A-2; LTR P-1 without medial gap; LTR P-3 length about 100% of P-2 length; typically appears uniformly black to pale brown but significant variations are known from southern Nevada and Arizona; tail musculature unicolored, bicolored, or irregularly blotched; sympatric: *Incilius alvaria*, *A. boreas*, *A. microscaphus*, *A. woodhousii* .... ***Anaxyrus punctatus* (Red-spotted Toad)**
Fig. 67A–C, Plate 7J

Key Middle Treefrog

I. Lengths of LTRs P-2 and P-3 about equal; lateral eyes might be interpreted as dorsal

Eastern two-thirds of Arizona and adjacent Utah and New Mexico; slow rocky streams; body depressed; A-2 gap about 15% of one side of LTR A-2; LTR P-1 without medial gap; LTR P-3 length 90–100% of P-2 length; uniformly brown or tan; sympatric: *Hyla wrightorum*, *Pseudacris maculata* ....
.... ***Hyla arenicolor* (Canyon Treefrog)**
Fig. 77A–C, Plate 8H

II. Length of LTR P-3 notably shorter than P-2 length; eyes definitely lateral

A. Band from northwestern to southeastern Arizona; montane nonflowing water; A-2 gap about 25% of one side of LTR A-2; LTR P-1 without medial gap; LTR P-3 length about 50% of P-2 length (based on *H. eximia*); jaw sheaths medium; sympatric: *Pseudacris maculata* ....
.... ***Hyla wrightorum* (Mountain Treefrog)**
Fig. 84A–B, Plate 9D

B. Central Arizona north to northeastern British Columbia east to Rocky Mountains; jaw sheaths medium; sympatric: *Hyla arenicolor*, *H. wrightorum* ....
.... ***Pseudacris maculate* (Boreal Chorus Frog)**
Fig. 94E

C. Southern California, Nevada and adjacent Arizona south into Baja California; usually nonflowing water; A-2 gap about 40% of one side of LTR A-2; LTR P-1 with medial gap; LTR P-3 about 60% of P-2 length based on *P. regilla*) jaw sheaths wide and robust; sympatric: none in group ....
.... ***Pseudacris hypochondriaca* (Baja California Treefrog)**
Fig. 92E, Plate 9J

[Central California, eastern Oregon and Idaho and western Montana; sympatric: none in group... *Pseudacris sierra* (**Sierran Treefrog**)]

D. South-central Arizona into Mexico; ephemeral nonflowing water; jaw sheaths narrow; sympatric: none in group .... *Smilisca fodiens* (**Lowland Burrowing Treefrog**)

Plate 10E

Key Middle Frog

I. Jaw sheaths wide, front surface of lower sheath convex; demarcation between keratinized (black) and nonkeratinized (white) parts of jaws abrupt and discrete; nostrils large, easily seen... *Lithobates pipiens* **group**

A. Exotic in Gila River drainage, Arizona; nonflowing water and slow streams; A-2 gap more than 150% of one side of LTR A-2; LTR P-1 with medial gap; LTR P-3 length about 80% of P-2 length; ventral marginal papillae large; oral disc less than 22% of body length; general coloration pale with well-defined reticulated pattern on tail; neuromasts obscure; belly musculature weakly developed; sympatric: *Lithobates yavapaiensis* .... *Lithobates berlandieri* (**Rio Grande Leopard Frog**)

Fig. 103A–B, Plate 10J

B. Southeastern Arizona; nonflowing water, often turbid; A-2 gap about 65% of one side of LTR A-2; LTR P-1 without medial gap; LTR P-3 length about 85–90% of P-2 length; ventral marginal papillae large; oral disc less than 22% of body length; uniformly pale coloration; neuromasts obscure; belly musculature prominently visible in preserved tadpole; sympatric: *Lithobates chiricahuensis* .... *Lithobates blairi* (**Plains Leopard Frog**)

Fig. 104C–E, Plate 10K

C. Central to southwestern Arizona and adjacent New Mexico, includes *L. subaquavocalis* populations; nonflowing water and slow streams; A-2 gap well more than 200% of one side of LTR A-2; LTR P-1 with medial gap; LTR P-3 length 60–70% of P-2 length; ventral marginal papillae medium; oral disc less than 22% of body length; general coloration dark, definite large blotches on posterior third of tail and ventrolaterally on body; neuromasts obscure; belly musculature easily visible in preserved tadpole; sympatric: *Lithobates blairi, L. "pipiens," L. yavapaiensis* .... *Lithobates chiricahuensis* (**Chiricahua Leopard Frog**)

Fig. 106B–C, Plate 11A

D. Most of Nevada and parts of Arizona, Idaho, and Washington; permanent and ephemeral nonflowing water and slow streams; A-2 gap about 200% of one side of LTR A-2; LTR P-1 with or without medial gap; LTR P-3 length 80–90% of P-2 length; ventral marginal papillae medium; oral disc less than 22% of body length; general coloration relatively uniform throughout or with subtle mottling and small contrasty tail marks; neuromasts obscure; belly musculature

barely visible through skin of preserved tadpole; sympatric: *Lithobates chir-icahuensis* .... **Lithobates pipiens** (**Northern Leopard Frog**)
Fig. 113D, G, Plate 11H

E. Southeastern Nevada to southern Arizona east to central Arizona south of Mogollon Rim and adjacent New Mexico; usually slow streams; A-2 gap 80–85% of one side of LTR A-2; LTR P-1 with medial gap; LTR P-3 length 75–80% of P-2 length; ventral marginal papillae medium; generally dark gray coloration formed of fine stippling and larger blotches common on tail, belly dark; oral disc 22% or more of body length; neuromasts easily visible on dark dorsum; posterior belly musculature partially visible; sympatric: *Lithobates berlandieri, L. chiricahuensis* ....
.... **Lithobates yavapaiensis** (**Lowland Leopard Frog**)
Plate 12A

[Extinct in Vegas Valley, Clark Co., Nevada but persists along the Mogollon Rim of Arizona; sympatric: *Lithobates chiricahuensis* ....
.... **Lithobates fisheri** (**Vegas Valley Leopard Frog**)]
Fig. 111A

[Dorsum uniformly bronzy to boldly mottled; southern Nevada and adjacent Arizona; sympatric: none in group ....
.... **Lithobates onca** (**Relict Leopard Frog**)]
Fig. 111B–C, Plate 11F

II. Jaw sheaths narrow to medium, front surface of lower sheath not strongly convex; demarcation between keratinized (black) and nonkeratinized (white) parts of jaws diffuse; nostrils small, difficult to see ....
.... **Lithobates catesbeianus** **group and relatives**

A. Scattered populations throughout much of the area between the Sierra Nevada–Cascade ranges and Rocky Mountains; larger, permanent nonflowing water; greenish to brownish dorsum with distinct black dots on body and dorsal fin; individuals smaller than 25 mm TL with dark ground color and dappling of gold iridophores; sympatric: *Lithobates clamitans, Rana luteiventris* .... **Lithobates catesbeianus** (**American Bullfrog**)
Fig. 105B, F, Plate 10L

B. Exotic in northwestern Montana, and near Salt Lake City, Utah; permanent, often swampy, nonflowing water; dorsum gray to brown and many small markings without discrete borders, distal third of tail often with contrasting marks; A-2 gap over 300% of one side of LTR A-2; LTR P-1 with medial gap; LTR P-3 length about 50% of P-2 length; sympatric: *Lithobates catesbeianus, Rana luteiventris* ....
.... **Lithobates clamitans** (**Green Frog**)
Fig. 107B, D–E, Plate 11B

C. Central British Columbia south to central Nevada and east to northeastern Colorado, disjunct populations in the southern part of range; dorsum uniform

or finely marbled; sympatric: *Lithobates catesbeianus, L. clamitans* ....
.... ***Rana luteiventris* (Columbia Spotted Frog)**
Fig. 121C, F–G

Key Central Toad

I. Length of LTR P-3 less than 40% of P-2 length

A. Southeastern Utah south into Mexico, east to central Oklahoma and Missouri and north to southeastern Alberta; ephemeral nonflowing water; A-2 gap 15–30% of one side of LTR A-2; LTR P-1 with medial gap; LTR P-3 length 30% or less of A-2; at least large individuals mottled with iridophores; long sloping snout in lateral view, broadly rounded in dorsal view; sympatric: *Anaxyrus speciosus* .... ***Anaxyrus cognatus* (Great Plains Toad)**
Plate 7D

B. Southeastern New Mexico east to northeastern Texas and south into Mexico; ephemeral nonflowing water; A-2 gap 60–85% of one side of LTR A-2; LTR P-1 without medial gap; LTR P-3 length about 60% of P-2 length; small specimens appear black, larger specimens tan to brassy; tail muscle crudely bicolored, with pale blotches within black portion, or pale with irregular dark stripe, sometimes with pale line through eye; snout curves sharply downward in lateral view, vaguely pointed in dorsal view; sympatric: *Anaxyrus cognatus* .... ***Anaxyrus speciosus* (Texas Toad)**
Fig. 68A–B, Plate 7L

II. Length of LTR P-3 greater than 40% of P-2 length

A. Dorsum of tail muscle distinctly banded

Eastern Texas and southern Louisiana, disjunct populations in west Texas and south-central Arkansas; ephemeral nonflowing water; A-2 gap about 50% of one side of LTR A-2; LTR P-1 with medial gap; LTR P-3 length about 50–70% of P-2 length; body appears black but at least large specimens with abundant iridophores; dorsum of tail muscle with definite white bands; snout slopes gradually in lateral view; sympatric: *Anaxyrus americanus, A. fowleri, A. houstonensis, A. woodhousii* .... ***Incilius nebulifer* (Gulf Coast Toad)**
Plate 8D

B. Dorsum of tail muscle uniform or at least not banded

1. East and north of a line connecting central Manitoba and northeastern Texas; usually ephemeral nonflowing water; A-2 gap 40–55% of one side of LTR A-2; LTR P-1 without medial gap; LTR P-3 length about 70–80% of P-2 length; appears black but abundant golden iridophores visible at slight magnification; iridophores often concentrated along top of tail muscle; white part of bicolored tail muscle about 25% of basal muscle height; snout moderate, about 1.5 eye diameters from front of eye to tip of snout; spiracle on longitudinal axis; snout sloping in lateral view; sympatric: *Anaxyrus fowleri, A. hemiophrys, A. woodhousii, Incilius nebulifer* ....
.... ***Anaxyrus americanus* (American Toad)**
Fig. 62B–C, Plate 7A

[Laramie Hills, southeastern Wyoming; sympatric: *Anaxyrus cognatus, A. punctatus, A. woodhousii* .... ***Anaxyrus baxteri*** (**Wyoming Toad**)]
Fig. 66C–D

[Eastern half of Alberta south and east to northeastern South Dakota; sympatric: *Anaxyrus americanus, A. woodhousii* ....
.... ***Anaxyrus hemiophrys*** (**Canadian Toad**)]
Fig. 66B, Plate 7G

[East-central Texas; sympatric: *Anaxyrus speciosus, A. woodhousii, Incilius nebulifer* .... ***Anaxyrus houstonensis*** (**Houston Toad**)]
Fig. 62D

2. East of a line connecting southeastern Texas and southeastern Iowa; usually ephemeral nonflowing water; A-2 gap about 50% of one side of LTR A-2; LTR P-1 without medial gap; LTR P-3 length about 60–65% of P-2 length; small specimens appear black, larger specimens with abundant silver to brassy iridophores that impart a frosted or mottled appearance; white part of bicolored tail about 50% of basal muscle height; snout curves sharply downward in lateral view; spiracle on longitudinal body axis; sympatric: *Anaxyrus americanus, A. woodhousii, Incilius nebulifer* ....
.... ***Anaxyrus fowleri*** (**Fowler's Toad**)
Fig. 65B, D, Plate 7F

3. Southwestern New Mexico north to central Montana and east to northeastern Missouri; ephemeral nonflowing water; A-2 gap 40–50% of one side of LTR A-2; LTR P-1 with medial gap; LTR P-3 length 50–70% of P-2 length; small specimens appear black but scattered golden iridophores visible at slight magnification; larger specimens with considerable iridophores on body and dorsum of tail muscle; sympatric: *Anaxyrus americanus, A. fowleri, A. "hemiophrys," A. houstonensis, Incilius nebulifer* ....
.... ***Anaxyrus woodhousii*** (**Woodhouse's Toad**)
Fig. 70B–C, Plate 8B

III. Lengths of LTR P-2 and P-3 about equal

A. Southern Rocky Mountains of New Mexico and southern Colorado east to eastern third of Texas; ephemeral nonflowing water and slow streams; A-2 gap about 13% of one side of LTR A-2; LTR P-1 without medial gap; LTR P-3 length about 100% of P-2 length; typically appears uniformly black to pale brown but significant variations are known from southern Nevada and Arizona; tail musculature unicolored, bicolored, or irregularly blotched; sympatric: *Rhinella marina* .... ***Anaxyrus punctatus*** (**Red-spotted Toad**)
Fig. 67A–C, Plate 7J

B. Southern Texas; nonflowing water; A-2 gap 30–40% of one side of LTR A-2; LTR P-3 length about 100% of P-2 length; entirely black, except bicolored tail (about 30% of basal tail height); snout slopes gradually in lateral view; sympatric: *Anaxyrus punctatus* .... ***Rhinella marina*** (**Cane Toad**)
Fig. 72A

Key Central Treefrog

I. Lengths of LTRs P-2 and P-3 about equal

   A. Trans-Pecos, Texas, and northeastern New Mexico; slow rocky streams; A-2 gap about 15% of one side of LTR A-2; LTR P-1 without medial gap; LTR P-3 length 90–100% of P-2 length; midventral marginal papillae biserial; fins low; uniformly tan to brown with numerous iridophores; larger tadpoles often with scattered black blotches; sympatric: none in group .... *Hyla arenicolor* **(Canyon Treefrog)**
Fig. 77A–C, Plate 8H

   B. Southern Texas into Mexico; nonflowing water; A-2 gap about 30% of one side of LTR A-2; LTR P-1 without medial gap; LTR P-3 length about 95% of P-2 length; marginal papillae biserial midventrally, multiserial laterally; jaws sheaths wide; sympatric: none in group .... *Smilisca baudinii* **(Mexican Treefrog)**
Fig. 95A–C, Plate 10D

II. Length of LTR P-3 notably shorter than P-2 length; pattern involves some distinctive component

   A. Disjunct areas in central Louisiana and central and southwestern Missouri; swampy habitats with emergent woody vegetation; A-2 gap about 20–25% of one side of LTR A-2; LTR P-1 without medial gap; LTR P-3 length about 60% of P-2 length; marginal papillae biserial midventrally; body black with pale silver to reddish marks from nostril to eye and between eyes, dorsum of tail muscle vividly banded; sympatric: *Hyla cinerea, H. "versicolor"* ....
....*Hyla avivoca* **(Bird-voiced Treefrog)**
Fig. 78A–B, Plate 8I

   B. Southern Texas north and east to southeastern Missouri and east to Mississippi River; permanent and ephemeral nonflowing water, usually with emergent vegetation; A-2 gap about 40% of one side of LTR A-2; LTR P-1 without medial gap; LTR P-3 length about 65% of P-2 length; marginal papillae biserial midventrally; throat pigmented, usually greenish ground color with speckled pattern, pale lines from eye to snout, often a yellowish tint in fins; sympatric: *Hyla avivoca, H. "versicolor"* .... *Hyla cinerea* **(Green Treefrog)**
Fig. 79A–B, D–E, Plate 8K

   C. Eastern half of Texas north to southern Manitoba and east to Mississippi River; permanent but usually ephemeral nonflowing water; A-2 gap about 35% of one side of LTR A-2; LTR P-1 without medial gap; LTR P-3 length 80–85% of P-2 length; dense patch of submarginal papillae ventrolaterally; varies from uniformly pale to boldly marked fins with reddish to orange background; sympatric: *Hyla avivoca, H. cinerea* ....
.... *Hyla versicolor* **(Gray Treefrog)**
Fig. 83I, L, Plate 9C

   [As for *H. versicolor*; sympatric: as for *H. versicolor* ....
.... *Hyla chrysoscelis* **(Cope's Gray Treefrog)**]
Fig. 83I, K, Plate 8J

III. Length of LTR P-3 notably shorter than P-2 length; pattern bland, uniform to nondescript

A. Southeastern Coastal Plain of Texas east to Mississippi River; ephemeral non-flowing water; A-2 gap about 30% of one side of LTR A-2; LTR P-1 with medial gap; LTR P-3 length about 60% of P-2 length; usually uniformly pale tan, sometimes with golden hue, no prominent markings; sympatric: *Hyla "versicolor," Pseudacris* .... ***Hyla squirella* (Squirrel Treefrog)**
Fig. 82B, D, Plate 9B

B. Eastern half of Texas north to southern Manitoba and east to Mississippi River; permanent but usually ephemeral nonflowing water; A-2 gap about 35% of one side of LTR A-2; LTR P-1 without medial gap; LTR P-3 length 80–85% of P-2 length; dense patch of submarginal papillae ventrolaterally; color varies from uniformly pale to boldly marked fins with reddish to orange background; sympatric: *Hyla squirella, Pseudacris clarkii, P. streckeri* .... ***Hyla versicolor* (Gray Treefrog)**
Fig. 83I, L, Plate 9C

[As for *H. versicolor*; sympatric: as for *H. versicolor* .... ***Hyla chrysoscelis* Cope's Gray Treefrog)]**
Fig. 83I, K, Plate 8J

C. Most of central Texas north to southern Kansas; ephemeral habitats; A-2 gap 40–45% of one side of LTR A-2; LTR P-1 with medial gap; LTR P-3 length about 50% of P-2 length; marginal papillae uniserial midventrally; sympatric: *Hyla squirella, H. "versicolor," Pseudacris streckeri, P. "triseriata"* .... ***Pseudacris clarkii* (Spotted Chorus Frog)**
Fig. 88A–C, Plate 9H

D. Eastern third of Texas north to south-central Nebraska, disjunct ranges in northeastern Arkansas and adjacent Missouri; ephemeral nonflowing water; A-2 gap about 20% of one side of LTR A-2; LTR P-1 with medial gap; LTR P-3 length 40–55% of P-2 length; considerably larger than sympatric conge-ners—up to 65 mm TL; sympatric: *H. "versicolor," Pseudacris clarkii, P. "triseriata"* .... ***Pseudacris streckeri* (Strecker's Chorus Frog)**
Fig. 93A, Plate 10B

E. Yukon Territory south and east to New Mexico and east to Mississippi River; ephemeral nonflowing water, often in grasslands; A-2 gap 20–30% of one side of LTR A-2; LTR P-1 without medial gap; LTR P-3 length 35–40% of P-2 length; midventral marginal papillae uniserial; sympatric: *Hyla squirella, H. "versicolor," Pseudacris clarkii, P. streckeri* .... ***Pseudacris "triseriata"* (Trilling Chorus Frogs)**
Fig. 94D, Plate 10C

[Eastern third of Texas north to northern Oklahoma, east to southern Mis-souri and south to coastal Mississippi; sympatric: *Hyla squirella, H. "versi-color," P. "triseriata"* .... ***Pseudacris fouquettei* (Cajun Chorus Frog)]**

[East-central Arizona and west-central Idaho to Northern Territories, east to central Ontario and south to southeastern Missouri, disjunct in southeastern Ontario and adjacent Québec; sympatric: *Hyla "versicolor"* ....
.... ***Pseudacris maculata* (Boreal Chorus Frog)**]
Fig. 94E

Key Central Frog

I. Jaw sheaths wide, front surface of lower sheath convex; demarcation between keratinized (black) and nonkeratinized (white) parts of jaws abrupt and discrete; nostrils large, easily seen.. ***Lithobates pipiens* group**

A. Irregular distribution from eastern Texas northeast to southeastern Iowa; non-flowing water, often ephemeral; A-2 gap 80–100% of one side of LTR A-2; LTR P-1 with medial gap; LTR P-3 length about 80–85% of P-2 length; dorsal fin with medium arch, extends to or anterior to plane of spiracle; white lip line usually absent, dorsum usually uniformly dark to uniformly pale, fins usually clear but may have bold marks; all three species are allopatric, but separating them from sympatric members of the *L. pipiens* group is highly subjective; sympatric: *Lithobates blairi, L. palustris, L. "pipiens"* ....
.... ***Lithobates areolatus* (Crawfish Frog)**
Plate 10I

B. Southwestern three-quarters of Texas; nonflowing water and slow streams; A-2 gap more than 150% of one side of LTR A-2; LTR P-1 with medial gap LTR P-3 length about 80% of P-2 length; ventral marginal papillae large; general coloration pale with well-defined reticulated pattern on tail; neuromasts obscure; belly musculature weakly developed; sympatric: *Lithobates blairi, L. pipiens, L. sphenocephalus* ....
.... ***Lithobates berlandieri* (Rio Grande Leopard Frog)**
Fig. 103A–B, Plate 10J

C. Much of Great Plains north of central Texas to southern Iowa; usually ephemeral, often turbid, nonflowing water; A-2 gap about 65% of one side of LTR A-2; LTR P-1 without medial gap; LTR P-3 length about 85–90% of P-2 length; ventral marginal papillae large; uniformly pale coloration; neuromasts obscure; belly musculature prominently visible in preserved tadpole; sympatric: *Lithobates areolatus, L. berlandieri, L. palustris, L. pipiens. L. sphenocephala* .... ***Lithobates blairi* (Plains Leopard Frog)**
Fig. 104C–E, Plate 10K

D. Southeastern Texas north and east to northeastern Missouri and east to Mississippi River; nonflowing water, often wooded; A-2 gap 80–95% of one side of LTR A-2; LTR P-1 with medial gap; LTR P-3 length 75–80% of P-2 length; dorsal fin originates anterior to tail-body junction and forms high arch; midventral marginal papillae large; white lip line faint to absent; dorsum purplish black freckled with paler pigment, tail fin uniformly speckled; sympatric: *Lithobates "areolatus," L. blairi, L. "pipiens"* ....
.... ***Lithobates palustris* (Pickerel Frog)**
Fig. 112A–B, Plate 11G

E. Central New Mexico north to northern Manitoba and east to Mississippi River; usually nonflowing water, often ephemeral; A-2 gap about 200% of one side of LTR A-2; LTR P-1 with or without medial gap; LTR P-3 length 80–90% of P-2 length; ventral marginal papillae medium; dorsal fin originates near tail-body junction and forms low to medium arch; white lip line present; color varies from uniformly dark to uniformly pale to considerable subtle mottling, fins clear to boldly marked; no accurate means of distinguishing between *L. areolatus* and *L. pipiens* groups; ephemeral and permanent nonflowing water and slow streams; sympatric: *Lithobates "areolatus," L. berlandieri, L. blairi, L. palustris* ....
.... ***Lithobates pipiens* (Northern Leopard Frog)**
Fig. 113D, G, Plate 11H

[Eastern third of Texas north and east to northern Missouri and east to Mississippi River; sympatric: *Lithobates areolatus, L. berlandieri, L. blairi, L. palustris* .... ***Lithobates sphenocephalus* (Southern Leopard Frog)**]
Fig. 113D, H–J, Plate 11J

II. Jaw sheaths narrow to medium, front surface of lower sheath not strongly convex; demarcation between keratinized (black) and nonkeratinized (white) parts of jaws diffuse; nostrils small, difficult to see ... ***Lithobates catesbeianus* group**

A. Eastern three-quarters of Texas north to central South Dakota and east to Mississippi River, disjunct populations in western part of area; usually larger, permanent nonflowing water; dorsum and dorsal fin with distinct black dots; individuals smaller than 25 mm TL with dark ground color and dappling of gold iridophores; sympatric: *Lithobates clamitans, L. grylio* ....
.... ***Lithobates catesbeianus* (American Bullfrog)**
Fig. 105B, F, Plate 10L

B. Eastern third of Texas north and east to Minnesota and east to Mississippi River; usually permanent, often swampy, nonflowing water; dorsum gray to brown and many small markings without discrete borders, distal third of tail often with contrasting marks; A-2 gap more than 300% of one side of LTR A-2; LTR P-1 with medial gap; LTR P-3 length about 50% of P-2 length; sympatric: *Lithobates catesbeianus, L. grylio* ....
.... ***Lithobates clamitans* (Green Frog)**
Fig. 107B, D–E, Plate 11B

C. Coastal Plain of southeastern Texas east to Mississippi River; larger, permanent nonflowing water; dorsum and fins faintly marbled and dorsal fin with longitudinal row of black dots; individuals smaller than 25 mm TL with maroon ground color and a pale band at midbody; A-2 gap more than 200% of one side of LTR A-2; LTR P-1 with medial gap; LTR P-3 length about 60% of P-2 length; sympatric: *Lithobates catesbeianus, L. clamitans* ....
.... ***Lithobates grylio* (Pig Frog)**
Fig. 108B–C, Plate 11C

Key Eastern Toad

I. Dorsum of tail muscle distinctly banded

   A. Coastal Plain from Mississippi River to southeastern Virginia; small, ephemeral, nonflowing water; body appears black, but large specimens have iridophores that produce subtle mottling; snout curves abruptly downward in lateral view; to 26 mm TL; sympatric: *Incilius nebulifer* ....
   .... ***Anaxyrus quercicus*** (**Oak Toad**)
   Plate 7K

   B. Southwestern Mississippi; ephemeral nonflowing water; A-2 gap about 50% of one side of LTR A-2; LTR P-1 with medial gap; LTR P-3 length about 50–70% of P-2 length; body appears black, but larger specimens have iridophore frosting; snout slopes gradually in lateral view; to 40 mm TL; sympatric: *Anaxyrus quercicus* .... ***Incilius nebulifer*** (**Gulf Coast Toad**)
   Plate 8D

II. Dorsum of tail muscle uniform or at least not banded, sides of tail muscle bicolored

   A. Throughout most of eastern North America above Fall Line; A-2 gap 40–55% of one side of LTR A-2; LTR P-1 without medial gap; LTR P-3 length about 70–80% of P-2 length; appears uniformly black but abundant golden iridophores visible at slight magnification, often concentrated along top of tail muscle; white part of bicolored tail muscle about 25% of basal muscle height; snout moderate, about 1.5 eye diameters from front of eye to tip of snout; spiracle on longitudinal axis; sympatric: *Anaxyrus fowleri, A. terrestris* ....
   .... ***Anaxyrus americanus*** (**American Toad**)
   Fig. 62B–C, Plate 7A

   B. Throughout most of eastern United States except southeastern Coastal Plain and Florida Peninsula; A-2 gap about 50% of one side of LTR A-2; LTR P-1 without medial gap; LTR P-3 length about 60–65% of P-2 length; smaller specimens appear black, larger specimens with abundant brassy iridophores that impart a frosted or mottled appearance; white part of bicolored tail about 50% of basal muscle height; snout bluntly rounded in lateral view; spiracle on longitudinal axis; sympatric: *Anaxyrus americanus, A. terrestris* ....
   .... ***Anaxyrus fowleri*** (**Fowler's Toad**)
   Fig. 65B, D, Plate 7F

   C. Coastal Plain from Mississippi River to southeastern Virginia; A-2 gap 65% of one side of LTR A-2; LTR P-1 without medial gap; LTR P-3 length about 80% of P-2 length; body dorsum uniformly dark with golden lines of iridophores extending diagonally backward from below eye toward midline (less visible at night); lower white part of bicolored tail 20–25% of basal muscle height; snout slopes gradually in lateral view, about 3 eye diameters from front of eye to tip of snout; spiracle below longitudinal axis; sympatric: *Anaxyrus americanus, A. fowleri, Rhinella marina* ....
   .... ***Anaxyrus terrestris*** (**Southern Toad**)
   Fig. 69A, Plate 8A

D. Exotic around Tampa Bay, Miami, and southern Florida; A-2 gap 30–40% of one side of LTR A-2; LTR P-3 length about 100% of P-2 length; entirely black, except bicolored tail (about 30% of basal tail height); snout slopes gradually in lateral view, about 2.5 eye diameters from front of eye to tip of snout; spiracle on longitudinal axis; sympatric: *Anaxyrus terrestris* ....
.... ***Rhinella marina*** **(Cane Toad)**
Fig. 72A

Key Eastern Treefrog

I. Pattern involves multiple contrasty bands on tail or a single contrasty band or saddle on dorsum of tail muscle
   A. Much of the Mississippi Embayment and east to southwestern South Carolina; swampy habitats with emergent woody vegetation; A-2 gap about 20–25% of one side of LTR A-2; LTR P-1 without medial gap; LTR P-3 length about 60% of P-2 length; marginal papillae biserial midventrally; throat darkly pigmented; body mostly black, dorsum of tail muscle with white, silver, or reddish bands that persist in preservative, a similar band between eyes and stripe from eye to adjacent naris usually disappear in preservative; sympatric: none with similar coloration .... ***Hyla avivoca*** **(Bird-voiced Treefrog)**
   Fig. 78A–B, Plate 8I

   B. Coastal Plain from Mississippi River to Delaware including most of peninsular Florida and north to southern Tennessee, disjunct in northern Mississippi and southwestern Kentucky; usually ephemeral nonflowing water; A-2 gap about 60% of one side of LTR A-2; LTR P-1 without medial gap; LTR P-3 length about 50% of P-2 length; jaw sheaths massive; marginal papillae uniserial midventrally; 25 mm TL or less: uniformly pale throughout except a single black band on tail muscle; larger sizes: fins clear during the day, dark at night; sympatric: *Hyla avivoca* .... ***Hyla gratiosa*** **(Barking Treefrog)**
   Fig. 81C–D, Plate 9A

II. Pattern involves contrasty stripes on tail or body
   A. Disjunct in New Jersey, North and South Carolina, and western panhandle of Florida; boggy, often ephemeral, habitats with emergent vegetation; A-2 gap about 65% of one side of LTR A-2; LTR P-1 without medial gap; LTR P-3 length 20–35% of P-2 length; marginal papillae biserial midventrally; throat not pigmented; body mostly lightly pigmented brown, sometimes with golden hue, lateral surface of tail muscle with either pale stripe bordered dorsally and ventrally by black or sometimes upper black stripe predominates, sometimes broken into blotches; sympatric: *Hyla avivoca, H. femoralis, Pseudacris brimleyi, P. ocularis*............................ ***Hyla andersonii*** **(Pine Barrens Treefrog)**
   Fig. 76A–B, Plate 8G

   B. Coastal Plain from Delaware to Mississippi River including the Mississippi Embayment and peninsular Florida; permanent and ephemeral nonflowing water, usually with emergent vegetation; A2 gap about 40% of one side of LTR A2; LTR P1 without medial gap; LTR P3 length about 65% of P2 length; marginal papillae biserial midventrally; throat pigmented, usually greenish

ground color with speckled pattern, pale lines from eye to snout, often a yellowish tint in fins; sympatric: none with similar coloration ....
.... *Hyla cinerea* (**Green Treefrog**)
Fig. 79A–B, D–E, Plate 8K

C. Coastal Plain from Mississippi River to southeastern Virginia; ephemeral, nonflowing water; A-2 gap about 20% of one side of LTR A-2; LTR P-1 without medial gap; LTR P-3 length about 90% of P-2 length; marginal papillae multiserial midventrally; throat not pigmented; body more or less uniformly brown to russet; basal two-thirds of tail muscle with prominent pale lateral stripe; fin areas adjacent to tail muscle usually lacking or with less dense aggregations of melanic blotches compared with remainder of fin; clear parts of fins usually reddish; geographical variations apparent—fins much higher, differently shaped, and more brightly colored at Tampa and Crestview, Florida with more prominent flagellum than near Gulfport, Mississippi; sympatric: *Hyla andersonii, Pseudacris brimleyi, P. ocularis* ....
.... *Hyla femoralis* (**Pine Woods Treefrog**)
Fig. 80B, Plate 8L

D. Coastal Plain from southeastern Virginia to coastal Georgia; wooded ephemeral habitats; A-2 gap about 50% of one side of LTR A-2; LTR P-3 length about 30% of P-2 length; marginal papillae uniserial midventrally; throat pigmented, often spotted; white ventral portion of tail coloration about 50% of basal muscle height; pale (silver to orange) line of iridophores from eye to tail junction (mostly disappears in preservative) sometimes extends onto tail to form pale stripe above black portion; sympatric: *Hyla andersonii, H. femoralis, Pseudacris ocularis* .... *Pseudacris brimleyi* (**Brimley's Chorus Frog**)
Fig. 86A, Plate 9F

E. Coastal Plain from western part of Florida panhandle to southeastern Virginia; grassy ephemeral habitats; A-2 gap 10–15% of one side of LTR A-2 LTR P-1 without medial gap; LTR P-3 length about 30% of P-2 length; marginal papillae biserial midventrally; throat lightly pigmented near oral disc; white, ventral portion of tail coloration about 25% of basal muscle height; pale (silver to orange) stripe from eyes to tail base may extend onto dorsolateral part of tail muscle as a pale stripe above dark area; dark tail banded dorsally, at least in smaller specimens; fins with abundant diffuse blotches; dorsum of large specimens with discrete black dots; fins of larger specimens with fairly uniform, diffuse pigmentation; sympatric: *Pseudacris brimleyi* ....
.... *Pseudacris ocularis* (**Little Grass Frog**)
Fig. 90A–B, Plate 9L

III. Pattern bland, uniform to nondescript without distinctive markings

A. Jaw sheaths notably wide and robust

1. Coastal Plain from Mississippi River to Delaware including most of peninsular Florida and north to southern Tennessee, disjunct in northern Mississippi and southwestern Kentucky; greater than 25 mm TL: A-2 gap about

60% of one side of LTR A-2; LTR P-1 without medial gap; LTR P-3 length about 50% of P-2 length; jaw sheaths massive; marginal papillae uniserial midventrally; 25 mm or less TL: uniformly pale with black band at about the middle of tail muscle; larger sizes: fins clear during the day, dark at night; sympatric: *Pseudacris ornata* ....
....*Hyla gratiosa* (**Barking Treefrog**)
Fig. 81C–D, Plate 9A

2. Coastal Plain from Mississippi River to southeastern Virginia except for most of Florida Peninsula; A-2 gap 30–40% of one side of LTR A-2; LTR P-1 without medial gap; LTR P-3 length about 50% of P-2 length; midventral marginal papillae uniserial; jaw sheaths massive; small specimens dark with bicolored tail muscle, older specimens nearly uniformly dark with minor mottling; sympatric: *Hyla gratiosa* ....
.... *Pseudacris ornata* (**Ornate Chorus Frog**)
Fig. 91B–C, Plate 10A

B. Jaw sheaths medium to narrow

1. Coastal Plain from Delaware to Mississippi River including the Mississippi Embayment and peninsular Florida; permanent and ephemeral, nonflowing water; A-2 gap about 40% of one side of LTR A-2; LTR P-1 without medial gap; LTR P-3 length about 65% of P-2 length; marginal papillae biserial midventrally; throat pigmented, usually greenish ground color with speckled pattern, pale lines from eye to snout, often a yellowish tint in fins; sympatric: *Hyla squirella, H. "versicolor," Pseudacris brachyphona, P. crucifer, P. nigrita, P. streckeri* ....
.... *Hyla cinerea* (**Green Treefrog**)
Fig. 79A–B, D–E, Plate 8K

2. Coastal Plain from Mississippi River to southeastern Virginia; A-2 gap about 30% of one side of LTR A-2; LTR P-1 with medial gap; LTR P-3 length about 60% of P-2 length; sympatric: *Hyla cinerea, H. "versicolor," Pseudacris crucifer, P. nigrita, P. "triseriata"* ....
.... *Hyla squirella* (**Squirrel Treefrog**)
Fig. 82B, D, Plate 9B

3. Throughout much of eastern United States except most of Florida Peninsula and northern Maine; A-2 gap about 35% of one side of LTR A-2; LTR P-1 without medial gap; LTR P-3 length 80–85% of P-2 length; marginal papillae biserial midventrally; dense patch of submarginal papillae ventrolaterally; color varies from uniformly pale to boldly marked fins with reddish to orange background; sympatric: *Hyla cinerea, H. squirella, Pseudacris brachyphona, P. crucifer, P. nigrita, P. "triseriata"* ....
.... *Hyla versicolor* (**Gray Treefrog**)
Fig. 83I, L, Plate 9C

[As for *H. versicolor*; sympatric: as for *H. versicolor* ....
.... ***Hyla chrysoscelis* (Cope's Gray Treefrog)**]
Fig. 83I, K, Plate 8J

4. Southeastern Pennsylvania south and west to northeastern Mississippi and central Alabama; A-2 gap 20–30% of one side of LTR A-2; LTR P-1 without medial gap; LTR P-3 length about 30–40% of P-2 length; midventral marginal papillae uniserial; throat not pigmented; sympatric: *Hyla cinerea, H. "versicolor," Pseudacris crucifer, P. "triseriata"* ....
.... ***Pseudacris brachyphona* (Mountain Chorus Frog)**
Plate 9E

5. Throughout much of eastern United States and Canada except for most of Florida Peninsula; ephemeral woodland habitats; A-2 gap about 30% of one side of LTR A-2; LTR P-1 without medial gap; LTR P-3 about 28% or less of LTR A-2; marginal papillae uniserial midventrally, sometimes with a medial gap in marginal papillae the width of LTR P-3, LTR P-3 sometimes absent; fins medium to tall, dorsum usually uniform, tail fin marks vary from nearly absent to fine speckles to large blotches with locality and size, dorsum of tail usually banded in younger stages, tail musculature never bicolored; throat speckled; sympatric: *Hyla cinerea, H. "versicolor," Pseudacris brachyphona, P. nigrita, P. streckeri, P. "triseriata"* ....
.... ***Pseudacris crucifer* (Spring Peeper)**
Fig. 89B–C, Plate 9I

6. Disjunct sites from southern to northern Illinois; A-2 gap about 20% of one side of LTR A-2; LTR P-1 with medial gap; LTR P-3 length 40–55% of P-2 length; up to 65 mm TL, considerably larger than sympatric congeners; sympatric: *Pseudacris maculata, P. triseriata* ....
.... ***Pseudacris illinoensis* (Illinois Chorus Frog)**
For similar species, see Fig. 93A, Plate 10B

7. Coastal Plain from southeastern Mississippi to southeastern North Carolina; A-2 gap 10–15% of one side of LTR A-2; LTR P-1 with medial gap; LTR P-3 length about 40% of P-2 length; marginal papillae uniserial midventrally; sympatric: *Hyla cinerea, H. "versicolor," Pseudacris crucifer, P. "triseriata"* .... ***Pseudacris nigrita* (Southern Chorus Frog)**
Fig. 94F, Plate 9K

8. Throughout most of North America east of Mississippi River except for much of Atlantic Coastal Plain, peninsular Florida, and higher elevations of northern Appalachian Mountains; A-2 gap 20–30% of one side of LTR A-2; LTR P-1 without medial gap; LTR P-3 length 35–40% of P-2 length; midventral marginal papillae uniserial; sympatric: *Pseudacris brachyphona, P. crucifer, P. nigrita, P. streckeri* ....
.... ***Pseudacris "triseriata"* (Trilling Chorus Frogs)**
Fig. 94D, Plate 10C

[Northeastern half of Mississippi east and north to central Pennsylvania; sympatric: *Hyla cinerea, Pseudacris crucifer, P. fouquettei, P. nigrita*.......
.... ***Pseudacris feriarum* (Upland Chorus Frog)**]

[Eastern third of Texas north to northern Oklahoma, east to southern Missouri, and south to coastal Mississippi; sympatric: *Pseudacris feriarum* ....
.... ***Pseudacris fouquettei* (Cajun Chorus Frog)**]
[Staten Island, New York south throughout Delmarva Peninsula; sympatric: *Hyla cinerea, Pseudacris crucifer* ....
.... ***Pseudacris kalmi* (New Jersey Chorus Frog)**]
[East-central Arizona and west-central Idaho to Northern Territories, east to central Ontario and south to southeastern Missouri, disjunct in southeastern Ontario and adjacent Québec; sympatric: none in group ....
.... ***Pseudacris maculata* (Boreal Chorus Frog)**]

Fig. 94E

Key Eastern Frog

I. Specimen 25 mm TL or less; fins clear; ground color dark

  A. Throughout eastern North America except southern third of Florida Peninsula; larger, nonflowing water; ground color dark with numerous specks and blotches of gold iridophores, black blood vessel along dorsum of tail muscle; sympatric: *Lithobates grylio, L. heckscheri, L. virgatipes* ....
.... ***Lithobates catesbeianus* (American Bullfrog)**

Fig. 105F

  B. Coastal Plain from Mississippi River to central South Carolina and entire Florida Peninsula; larger nonflowing water, hides in bottom debris; ground color dark with diffuse gold band on body; A-2 gap more than 200% of one side of LTR A-2; LTR P-1 with medial gap; LTR P-3 length about 60% of P-2 length; sympatric: *Lithobates catesbeianus, L. heckscheri, L. virgatipes* ....
.... ***Lithobates grylio* (Pig Frog)**

Fig. 108B

  C. Coastal Plain from southern Mississippi to southern North Carolina and entire Florida Peninsula; slow streams and larger, usually permanent, nonflowing water, stationary schools when small, mobile schools when larger; ground color black with bright golden band on body; A-2 gap about 150% of one side of LTR A-2; LTR P-1 with medial gap; LTR P-3 length about 90% of P-2 length; sympatric: *Lithobates catesbeianus, L. grylio, L. virgatipes* ....
.... ***Lithobates heckscheri* (River Frog)**

Fig. 109B

  D. Atlantic Coastal Plain from northeastern Florida to New Jersey; swampy nonflowing water; ground color dark to maroon with fine speckling throughout, longitudinal row of black dots in dorsal fin; sympatric: *Lithobates catesbeianus, L. grylio, L. heckscheri* .... ***Lithobates virgatipes* (Carpenter Frog)**

Fig. 116B

II. Specimen greater than 25 mm TL; coloration distinctive; fins often with some pattern

  A. Throughout most of eastern North America except southern third of Florida Peninsula; larger, nonflowing water; dorsum green to brown with distinct

black dots on dorsum and dorsal fin; belly opaque white; sympatric: *Lithobates grylio, L. heckscheri, L. okaloosae, L. virgatipes* ....
.... *Lithobates catesbeianus* (**American Bullfrog I**)
Plate 10L

B. Probable distinct taxon presently known from Tampa Bay north to near Tallahassee but likely ranges well north and perhaps west along Coastal Plain; often smaller nonflowing water; body uniformly bright green, belly bright yellow; dorsum and fins lack black dots and most other contrasting marks; sympatric: *Lithobates grylio, L. heckscheri*, perhaps others ....
.... *Lithobates catesbeianus* (**American Bullfrog II**)
Fig. 105H

C. Coastal Plain from Mississippi River to central South Carolina and entire Florida Peninsula; larger nonflowing water, hides in bottom debris; ground color brown to maroon, dorsum and fins mottled, longitudinal row of black dots in dorsal fin formed by melanophores around neuromasts; A-2 gap over 200% of one side of LTR A-2; LTR P-1 with medial gap; LTR P-3 length about 60% of P-2 length; sympatric: *Lithobates catesbeianus, L. heckscheri, L. okaloosae, L. virgatipes* .... *Lithobates grylio* (**Pig Frog**)
Fig. 108C, Plate 11C

D. Coastal Plain from southern Mississippi to southern North Carolina and entire Florida Peninsula; slow streams and larger, usually permanent, nonflowing water; ground color gray to black, finely speckled, body band usually visible; fins opaque white rimmed in black; dorsum of tail muscle dark; A-2 gap about 150% of one side of LTR A-2; LTR P-1 with medial gap; LTR P-3 length about 90% of P-2 length; sympatric: *Lithobates catesbeianus, L. grylio, L. okaloosae, L. virgatipes* .... *Lithobates heckscheri* (**River Frog**)
Fig. 109D, Plate 11D

E. Okaloosa and Santa Rosa cos., western Florida; small, swampy streams; notable silver markings ventrolaterally, ground color brownish; sympatric: *Lithobates catesbeianus, L. grylio, L. heckscheri, L. virgatipes* ....
.... *Lithobates okaloosae* (**Florida Bog Frog**)
Fig. 110A, C, Plate 11E

F. Atlantic Coastal Plain from northeastern Florida to New Jersey; swampy nonflowing water; ground color brown to maroon, dorsum and fins mottled, longitudinal row of black dots in dorsal fin formed by melanophores around neuromasts; A-2 gap more than 300% of one side of LTR A-2; LTR P-1 with medial gap; LTR P-3 length about 75% of P-2 length; sympatric: *Lithobates catesbeianus, L. grylio, L. heckscheri* ....
.... *Lithobates virgatipes* (**Carpenter Frog**)
Plate 11L

III. Specimen greater than 25 mm TL; coloration nondescript and bland, uniform to nondescript
    A. Throughout much of eastern North America except most of Florida Peninsula; nonflowing water, often swampy; dorsum gray to brown and many small

markings without discrete borders, distal third of tail often with contrasting marks; dorsal fin low; A-2 gap more than 300% of one side of LTR A-2; LTR P-1 with medial gap; LTR P-3 length about 50% of P-2 length; sympatric: *Lithobates catesbeianus, L. okaloosae, L. septentrionalis* ....
.... ***Lithobates clamitans* (Green Frog)**
Fig. 107D–E, Plate 11B

B. Okaloosa and Santa Rosa cos., western Florida; small, swampy streams; brownish ground color, notable silvery blotches ventrolaterally; sympatric: *Lithobates catesbeianus, L. clamitans* ....
.... ***Lithobates okaloosae* (Florida Bog Frog)**
Fig. 110A, C, Plate 11E

C. Northeastern Minnesota east and north to Labrador and south to northern New York; streams, lakes, and swampy nonflowing water; dorsal fin medium; tail with pinkish spots; A-2 gap more than 200% of one side of LTR A-2; LTR P-1 with medial gap; LTR P-3 length about 70% of P-2 length; sympatric: *Lithobates catesbeianus, L. clamitans* ....
.... ***Lithobates septentrionalis* (Mink Frog)**
Plate 11I

TAXONOMIC ACCOUNTS

Bufonidae (Toads)

Toads are introduced in Hawaii and occur naturally throughout the North American mainland, usually in nonflowing water at ephemeral sites; LTRF usually 2/3, 2/2 in 2 cases, all rows uniserial, rows A-1 and A-2 about equal in length; small marginal papillae with wide dorsal and ventral gaps, LTR P-3 about equals the width of ventral gap in marginal papillae; submarginal papillae sparse; oral disc emarginate, folds into roughly a rectangular shape at rest; finely serrate jaw sheaths narrow to medium; vent medial; eyes dorsal; spiracle sinistral; dorsal fin low with rounded tip, originates near tail-body junction; body depressed to slightly globular with rounded snout; nonornamented nostrils notably large and not round; digits without toe pads; most commonly uniformly dark, although larger tadpoles often patterned; metamorph coloration resembles that of adults; to 55.0 mm TL, usually smaller; small pigmented eggs usually in strings in ephemeral sites; 3 genera, 19 species. The tadpoles of *Anaxyrus* (19 species), *Incilius* (2 species), and *Rhinella* (1 species) are not distinguishable as genera.

Anaxyrus americanus (American Toad) and *A. houstonensis* (Houston Toad)   Fig. 62; PL7A

IDENTIFICATION   LTRF 2/3; A-2 gap 40–55% of one side of LTR A-2; LTR P-1 without medial gap; LTR P-3 length 75–85% of P-2 length; appears black throughout ontogeny but considerable gold iridophores obvious at slight

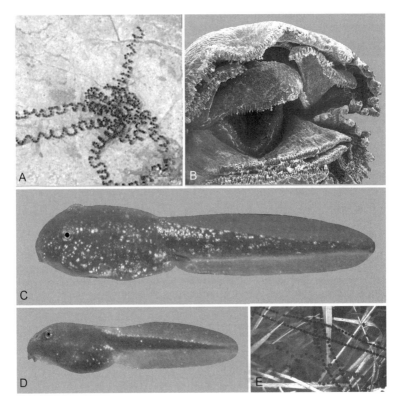

Figure 62. **Anaxyrus americanus and A. houstonensis.** (A–C) *A. americanus*: (A) egg string (Union Co., IL; JCM), (B) oral apparatus (Cannon Co., TN; LOT), (C) tadpole (24.0 mm TL, stage 28, Will Co., IL; MR), (D–E) *A. houstonensis*: (D) tadpole (15.0 mm TL, stage 30, Bastrop Co., TX; RA), (E) egg string (Bastrop Co., TX; JF). See also Plate 7A.

magnification, concentrations along dorsum of tail sometimes obvious; tail muscle usually distinctly bicolored, lower quarter of tail muscle white; 20.0–28.0 mm TL/stage 36. *A. houstonensis* similar, uniformly dark but perhaps less iridophore flecking and smaller amount of white on venter of tail muscle than *A. americanus*; about 20.0 mm TL/stage 25.

NATURAL HISTORY American Toads oviposit between January and June in temporary pools and shallow backwaters of permanent lakes and ponds. Clutch size varies between 1840 and 20,600 brown to black ova (OD 1.0–2.0 mm, 2.9–6.8 ova/cm) laid as twin, loosely spiraled, tubes (ED 2.9–15.0 mm) that are distinct but flimsy. The strings of eggs measure between 6 and 20 m; they have a bilayered tube with straight margins and partitions between successive ova. The ova may be uni-, bi-, or triserial within the tube; they hatch in 3–12 days. Tadpoles grow to 28.0 mm TL and metamorphose in 1.3–2.5 months at 6.0–12.0 mm SVL. Metamorphs reach sexual maturity in 2–4 years.

RANGE *A. americanus* is widely distributed in eastern North America from the maritime provinces of Canada to central Oklahoma and south to Louisiana; this species is absent from most of the Gulf and Atlantic Coastal Plains. *A. houstonensis* is restricted to a small area in east-central Texas.

CITATIONS ***Development/Morphology***: Altig and Pace 1974, Bragg 1947a, Bresler and Bragg 1954, L. E. Brown 1973, L. E. Brown et al. 1984, Gosner 1959, Hinckley 1881, Trauth et al. 2004, Tubbs et al. 1993. ***Reproductive Biology***: Bragg 1958a, Collins and Wilbur 1979, Dodd 2004, N. L. Jackson 1989, Kennedy 1961, Komoroski et al. 1998, Livezey and Wright 1947, Parmalee et al. 2002, Quinn and Mengden 1984, G. R. Smith and Rettig 1998, Trauth et al. 1990. ***Ecobehavior***: Ahlgren and Bowen 1991, Alford 1989, Arendt 2003, Beiswenger 1975, 1977, Beiswenger and Test 1966, Black 1969, 1971, Bragg 1947a, 1952, 1955a, 1958b, 1961, Breden and Kelly 1982, Brockelman 1969, Brodie and Formanowicz 1987, Brodie et al. 1978, Drake et al. 2007, Dupré and Petranka 1985, Formanowicz and Brodie 1982, Gee and Waldick 1995, Gomez-Mestre et al. 2006, Gosner and Black 1958a, Greuter and Forstner 2003, Hamel 2009, Hoff and Wassersug 1985, Holomuzski 1997, S. P. Lawler and Morin 1993, Licht 1967, Pearman 1995, Petranka 1989a, Petranka and Hayes 1998, J. M. L. Richardson 2001, Rot-Nikcevic et al. 2005, Seale 1980, Seale and Beckvar 1980, Sherman and Levitis 2003, Taigen and Pough 1981, C. N. Taylor et al. 2004, Test and McCann 1976, Thomas and Allen 1997, Touchon et al. 2006, Tumlison and Trauth 2006, Van Buskirk 1988, Volpe 1953, Voris and Bacon 1966, Waldman 1981, 1982b, 1985a–b, Waldman and Adler 1979, B. Walters 1975, Wassersug 1989, Wassersug and Feder 1983, Wassersug and Hoff 1985, Werner and Glennemeier 1999, Wilbur 1977b, Wilbur and Alford 1985, A. H. Wright 1910.

Anaxyrus boreas (Western Toad), *A. exsul* (Black Toad), and *A. nelsoni* (Amargosa Toad)   Fig. 63; PL 7B

IDENTIFICATION LTRF 2/3; A-2 gap about 36% of one side of LTR A-2; LTR P-1 without medial gap; LTR P-3 length about 85% of P-2 length. *A. boreas*: variable, but typically appears uniformly black with slightly paler areas ventrolaterally, slight ventral and ventrolateral bronze cast sometimes occurs in larger tadpoles; tail fin clear to dusky, tail muscle usually bicolored; tadpoles from clear, rocky streams often are mottled brown. *A. exsul*: appears black but has extensive sprinkling of iridophores. *A. nelsoni*: appears black but slight magnification reveals a uniform peppering of silverish iridophores and small white blotches common on sides of tail muscle; 35.0–55.0 mm TL/ stage 36.

NATURAL HISTORY A. *boreas* oviposits from January to August in shallow water of temporary and permanent ponds, including the slow parts of low-gradient streams, often immediately after snow melt. Desert populations breed in the spring or opportunistically with rain events. Clutches consist of 1200–18,000

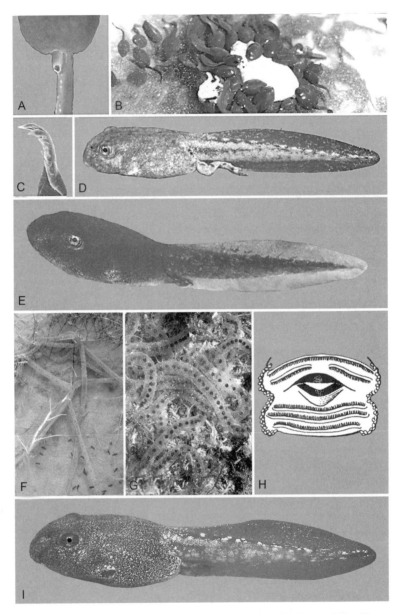

Figure 63. **Anaxyrus boreas and relatives**. (A) medial vent tube of *A. boreas* (RA), (B) group of tadpoles of *A. boreas* eating a dead adult toad (Clackamas Co., OR; WPL), (C) head of a single labial tooth of *A. boreas* viewed from the side (RA), (D) tadpole of *A. boreas* (32.0 mm TL, stage 37, Santa Barbara Co., CA; RA), (E) tadpole of *A. exsul* (31.0 mm TL, stage 34, Inyo Co., CA; RWH), (F–G) egg strings: (F) *A. nelsoni* (Nye Co., NV; RA) and (G) *A. boreas* (Sublette Co., WY; DWZ), (H) oral apparatus of *A. boreas* (modified from Stebbins 1985), (I) tadpole of *A. nelsoni* (37.0 mm TL, stage 38; Nye Co., NV; RA). See also Plate 7B.

black and white ova (OD 1.5–2.0 mm) strung out in a bilayered tube (4.8–5.3 mm diameter, 2 jelly layers). The ova are uniserial or staggered, sometimes appearing triserial, and separated by partitions between successive ova or every other one. Data provided by Savage and Schuierer (1961) differ from those of Karlstrom and Livezey (1955), Livezey and Wright (1947), Stebbins (1951), and Storer (1925). Eggs hatch in 3–10 days at about 10.0 mm TL, and depending on locality metamorphosis occurs in 1–3 months at 9.0–25.0 mm SVL. Some larvae may overwinter at high-elevation sites. Metamorphosis may occur en masse, and metamorphs sometimes bask in piles on the bank. The typical benthic *B. boreas* tadpole is larger than most congeners and often swims in midwater; after spending daylight hours crowded together in shallow water of the margins of lakes at high elevations, they sometimes move in a long, meandering file to deeper water at night.

Females of *A. exsul* oviposit between February and August, but most often in March–April, or opportunistically with sufficient rain in spring runs, sloughs, and adjacent pools. They may be active only during the day. Clutches consist of about 16,000 pigmented ova (OD 1.2–1.7 mm) arranged uni- or biserially in a bilayered tube (4.0–6.9 mm diameter, 5.2–20.8 ova/cm, mean = 10.8) laid as strings. Eggs hatch in 3–10 days, and tadpoles grow to 35.0 mm TL. Black Toad larvae metamorphose in 3.0–6.4 months at 10.0–19.0 mm SVL; recent metamorphs are common in June. They grow to 22.0–33.0 mm SVL by November, and resemble those of *A. boreas*, becoming darker with age.

Females of *A. nelsoni* oviposit between March and August or opportunistically with rains in pools along the Amargosa River and in adjacent marshy spring runs. The pigmented ova (OD 1.3–1.8 mm) are arranged uniserially in a bilayered tube 4.3–5.0 mm diameter.

RANGE *A. boreas* occurs in much of the area west of the crest of the Rocky Mountains from southern Alaska to northen Baja California; it is absent from most of Arizona, New Mexico, and the drier portions of Utah and western Colorado. *A. exsul* is a desert relict restricted to about 10,000 m^2 of marshy habitat in Deep Springs Valley, Inyo Co., California. *A. nelsoni* is also a desert relict and occurs in the upper drainage of the Amargosa River, Nye Co., Nevada.

CITATIONS *Development/Morphology*: Altig and Pace 1974, Grismer 2002, Livezey 1960. *Reproductive Biology*: Livezey and Wright 1947, G. S. Myers 1942, Savage and Schuierer 1961, Schuierer 1962. *Ecobehavior*: Brattstrom 1962, Hews 1988, Hews and Blaustein 1985, Huey 1980, J. E. Johnson et al. 2007, M. S. Jones et al. 1999, Kiesecker and Blaustein 1997, Livo and Lambert 2001, Pearl 2000, Pearl and Hayes 2002, J. A. Peterson and Blaustein 1992, Scherff-Norris and Livo 1999, Vonesh and de la Cruz 2002.

Anaxyrus canorus (Yosemite Toad)   Fig. 64; PL 7C

IDENTIFICATION LTRF 2/3; A-2 gap about 30% of one side of LTR A-2; LTR P-1 usually without a medial gap; LTR P-3 length about 45–50% of

Figure 64. ***Anaxyrus canorus.*** (A) egg string visible among organic material between arrow points (Mono Co., CA; RLG), (B) tadpole (26.0 mm TL, stage 30, Alpine Co., CA; RA), (C) tadpoles in a meadow rivulet at (D) Tyron Meadows (2430 m, Alpine Co., CA; JET). See also Plate 7D.

P-2 length; uniquely entirely sooty black, including fins; pronounced sexual color dimorphism of adults not apparent in metamorphs; 25.0–30.0 mm TL/ stage 36.

NATURAL HISTORY Yosemite Toads oviposit from May to August, usually in very shallow water of vegetated pools and slow streams or rivulets in alpine meadows and lake margins above 1460 m. The upper temperature maximum for ova is 31 °C, and tadpoles can tolerate temperatures to 36–38 °C. Clutches consist of 1500–2000 pigmented ova (OD 1.7–2.7 mm) uniserially arranged in a bilayered tube (3.7–4.6 mm diameter) with partitions between the ova. Strings often are short and deposited as a radiating group. Eggs hatch in 3–12 days, and tadpoles metamorphose in 6–10 weeks at about 10.0 mm SVL. The tadpoles are benthic and often aggregate in the presumably warmer swallow water microhabitats; individuals often swim well off the bottom. Yosemite Toads reach sexual maturity in 3–6 years.

RANGE *A. canorus* is known from localities above 1460 m in the central Sierra Nevada from Eldorado to Fresno cos., California.

CITATIONS ***General:*** Karlstrom 1973. ***Development/Morphology***: Karlstrom 1962. ***Reproductive Biology:*** Livezey and Wright 1947, Karlstrom and Livezey 1955. ***Ecobehavior:*** Brattstrom 1962, Chan 2001, Mullally 1953.

Anaxyrus cognatus (Great Plains Toad)   PL 7D

IDENTIFICATION LTRF 2/3; A-2 gap 15–30% of one side of LTR A-2; LTR P-1 with medial gap; LTR P-3 length about 30% of P-2 length; snout slopes gradually in lateral view; generally pale brassy to brownish, yellowish to pinkish ventrolaterally; to 31.0 mm TL/stage 36.

NATURAL HISTORY Female Great Plains Toads oviposit explosively from March to September in temporary rain-filled pools in open areas; some females lay 2 clutches/year. A clutch consists of 1340–45,000 black and white ova (OD 1.2 mm, ED 1.7–2.6 mm), arranged in a bilayered (2.1–2.7 mm diameter) and scalloped (i.e., slight indentation between successive eggs accompanied by a partition) tube. Eggs hatch in 2–7 days and tadpoles metamorphose in 2.4–6.4 months at 10.0–25.0 mm SVL. Metamorphs reach sexual maturity in 2–5 years.

RANGE *A. cognatus* is known from southeastern Alberta south to western Texas and west through southern Arizona to southeastern California, southern Nevada and Utah, and south into the Sonoran and Chihuahuan deserts of northern Mexico.

CITATIONS *General*: Krupa 1990. *Development/Morphology*: Bragg 1947b, Bresler 1954, Grismer 2002, H. M. Smith 1946. *Reproductive Biology*: Bragg 1936, 1937a–b, Bragg and Bresler 1952, Krupa 1986, 1988, 1994, Livezey and Wright 1947, Parmalee et al. 2002. *Ecobehavior:* Bragg 1940b, 1947a.

Anaxyrus debilis (Green Toad) and *A. retiformis* (Sonoran Green Toad)   PL 7E

IDENTIFICATION LTRF 2/2; A-2 gap about 30% of one side of LTR A-2; LTR P-1 with a medial gap; dorsum with subtle mottling and dense iridophores; dorsum of tail muscle banded; to 20.0 mm TL/stage 36.

NATURAL HISTORY Green toads oviposit between March and September in temporary pools formed by summer rains. Clutches consist of very lightly pigmented ova laid as flimsy, fragile, short strings (OD 0.9–1.1 mm, 1287/clutch) on the bottom or attached to the bases of vegetation. The tube is easily overlooked and may fall apart to produce what appear to be single eggs. Eggs hatch at 3.1–3.4 mm TL in 1 day at 33.1 °C and 5.6 days at 18.2 °C. Tadpoles grow to 17.0–21.0 mm TL in 1.1–2.9 weeks and metamorphose in 3–4 weeks at 8.0–11.0 mm SVL. The benthic, very cryptically colored, small tadpoles do not aggregate even when numerous and often swim directly along the bottom when frightened.

RANGE *A. debilis* is known from localities in central Texas, southwestern Oklahoma, and western Kansas south and west to southeastern Arizona and adjacent Mexico. *A. retiformis* is restricted to south-central Arizona and adjacent Sonora, Mexico.

CITATIONS *General:* Hulse 1978. *Development/Morphology*: Bragg 1955c, Zweifel 1970. *Reproductive Biology:* Livezey and Wright 1947, Taggart 1997, Zweifel 1968d.

Anaxyrus fowleri (Fowler's Toad)   Fig. 65; PL 7F

IDENTIFICATION LTRF 2/3; A-2 gap 40–65% of one side of LTR A-2; LTR P-1 without medial gap; LTR P-3 length 60–70% of P-2 length; small individuals dark to black, larger ones almost uniformly bronzy to somewhat mottled; tail musculature bicolored, with lower half white; to 30.0 mm TL/stage 36.

NATURAL HISTORY Fowler's Toads oviposit from March to July, in temporary pools and backwaters of permanent sites. Clutches consist of 2000–15,600 pigmented ova arranged in a unilayered tube (1.2–3.4 mm diameter, 7.3–8.0 ova/cm) and laid as strings. Eggs hatch in about a week; tadpoles metamorphose in 1–2 months at 8.0–14.0 mm SVL.

RANGE *A. fowleri* occurs through most of eastern North America from extreme southern Canada and the Great Lakes Region south to the Gulf of Mexico and west to eastern Oklahoma and Texas; it is absent from the southeastern Coastal Plain and peninsular Florida.

CITATIONS ***Development/Morphology***: Bragg 1947a, Dodd 2004, Gosner and Black 1954, Hinckley 1881, 1882b, Nichols 1937, Trauth et al. 2004, Volpe 1956. ***Reproductive Biology***: Livezey and Wright 1947, Parmalee et al. 2002, Trauth et al. 1990. ***Ecobehavior***: Alford and Harris 1988, Altig et al. 1975, Altig and McDearman 1975, Bragg 1947a, 1955a, Breden 1988, Gosner and Black 1958a, S. P. Lawler 1989, Morgan 1891, Schorr et al. 1990, Seale and Beckvar 1980, C. L., Vences et al. 2003, Volpe 1953, 1959a, Wassersug et al. 1981, Wilbur 1987, Wilbur et al. 1983.

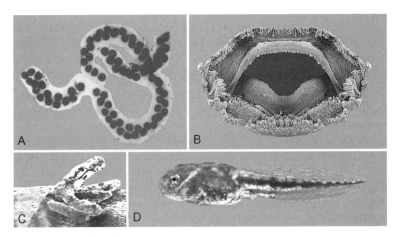

Figure 65. **Anaxyrus fowleri.** (A) part of an egg string (Oktibbeha Co., MS; RA), (B) SEM of the oral apparatus (MPM), (C) external gills at stage 21 (MPM), (D) tadpole (23.0 mm TL, stage 35, Okaloosa Co., FL; RA). See also Plate 7F.

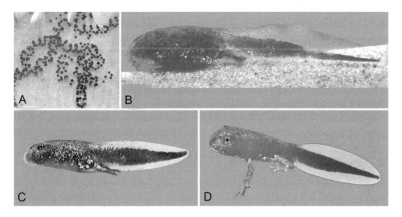

Figure 66. **Anaxyrus baxteri and A. hemiophrys.** (A) egg string of *A. baxteri* (culture, Laramie Co., WY; DP), (B) tadpole (25.0 mm TL, stage 35, Grand Forks Co., ND; RA) of *A. hemiophrys*, (C) tadpole (18.0 mm TL, stage 36; culture, Laramie Co., WY; RA) of *A. baxteri*, (D) metamorph of *A. baxteri* (11.0 mm TL, stage 42; RA). See also Plate 7G.

Anaxyrus hemiophrys (Canadian Toad) and *A. baxteri* (Wyoming Toad)   Fig. 66; PL 7G

IDENTIFICATION LTRF 2/3; similar to *A. americanus*; P-1 with medial gap; dorsum dark with pale blotches of various sizes on tail muscle and paler venter, especially anteriorly; about 25.0 mm TL/stage 26.

RANGE *A. baxteri* is restricted to the Laramie Hills in southeastern Wyoming. *A. hemiophrys* is found in the Northwest Territories south through eastern Alberta, most of Saskatchewan and southwestern Manitoba to northern Montana, northeastern North Dakota, western Minnesota, and northeastern South Dakota.

NATURAL HISTORY Canadian Toads oviposit from May to August in temporary pools, shallows of lakes, and slow-flowing streams. The ova are pigmented and laid as strings (clutch to 7000) and hatch in 3–12 days. Tadpoles metamorphose in 1.4–3.0 months at 7.0–14.0 mm SVL.

CITATIONS *Reproductive Biology:* Livezey and Wright 1947, Parmalee et al. 2002.

Anaxyrus microscaphus (Arizona Toad) and *A. californicus* (Arroyo Toad)   PL 7H,I

IDENTIFICATION LTRF 2/3; A-2 gap about 35% of one side of LTR A-2; LTR P-1 with medial gap; LTR P-3 length about 60% of P-2 length. *A. californicus*: medium bronzy brown; paler patch laterally; tail banded dorsally and with some lateral reticulations; 33.0–40.0 mm TL/stage 36. *A. microscaphus*: almost uniformly bronzy brown; belly white; tail muscle with a pattern of dark reticulations.

NATURAL HISTORY These toads occur primarily in riparian habitats, and their eggs and larvae can be found in the shallow parts of slow reaches of sandy or rocky-bottomed rivers, streams, washes, and arroyos, and occasionally in adjacent nonflowing water. Females lay 2013–10,368 ova/clutch (OD 1.2–1.7 mm, ED 3.3–4.2 mm) from January through July, extruding them in a relatively inelastic, unilayered tube (3.3–6.1 mm diameter). The strings of multiserial eggs are about 3.0–10.6 m long, have about 16.8 ova/cm., and are without intervening partitions. Eggs hatch in 3–6 days at 12–15 °C and the larvae are at 4.0 mm; they remain on egg jelly for 5–6 more days. Larvae grow to 34.0–40.0 mm TL and undergo metamorphosis in 2.2–2.7 months (April–October) at 9.0–15.0 mm SVL. These quiescent, benthic, nonsocial tadpoles prefer open sand or gravel substrates and sometimes escape by swimming at an angle toward the surface before returning to the bottom in deeper water.

RANGE *A. californicus* is restricted to the Coast Range of California from near Monterey south into Baja California, Mexico. *A. microscaphus* is known from several disjunct populations in southern Utah, adjacent Nevada and in a line southeasterly across Arizona into New Mexico and mountains of northwestern Mexico.

CITATIONS *General:* A. H. Price and Sullivan 1988. *Development/Morphology*: Grismer 2002, Sweet 1991, 1993. **Reproductive Biology**: Dahl et al. 2000, Livezey and Wright 1947, Sweet 1991, 1993. *Ecobehavior:* Sweet 1991.

Anaxyrus punctatus (Red-spotted Toad)   Fig. 67; PL 7J

IDENTIFICATION LTRF 2/3; A-2 gap less than 20% of one side of LTR A-2; LTR P-1 without medial gap; LTR P-3 length 85% or greater of P-2 length; typically described as being black, but members of some populations are brown to mottled and begin to attain metamorphic colors well before metamorphosis; sometimes deviates strongly from published descriptions; 23.0–30.0 mm TL/ stage 36.

NATURAL HISTORY Female Red-spotted Toads oviposit between March and September in oases, temporary pools, and slow parts of desert streams and spring runs, often in open areas. Between 30 and 5000 darkly pigmented ova (OD 1.0–1.3 mm, ED 3.2–4.0 mm, 1 jelly layer) are laid as single eggs that are weakly adherent; they may group together in a clump or form a linear series. The eggs hatch in 3 days and larvae metamorphose in 1.3–1.9 months at about 9.0–11.0 mm SVL. Tadpoles show considerable geographic variation in coloration, from uniformly dark to mottled, throughout their range.

RANGE *A. punctatus* occurs from the arid regions of southeastern California to central Texas and south in Baja California and mainland Mexico.

CITATIONS *General:* Korky 1999. *Development/Morphology*: Gosner 1959, Grismer 2002, Livo 2000, Storer 1925. *Reproductive Biology:* Livezey and Wright 1947. *Ecobehavior:* Livo and Kondratieff 2000, Strecker 1926b.

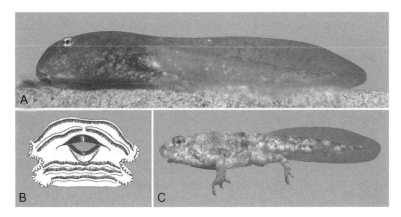

Figure 67. **Anaxyrus punctatus.** (A) tadpole (28.0 mm TL, stage 34, Maricopa Co., AZ; RA), (B) oral apparatus (modified from Stebbins 1985), (C) metamorph (28.0 mm TL, stage 43, Pinal Co., AZ; RA). See also Plate 7J.

Anaxyrus quercicus (Oak Toad)   PL 7K

IDENTIFICATION LTRF 2/3; A-2 gap about 50% of one side of LTR A-2; LTR P-1 with medial gap; LTR P-3 length 50–70% of P-2 length; marginal papillae often do not extend to the lateral tips of row P-3; medium brown dorsally; darker mottling ventrolaterally; tail banded dorsally, sometimes with darker marks laterally; to 26.0 mm TL/stage 36.

NATURAL HISTORY Oak Toads oviposit from April to October in shallow, often grassy, temporary sites coincidental with the warm summer rains. Between 300 and 700 black ova (OD 0.8–1.4.0 mm, ED 1.2–1.6.0 mm, 1 jelly layer) form a clutch, and eggs are laid as short strings of 2–8 ova, with several strings sometimes extending from a common attachment point. The tubular jelly is thin, transparent and inconspicuous, and somewhat constricted between successive ova. Eggs hatch in 24 h to 3.5 days, and larvae grow to 10.0–19.0 mm TL. Metamorphosis is in 4.0–6.3 weeks at 7.0–9.0 mm SVL.

The small, strongly benthic tadpoles occur in shallow, temporary pools and seldom aggregate. As a result the tadpole populations seem sparse relative to what one might expect based on the calling intensity of males at a pool (see *Hyla cinerea*), and these tadpoles often escape notice because of their small size, cryptic coloration, and relative quiescence.

RANGE *A. quercicus* is found along the Gulf and Atlantic coastal plains from southeastern Virginia to eastern Louisiana.

CITATIONS *General:* Ashton and Franz 1979. *Development/Morphology*: Altig and Pace 1974, Volpe and Dobie 1959. *Reproductive Biology:* Ashton and Ashton 1988, Hamilton 1955, A. H. Wright 1932. Livezey and Wright 1947.

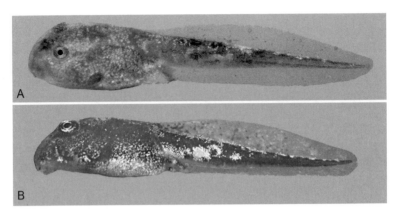

Figure 68. **Anaxyrus speciosus.** Tadpoles: (A) 17.0 mm TL, stage 31, tilted slightly to left, (B) 22.0 mm TL, stage 34, tilted slightly to right (Eddy Co., NM; RA). See also Plate 7L.

Anaxyrus speciosus (Texas Toad)   Fig. 68; PL 7L

IDENTIFICATION LTRF 2/3; A-2 gap about 60–85% of one side of LTR A-2; LTR P-1 without medial gap; LTR P-3 length about 60% of P-2 length; dorsum varies from dark to pale and typically paler colored than most bufonid tadpoles; belly white; sides of tail muscle blotched; minor dark marks in dorsal fin; 20.0–28.0 mm TL/stage 36.

NATURAL HISTORY *Anaxyrus speciosus* breeds in temporary rain pools from April to September. The brown and yellow ova (OD 1.2–1.6 mm, ED 1.8–2.4 mm, 1 jelly layer) are deposited in slightly scalloped tubes (4.4–6.8 ova/cm) that are laid as coiled strings. Eggs hatch in 2 days, and the tadpoles metamorphose in 1.3–2.0 months at about 12.0 mm SVL.

RANGE *A. speciosus* is known from the western two-thirds of Texas and southeastern New Mexico, southern Oklahoma, and the adjacent Mexican states of Chihuahua, Coahulia, Nuevo León, and Tamaulipas.

CITATIONS *Development/Morphology*: Bragg 1947a–b. *Reproductive Biology:* Livezey and Wright 1947. *Ecobehavior:* Bragg 1947a, Licht 1967.

Anaxyrus terrestris (Southern Toad)   Fig. 69; PL 8A

IDENTIFICATION LTRF 2/3; A-2 gap about 65% of one side of LTR A-2; LTR P-1 without medial gap; LTR P-3 length about 80% of P-2 length; usually appears black with a diagonal golden band extending posterolaterally from near eye, more visible at night; to 35.0 mm TL/stage 36.

NATURAL HISTORY Southern Toads oviposit between January and October in ditches, shallow swamp pools, farm ponds, isolated wetlands, and the shallows of more permanent bodies of water. Between 2500 and 8000 pigmented ova (OD 1.0–1.5 mm, ED 2.2–4.6 mm, 2 jelly layers, 2.3–2.6 ova/cm) are laid as coiled

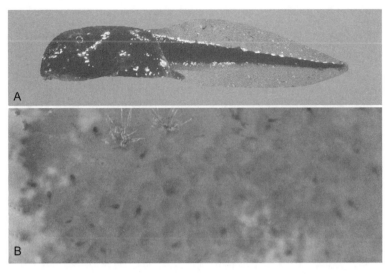

Figure 69. **Anaxyrus terrestris.** (A) tadpole (26.0 mm TL, stage 35, Hancock Co., MS; RA), (B) hexagonal and pentagonal feeding depressions in the flocculent sediment of a temporary pool (i.e., "tadpole nests or holes"), 1–2 tadpoles rest in some pits (RA). See also Plate 8A.

strings without partitions between the ova. Eggs hatch in 2–4 days and larvae grow to 25.0 mm TL. Metamorphosis is in 1.0–1.8 months at 6.0–11.0 mm SVL.

RANGE *A. terrestris* occurs in the Gulf and Atlantic coastal plains from southeastern Virginia to southeastern Louisiana and north through Mississippi and Alabama.

CITATIONS *Development/Morphology*: Altig and Pace 1974, Gosner 1959, Volpe 1953, 1959a–b. *Reproductive Biology:* Livezey and Wright 1947, A. H. Wright 1932. *Ecobehavior:* Alford 1986, Babbitt 1995, Babbitt and Jordan 1996, Beck and Congdon 2003, Bragg 1955a, Fedewa 2006, Gosner and Black 1958a, Jensen 1996, J. M. L. Richardson 2001, K. G. Smith 2005a, 2006, Travis and Trexler 1986, Volpe 1953, Wilbur 1987, Wilbur et al. 1983.

Anaxyrus woodhousii (Woodhouse's Toad)   Fig. 70; PL 8B

IDENTIFICATION LTRF 2/3; A-2 gap about 40–50% of one side of LTR A-2; LTR P-1 with medial gap; LTR P-3 length 50–70% of P-2 length; extremely variable throughout the large range, grossly appears dark, especially in smaller individuals, but abundant iridophores arranged in blotches or evenly distributed are visible at slight magnification; venter gray to white; to 37.0 mm TL/ stage 36.

NATURAL HISTORY Females of Woodhouse's Toad oviposit from February to September in all types of nonflowing water (lakes, pools, irrigation ditches), often after spring and summer rains; sometimes in quiet waters of streams. Clutch

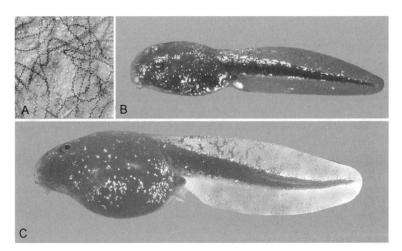

Figure 70. **Anaxyrus woodhousii.** (A) egg string (Benton Co., WA; CCC), (B) tadpole (23.0 mm TL, stage 30, Maricopa Co., AZ; RA), (C) tadpole (15.0 mm TL, stage 27, Benton Co., WA; WPL). See also Plate 8B.

size large, up to 28,000 uni- or biserial ova deposited in a unilayered, distinct and firm tube with 6.8–10.0 ova/cm. Ova black and gray (OD 1.0–1.5 mm, ED 2.6–4.6 mm). Tadpoles grow to 20.0–30.0 mm TL and metamorphose in 1.2–2.0 months at 10.0–15.0 mm SVL.

RANGE Disjunct populations of *A. woodhousii* occur in southeastern Washington and eastern Oregon and adjacent Idaho. It is also distributed widely through western states from southeastern California east to Missouri and Louisiana and north to Montana and North Dakota and south into central Mexico.

CITATIONS **Development/Morphology**: Altig and Pace 1974, Gosner 1959, Grismer 2002, V. O. Johnson 1939, Youngstrom and Smith 1936. **Reproductive Biology:** Livezey and Wright 1947, Parmalee et al. 2002. **Ecobehavior:** Bragg 1940a, 1955a, Krupa 1995, Licht 1967, Swart and Taylor 2004, Verma and Pierce 1994, Volpe 1953, Wassersug and Seibert 1975, Woodward 1982a, 1987b.

Incilius alvaria (Sonoran Desert Toad)   Fig. 71; PL 8C

IDENTIFICATION LTRF 2/3(1); A-2 gap about 55% of one side of LTR A-2; LTR P-1 without medial gap; LTR P-3 length about 65% of P-2 length; uniformly pale tan, abundant iridophores give body brassy appearance, fins clear, tip sometimes weakly spotted; venter transparent; about 57.0 mm TL/stage 36.

NATURAL HISTORY The Sonoran Desert Toad is an explosive breeder from May to September in temporary desert pools and more permanent water of springs and streams usually after moderate seasonal rains. Between 7500 and

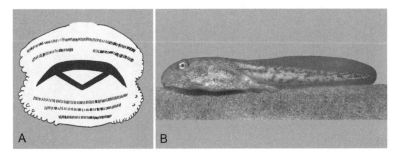

Figure 71. **Incilius alvaria.** (A) schematic drawing of oral apparatus (modified from Altig 1971), (B) Tadpole (30.0 mm TL, stage 34, Pima Co., AZ; RA). See also Plate 8C.

8000 brown and tan ova are laid as uniserial strings in unilayered tube (OD 1.1–1.7 mm, ED 2.1–2.3 mm, 4.8–11.2 ova/cm) without partitions between ova or scalloping of string margins. Tadpoles grow to about 57.0 mm TL and metamorphose in about 30 days; they are poorly known and may occur in deeper water than most bufonid tadpoles.

RANGE *I. alvaria* occurs in southeastern California, the southern third of Arizona, southwestern New Mexico, and south into Mexico.

CITATIONS *General:* Fouquette 1970. *Development/Morphology*: Altig 1971, Grismer 2002. *Reproductive Biology:* Livezey and Wright 1945, 1947, Savage and Schuierer 1961. *Ecobehavior:* Ruthven 1907.

Incilius nebulifer (Gulf Coast Toad)   PL 8D

IDENTIFICATION LTRF 2/3; A-2 gap about 50% of one side of LTR A-2; LTR P-1 with medial gap; LTR P-3 length 50–70% of P-2 length; small individuals appear uniformly dark; large individuals are mottled or frosted with abundant iridophores; dorsum of tail muscle prominently banded at all stages; 30.0–40.0 mm TL/stage 36.

NATURAL HISTORY Female *Incilius nebulifer* oviposit from March to September in pools, shallow parts of ponds and slow streams, and perhaps brackish water. The ova are pigmented (OD about 1.2 mm, ED 2.8–3.2 mm, 2 jelly layers), laid as strings (2.3–3.3 to 8.3–9.0 ova/cm), and commonly biserial. Eggs hatch in 1.5–2.0 days; tadpoles grow to 25.0 mm TL and metamorphose in 2.9–4.0 weeks at 7.0–12.0 mm SVL. The tadpoles are strictly benthic and do not aggregate as much as some other bufonid tadpoles; they usually swim directly across the bottom for relatively short distances.

RANGE *I. nebulifer* occurs from central Texas east to southwestern Mississippi and south into Mexico.

CITATIONS *General:* Porter 1970. *Development/Morphology*: Limbaugh and Volpe 1957, Savage 1980, Volpe 1956, 1957a. *Reproductive Biology:* Livezey and Wright 1947. *Ecobehavior:* Licht 1968, Volpe 1957a, 1959b.

Figure 72. **Rhinella marina.** (A) tadpole (preserved, 28.0 mm TL, stage 34, Veracruz, Mexico; RA).

Rhinella marina (Cane Toad)   Fig. 72

IDENTIFICATION LTRF 2/3; A-2 gap about 15–25% of one side of LTR A-2; LTR P-1 with medial gap; LTR P-3 length about 100% of P-2 length; uniformly dark, sometimes with bicolored tail muscle; fins clear; about 30.0 mm TL at stage 36.

NATURAL HISTORY Cane Toads oviposit opportunistically throughout the year after rains and usually in shallow, often temporary ponds; in Hawaii they also breed in mountain streams. The darkly pigmented ova (4240–36,100) are laid as strings and hatch in 48–72 h, and the tadpoles undergo metamorphosis in 45–55 days at 11.0–12.0 mm SVL. Tadpole densities of 15–61/m^2 have been reported, and some populations may breed in saltwater up to 5–10 ppt.

RANGE *R. marina* occurs naturally in extreme southern Texas and ranges through Central America into South America. This toad is also exotic on all major islands of Hawaii and in the southern half of peninsular Florida.

CITATIONS *General:* Easteal 1986. *Development/Morphology*: Altig and Pace 1974, Breder 1946, Ford and Scott 1996, Kenny 1969, Nokhbatolfoghahai and Downie 2007, Savage 1960, 1980, *Reproductive Biology:* Livezey and Wright 1947. *Ecobehavior:* Crossland et al. 2011, Floyd 1985, K. L. Lawler and Hero 1997, Mares 1972, K. G. Smith 2005a, Valerio 1971.

Dendrobatidae (Dart-poison Frogs)

This neotropical exotic occurs on Maui and Oahu, Hawaii; in phytotelms; LTRF 2/3, all rows uniserial, rows A-1 and A-2 about equal in length; small marginal papillae, with wide dorsal gap; submarginal papillae sparse; oral disc not emarginate; jaw sheaths massive and prominently serrate; vent medial; eyes dorsal; spiracle sinistral; dorsal fin low with rounded tip, originates near tail-body junction; body depressed with broadly rounded snout; nostrils small, without ornamentation; digital pads visible by stage 36; uniformly shiny dark brown to black; metamorph coloration resembles that of adults; 10.0–26.0 mm TL/stage 36; 1 genus, 1 species. *Dendrobates*: as for family.

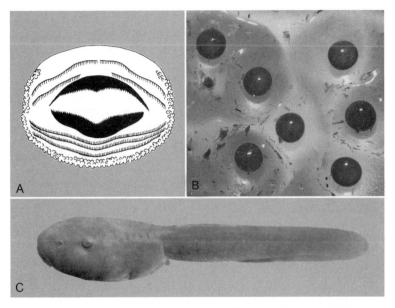

Figure 73. **Dendrobates auratus.** (A) oral apparatus (modified from Silverstone 1975), (B) eggs of similar species (*D. leucomelas*, TRK), (C) tadpole (29.0 mm TL, stage 30, culture; GG).

Dendrobates auratus (Green and Black Dart-poison Frog)   Fig. 73

IDENTIFICATION A-2 gap about 50% of one side of LTR A-2; LTR P-1 without medial gap; LTR P-3 length 100% of P-2 length; as for family.

NATURAL HISTORY Females of *D. auratus* oviposit 4–7 nonpigmented ova in a terrestrial clump, tended by the male. Eggs hatch in 1.4–1.9 weeks and the male transports the tadpoles on his back to phytotelms where they feed and grow. Tadpoles metamorphose in 1.4–2.0 months. This carnivorous tadpole has formidable mouthparts, and typically only one tadpole occurs in each tree hole or bromeliad cistern.

RANGE The Green and Black Dart-poison Frog is exotic on Maui and Oahu, Hawaii.

CITATIONS *Development/Morphology*: Savage 1968, 1980, Silverstone 1975. *Reproductive Biology:* Dunn 1941, K. D. Wells 1978. *Ecobehavior:* Dunn 1941, T. H. Eaton 1941, Lannoo et al. 1987.

Hylidae (Treefrogs and relatives)

Treefrogs and their relatives occur throughout mainland North America; most commonly occur in nonflowing water as benthic or pelagic inhabitants, rarely benthic in flowing water; LTRF 2/2 (*Acris*), 2/3 (common), or 2/4 (*Osteopilus*), all rows uniserial, rows A-1 and A-2 about equal in length; small marginal papillae uni- to triserial at various areas with wide dorsal gap, rarely a narrow ventral gap; submarginal papillae usually sparse, sometimes abundant ventrolaterally; oral disc not emarginate, anterior labium folds downward to form a roughly triangular-shaped disc at rest; finely or coarsely serrate jaw sheaths narrow to wide; vent dextral; eyes usually lateral, dorsal in *Acris*; spiracle sinistral; dorsal fin low to tall, originates at tail-body junction to well forward on body; body shape compressed, globular to depressed; nostrils small and round, without ornamentation; digital pads visible by stage 36; coloration variable, patterns range from uniform or striped to banded; metamorph coloration resembles that of adults; 30.0–70.0 mm TL/stage 36; 5 genera, 35 species; the tadpoles of *Hyla*, *Pseudacris*, and *Smilisca* are not distinguishable as genera. *Acris* (3 species): LTRF 2/2; eyes dorsal. *Hyla* (11 species), *Pseudacris* (18 species), and *Smilisca* (2 species): LTRF typically 2/3, rarely 2/2; eyes lateral. *Osteopilus* (1 exotic species): LTRF 2/4; eyes lateral.

Acris blanchardi (Blanchard's Cricket Frog) and *A. crepitans* (Northern Cricket Frog)   Fig. 74; PL 8E

IDENTIFICATION LTRF 2/2; A-2 gap greater than 100% of one side of LTR A-2; LTR P-1 without medial gap; spiracular tube short; dorsum medium brown with some faint mottling; darker mottling ventrolaterally; tail muscles sometimes banded; tail tip usually black; tadpoles with large hind legs: toes webbed about three-quarters of toe length; to 36.0 mm TL/stage 36.

NATURAL HISTORY Females of *A. crepitans* and *A. blanchardi* have ovulated by the time they arrive at a breeding site. Eggs are deposited between March and September in ponds and lake margins, sometimes in shallow streams but usually among vegetation. Females may produce multiple clutches/season, each of which may include 174–431 pigmented ova (OD about 1.0 mm, ED 2.3–5.0 mm, 1 or 2 jelly layers). Literature reports indicate that the ovipositional mode may vary from single eggs attached to vegetation or placed on the bottom, to groups of eggs, to a floating film of 6–15 ova/group. Tadpoles grow to 30.0–50.0 mm TL and frequently have a black tail tip. Metamorphosis is in 4.1–12.9 weeks at 10.0–15.0 mm SVL.

The benthic tadpoles usually occur in 2–4 cm of water at the very margins of lakes and ponds among debris below emergent vegetation; in this way these cryptic, inactive tadpoles commonly occur with predatory fishes.

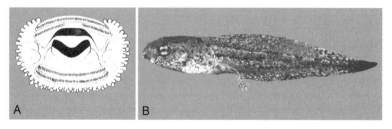

Figure 74. ***Acris crepitans.*** (A) oral apparatus (modified from Orton 1952), (B) tadpole (36.0 mm TL, stage 36, Pope Co., IL; MR). See also Plate 8E.

RANGE A. *blanchardi* is broadly distributed from central Nebraska in the west to Ohio in the east and south through most of Texas with extensions into eastern Colorado and southeastern New Mexico along the Pecos River. *A crepitans* is found in most of the remainder of eastern North America from New York southwest to Louisiana. It is absent from the Atlantic Coastal Plain and Peninsular Florida.

CITATIONS ***Development/Morphology***: Gosner and Black 1957a, Regan 1969. ***Reproductive Biology:*** Collins and Wilbur 1979, Livezey 1950, Livezey and Wright 1947, Parmalee et al. 2002, Trauth et al. 1990, 2004. ***Ecobehavior:*** Bragg 1947a, J. P. Caldwell 1982, Gosner and Black 1957b, L. M. Johnson 1991.

Acris gryllus (Southern Cricket Frog)   Fig. 75; PL 8F

IDENTIFICATION LTRF 2/2; A-2 gap about 100% of one side of LTR A-2; LTR P-1 without medial gap; spiracular tube long; dorsum uniformly brown to dark; dorsum of tail muscle sometimes banded; tail tip usually black; tadpoles with large hind legs: toes webbed about half of toe length; 35.0–45.0 mm TL/stage 36.

NATURAL HISTORY Females of *A. gryllus* oviposit in any month especially in the Florida, but breeding peaks from April to October around ponds and lake margins. Between 75 and 250 brown ova (OD 0.9–1.0 mm, ED 2.4–3.6 mm, 1 jelly layer) are laid as single eggs or small groups on the pond bottom or on vegetation; sometimes they are in small masses. They hatch in 3–4 days, and

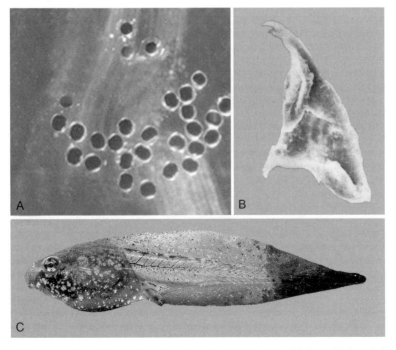

Figure 75. ***Acris gryllus.*** (A) nonadherent, single eggs (Noxubee Co., MS; RA), (B) lateral view of single labial tooth (RA), (C) tadpole (28.0 mm TL, stage 32, Bryan Co., GA; DJS). See also Plate 8F.

tadpoles grow to 42.0 mm TL and metamorphose in 1.3–2.0 months at 9.0–15.0 mm SVL. Toe pads visible by stage 36. Their habitat and behavior are similar to those of *A. crepitans* but often in more acidic waters.

RANGE Southern Cricket Frogs occur throughout the Atlantic Coastal Plain from southeastern Virginia to Florida and west through most of southern Georgia, southern and western Alabama, and Mississippi and into the eastern edge of Louisiana and southwestern Tennessee.

CITATIONS *Development/Morphology*: Altig and Pace 1974, Strauss and Altig 1992. *Reproductive Biology:* Livezey and Wright 1947, A. H. Wright 1932. *Ecobehavior:* Abbott 1882, Altig et al. 1975, Altig and McDearman 1975, J. M. L. Richardson 2001, Schorr et al. 1990, Turnipseed and Altig 1975.

Hyla andersonii (Pine Barrens Treefrog)   Fig. 76; PL 8G

IDENTIFICATION LTRF 2/3; midventral marginal papillae biserial; dorsum pale, olive or often golden; belly greenish yellow; tail muscle striped, perhaps more distinct in northern populations than those further south, or banded and fins often with well-defined blotches and spots: 25.0–38.0 mm TL/stage 36.

NATURAL HISTORY Female Pine Barrens Treefrogs oviposit from late April through September and place their eggs in small, shallow backwaters of acidic bogs, ponds, seepages, and swamps. A clutch consists of 500–1000 brown ova (OD 1.2–1.4 mm, ED 3.5–4.0 mm, 2 jelly layers) laid as singles or small, loosely adherent clumps attached to vegetation, on the bottom or floating at the surface. Eggs hatch in 3–4 days and tadpoles grow to 35.0 mm TL in 1.6–2.5 months. Metamorphosis occurs at 11.0–15.0 mm SVL.

At least in Florida, this tadpole often occurs in bodies of water smaller than one would expect to find hylid tadpoles; footprint depressions in a boggy site are sufficient. The benthic tadpoles seem behaviorally inactive compared with those of most congeners (see *Hyla avivoca*). At some sites, these tadpoles and those of *Hyla squirella* have a rather uniform, golden metallic luster. The significance of this coloration and if and how it is induced or formed are not known.

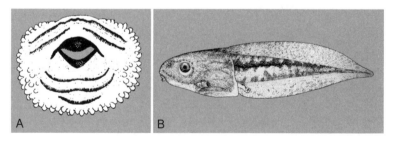

*Figure 76. **Hyla andersonii.** Tadpole: (A) oral apparatus, (B) lateral view (30.0 mm TL, stage 33; Ocean Co., NJ; modified from Noble and Noble 1923). See also Plate 8G.

RANGE *Hyla andersonii* occurs in three widely separated disjunct populations in the Pine Barrens of New Jersey, in central North and South Carolina, and in the western part of the Florida panhandle.

CITATIONS **General:** Gosner and Black 1967. ***Development/Morphology***: Gosner 1959, Gosner and Black 1957a, Noble and Noble 1923. ***Reproductive Biology:*** Livezey and Wright 1947, Means and Longden 1976, A. H. Wright 1932. ***Ecobehavior:*** Gosner and Black 1957b, S. P. Lawler 1989, Pehek 1995.

Hyla arenicolor (Canyon Treefrog)   Fig. 77; PL 8H

IDENTIFICATION LTRF 2/3; A-2 gap about 30% of one side of LTR A-2; LTR P-1 with medial gap; LTR P-3 length about 90% of P-2 length. Eyes lateral but sometimes evaluated as dorsal, these tadpoles are commonly misidentified as bufonids or ranids at younger stages; appear black when small, grayish to brown when older; quite variable in coloration, body shape and fin height; to 45.0 mm TL/stage 36.

NATURAL HISTORY Canyon Treefrogs oviposit from March to August along streams and in stream pools. Several hundred pigmented ova (OD 1.8–2.4 mm, ED 3.8–5.0 mm, 1 jelly layer) are laid as small, adherent clumps or as singles, usually on the bottom or attached to vegetation. Metamorphosis occurs in 1.3–2.5 months at 14.0–28.0 mm SVL.

RANGE Populations of *Hyla arenicolor* are known from desert streams throughout much of Arizona, the western half of New Mexico, and the southern third of Utah, with scattered populations in eastern New Mexico, southeastern Colorado and Trans Pecos, Texas and south into northwestern Mexico. **Citations:**

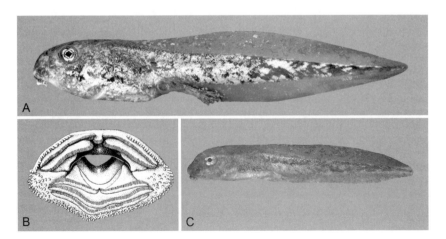

Figure 77. **Hyla arenicolor.** (A) tadpole (36.0 mm TL, stage 36, Cochise Co., AZ; MR), (B) oral apparatus (modified from Duellman 1970), (C) tadpole (28.0 mm TL, stage 32, Jeff Davis Co., TX; RA). See also Plate 8H.

Development/Morphology: Zweifel 1961. ***Reproductive Biology:*** Livezey and Wright 1947.

Hyla avivoca (Bird-voiced Treefrog)   Fig. 78; PL 8I

IDENTIFICATION   LTRF 2/3; A-2 gap 20–25% of one side of LTR A-2; LTR P-1 without medial gap; LTR P-3 length about 60% of P-2 length; perhaps the most spectacular tadpole of North America—black body with dorsum of tail muscle banded, a silver or reddish band extends between the eyes and similar stripes are present from eye to naris; fins clear with fine black reticulations. This pattern appears early in embryology and remains throughout ontogeny; 30.0–40.0 mm TL/stage 36.

NATURAL HISTORY   Females of *H. avivoca* breed in the late spring to summer months (March–August) in acidic, cypress-tupelo gum swamps associated with large streams and rivers. Between 400 and 800 brown ova (OD 0.8–1.4 mm, ED 4.0–5.8 mm, 2 jelly layers) are laid in small films or clusters that sink to the bottom or become attached to vegetation. The eggs hatch in 1.5–4.0 days and tadpoles grow to 32.0 mm TL. They metamorphose in 4.1–5 weeks at 90.0–13.0 mm SVL.

The benthic tadpoles occur solitarily and rest quietly for long periods of time, usually among organic litter adjacent to shore; at night they spend long quiescent

Figure 78. ***Hyla avivoca.*** (A) tadpole (39.0 mm TL, stage 34, Pope Co., IL; MR), (B) Tadpole (28.0 mm TL, stage 34, Noxubee Co., MS; RA), (C) Typical southeastern swamp habitat where *H. avivoca* is sympatric with sirenids, amphiumids, *Hyla cinerea*, and *Lithobates clamitans*. See also Plate 8I.

periods in very shallow water with the snout tipped to the surface. RA has seen only one incidence of tadpoles not in deep shade, in water deeper than 5 cm, and acting as a group. In this case, the tadpoles were in a group on the bottom in about 30 cm of clear water, and individuals were making frequent trips to the surface to gulp air. It is almost a rule that *H. avivoca* occurs only where emergent, woody vegetation is present.

RANGE The Bird-voiced Treefrog occurs in disjunct populations in Arkansas and Louisiana west of the Mississippi River and throughout much of the Mississippi Embayment and east through southern Alabama, northwestern Florida, and central Georgia to southwestern South Carolina.

CITATIONS *General:* P. W. Smith 1966b. *Development/Morphology*: Altig and Pace 1974, Fortman and Altig 1973, Hellman 1953, Trauth et al. 2004, Volpe et al. 1961. *Reproductive Biology:* Livezey and Wright 1947, Redmer 1998, Trauth and Robinette 1990. *Ecobehavior:* M. V. Parker 1951, Redmer et al. 1999, Turnipseed and Altig 1975.

Hyla cinerea (Green Treefrog)   Fig. 79; PL 8K

IDENTIFICATION LTRF 2/3; A-2 gap about 35–60% of one side of LTR A-2; LTR P-1 without medial gap; LTR P-3 length about 30–45% of P-2 length; midventral marginal papillae biserial; body dark or greenish with yellowish or yellowgreen tinge and with pale lines converging from eyes to nostrils, dorsum of at least larger tadpoles with freckled pattern; throat speckled diffusely; 34.0–60.0 mm TL/stage 36.

NATURAL HISTORY Females of the Green Treefrog oviposit from April to September in ponds, pools and swamps. Between 500 and 1000 black to brown ova (OD 0.8–1.6 mm, ED 3.5–4.0 mm, inner layer 2.2–3.4, 2 jelly layers), are laid as a film; they sometimes sink and appear as small, poorly defined clumps, often

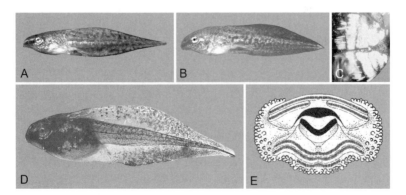

Figure 79. **Hyla cinerea.** (A) tadpole (42.0 mm TL, stage 35, Pope Co., IL; MR), (B) tadpole (37.0 mm TL, stage 37, Hancock Co., MS; RA), (C) rectus abdominus in belly, posterior to right (RA), (D) tadpole (34.0 mm TL, stage 30, Conway Co., AR; SET), (E) oral apparatus (modified from Orton 1952). See also Plate 8K.

upon vegetation near the surface. A single female may lay multiple clutches of 5–40 eggs/group. Eggs hatch in 2–4 days at 4.5–5.5 mm TL, and the tadpoles grow to 49.0–60.0 mm TL. Metamorphosis occurs in 4–9.3 weeks at 12.0–17.0 mm SVL. RA and others have noticed that tadpoles of *Hyla cinerea* commonly are sparse at sites where many males call, and an explanation for the difference is not known.

RANGE Populations of Green Treefrogs are common on the Coastal Plain from Delaware to southern Texas including the Mississippi Embayment and peninsular Florida. In some mid-Atlantic states (e.g., North Carolina) they seem to be expanding their range into the Piedmont.

CITATIONS *Development/Morphology*: Altig 1972, Altig and Pace 1974, Fortman and Altig 1973, Gosner 1959, Trauth et al. 2004. *Reproductive Biology:* Livezey and Wright 1947, Perrill and Daniel 1983, Trauth et al. 1990, A. H. Wright 1932. *Ecobehavior:* Faragher and Jaeger 1998, Gunzburger and Travis 2004, 2005a–b, Leips et al. 2000, Redmer et al. 1999, J. M. L. Richardson 2001, 2002a–b, Roth and Jackson 1987, Schorr et al. 1990, Turnipseed and Altig 1975.

Hyla femoralis (Pine Woods Treefrog)   Fig. 80; PL 8L

IDENTIFICATION LTRF 2/3; A-2 gap about 20% of one side of LTR A-2; LTR P-1 without medial gap; LTR P-3 length about 50–70% of P-2 length; tail with distinct flagellum; body more or less uniformly brown or russet; basal two-thirds of tail muscle with prominent pale lateral stripe that remains in preservative; fin areas adjacent to tail muscle usually lacking or with less dense aggregations of melanic blotches compared with remainder of fin; clear parts of fins usually reddish; 30.0–35.0 mm TL/stage 36.

NATURAL HISTORY Females of *H. femoralis* oviposit between March and October usually in temporary pools in pine flatwood forests and often in ditches in areas in early succession. Clutch size ranges from 200 to 2000 brown ova (OD 0.8–0.9 mm, ED 4.0–8.0 mm, 2 jelly layers) laid as a cluster or as film, often in ovipositional rafts of about 100 eggs. They hatch in 3 days; tadpoles grow to 33.0–36.0 mm TL and metamorphose in 1.2–2.5 months at 8.0–15.0 mm SVL.

RANGE Pine Woods Treefrogs occur on the Coastal Plain from central Virginia to southeastern Louisiana and through most of peninsular Florida.

CITATIONS *General:* R. L. Hoffman 1988. *Development/Morphology*: Altig 1972, Altig and Pace 1974, Fortman and Altig 1973, Gosner 1959. *Reproductive Biology:* Livezey and Wright 1947, A. H. Wright 1932. *Ecobehavior:* S. A. Johnson 1996, Leips and Travis 1994, McCoy 2007, J. C. Mitchell 1986, J. M. L. Richardson 2001, Travis 1980b, Warner et al. 1993, Wilbur 1982.

Hyla gratiosa (Barking Treefrog)   Fig. 81; PL 9A

IDENTIFICATION LTRF 2/3; A-2 gap about 55% of one side of LTR A-2; LTR P-1 without medial gap; LTR P-3 length about 50% of P-2 length; jaw sheaths

Figure 80. **Hyla femoralis.** (A) adherent egg film (Okaloosa Co., FL; rafts resulting from 5 or more ovipositional bouts delimited by lines of white dots; RA), (B) tadpole (29.0 mm TL, stage 34; Evans Co., GA; DJS). See also Plate 8L.

Figure 81. **Hyla gratiosa.** (A) labial teeth in situ (modified from Altig 1973), (B) fibers of the mandibulo-labialis muscle extending to the first and second upper labial tooth rows (RA), (C) metamorph (42.0 mm TL, stage 42, Liberty Co., GA; DJS), (D) tadpole (29.0 mm TL, stage 25, Liberty Co., GA; DJS) showing dark tail saddle (= single band) present in small specimens. See also Plate 9A.

massive. Tadpoles less than 30 mm TL: lightly colored with nonpigmented fins and tail muscle except for a single black band at midlength of tail muscle. Larger tadpoles: overall uniformly dark, punctate melanophores on the anterior third of the tail and stellate cells on the posterior two-thirds of the tail; stellate ones disperse at night and impart an oily, black color to the tail; uniformly silver iris gains brownish maroon pigment late in ontogeny; line extending from eye to naris formed by lack of pigment in subintegumentary layers; 45.0–70.0 mm TL/ stage 36.

NATURAL HISTORY   Barking Treefrogs oviposit from March to October, in ponds and pools in shallow wetlands often following rains. Clutches vary in size from 800 to 4000 pigmented ova (OD 1.0–1.8 mm, ED 2.3–5.0 mm, 1 jelly layer) that are laid as loose and glutinous single eggs on pond bottoms or as an indistinct group draped over vegetation, usually in ponds that lack fishes. Eggs hatch in 2–3 days, and tadpoles grow to 70.0 mm TL in 1.2–5.3 months. Metamorphosis is at 14.0–28.0 mm SVL, and metamorphs are usually uniformly green.

The nektonic tadpoles commonly occur in deep, often clear water, and smaller tadpoles tend to stay in areas of dense plant cover. Larger tadpoles frequent the deepest water that is clear of emergent vegetation, and their tails turn very black. Especially at night, tadpoles often escape at an angle to the surface before descending again to midwater or the bottom; they often swing momentarily head-upward at the end of a sprinting swim.

RANGE   The distribution of *Hyla gratiosa* is continuous on the Coastal Plain from North Carolina to southeastern Louisiana including most of peninsular Florida and north into southern Tennessee. Disjunct populations are also known from Delaware and adjacent Maryland, southeastern Virginia, southwestern Kentucky, and adjacent northwestern Tennessee.

CITATIONS   *Development/Morphology*: Altig 1972, Altig and Pace 1974, Fortman and Altig 1973, Gosner 1959, Strauss and Altig 1992. *Reproductive Biology:* Delis and Summers 1996, Gunzburger and Travis 2007, Livezey and Wright 1947, Perrill and Daniel 1983, A. H. Wright 1932. *Ecobehavior:* J. P. Caldwell et al. 1980, Leips and Travis 1994, Leips et al. 2000, J. M. L. Richardson 2001, Travis 1980a, 1981b, 1983a–b, 1984, Travis et al. 1985a, Warner et al. 1993, Wilbur 1982.

Hyla squirella (Squirrel Treefrog)   Fig. 82; PL 9B

IDENTIFICATION   LTRF 2/3; A-2 gap 55–75% of one side of LTR A-2; LTR P-1 without medial gap; LTR P-3 length about 40–45% of P-2 length; easily confused with several other species; 30.0–42.0 mm TL/stage 36.

NATURAL HISTORY   Squirrel Treefrogs oviposit from March to December in temporary and permanent ponds, pools and other shallow-water habitats, often opportunistically with rains. Clutch size varies between 800 and 1000 brown ova (OD 0.8–1.0 mm, ED 1.4–2.0 mm, 2 jelly layers) that are laid as single eggs on

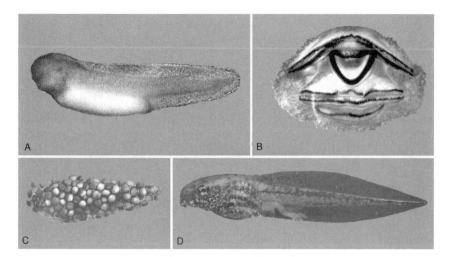

Figure 82. **Hyla squirella.** (A) embryo (7.0 mm TL, stage 21), (B) oral apparatus (Oktibbeha Co., MS; RA), (C) recently oviposited eggs (lab origin; prior to full jelly hydration and ovum rotation, group soon fell apart into single eggs on the bottom; St. Martin Parish, LA; DLD), (D) tadpole (31.0 mm TL, stage 36, Oktibbeha Co., MS; RA). See also Plate 9B.

pond bottoms or as small irregular clumps attached to vegetation. Eggs hatch in 1.5–2 days, and tadpoles grow to 32.0–38.0 mm TL, undergoing metamorphosis in 1.2–2.0 months at 11.0–13.0 mm SVL. The Squirrel Treefrog is an opportunistic, ruderal species that commonly breeds in temporary sites in disturbed areas and then disappears as succession proceeds.

RANGE Populations of Squirrel Treefrogs occupy the Coastal Plain from southeastern Virginia south through peninsular Florida and west to southeastern Texas.

CITATIONS *General:* Martof 1975c. *Development/Morphology*: Altig 1972, Altig and Pace 1974, Fortman and Altig 1973, Webb 1965. *Reproductive Biology:* Livezey and Wright 1947, A. H. Wright 1932. *Ecobehavior:* Babbitt and Tanner 1997, Beck 1997, J. M. L. Richardson 2001, Walls et al. 2002.

Hyla chrysoscelis (Cope's Gray Treefrog) and *H. versicolor* (Gray Treefrog)   Fig. 83; PL 8J, 9C

IDENTIFICATION LTRF 2/3; A-2 gap about 35–45% of one side of LTR A-2; LTR P-1 without medial gap; LTR P-3 length about 55–75% of P-2 length; Travis's (1981a) report of 4 lower tooth rows seems unlikely; among sites and throughout the ranges, the fin heights and coloration are quite variable; under certain circumstances orange to red pigment appears on the fins and almost always late in ontogeny; besides the predatory effects, what appears to be geographical variation in

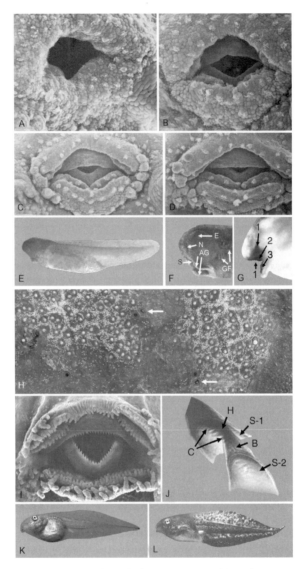

Figure 83. **Hyla chrysoscelis and Hyla versicolor.** Development of the oral apparatus at stages (A) 21, (B) 23, (C) 24, (D) 25 (modified from Thibaudeau and Altig 1988), (E) embryo (6.0 mm TL, stage 20; RA), (F) head of an embryo (AG = adhesive glands, E = eye placode, GF = gills, N = narial pit, S = stomodeum; AMR), (G) oral apparatus in lateral view (1 down-arrow = upper jaw sheath viewed through translucent tissue of snout, 1 up-arrow = edge of upper jaw sheath, 2 = lower jaw sheath, 3 = lower tooth rows; Oktibbeha Co., MS; RA), (H) two egg rafts with entrapped bubbles (arrow = interspersed eggs of *Gastrophryne carolinensis*), (I) SEM of the oral apparatus in face view (GT), (J) two isolated labial teeth (B = body of replacement tooth, C = cusps of replacement tooth, H = head of replacement tooth, S-1 = sheath of erupted tooth, S-2 = sheath of replacement tooth; RA), (K–L) tadpoles: (K) *H. chrysoscelis* (21.0 mm TL, stage 33, Hancock Co., MS; RA), (L) *H. versicolor* (42.0 mm TL, stage 30, Will Co., IL; MR). See also Plates 8J and 9C.

the propensity to form these colored fins needs to be documented; to 50.0 mm TL/ stage 36.

NATURAL HISTORY The Gray Treefrogs oviposit between March and October in ephemeral wetlands, roadside ditches, and the edges of permanent ponds. Clutch size is between 1400 and 2000 pigmented ova (OD 1.1–1.2 mm, ED 4.0–6.0 mm, 2 jelly layers) that are laid in separate rafts of 6–60 eggs that hatch in 3–6 days. Tadpoles grow to 42.0–50.0 mm TL and metamorphose in 3.4–9.3 weeks at 13.0–20.0 mm SVL.

At northern sites this treefrog often breeds in permanent water, but recently disturbed sites often are used in southern areas. These frogs have been the subjects of studies on predator-induced changes; reddish to yellow coloration in the fin and height of tail increases in the presence of competitors and predators.

RANGE The composite range of *H. versicolor* and *H. chrysoscelis* includes most of eastern North America from southern Manitoba south to southern Texas and east to the Atlantic Coast except for peninsular Florida.

CITATIONS *Development/Morphology*: Altig and Pace 1974, Bragg 1947a, Dodd 2004, Fortman and Altig 1973, Gosner 1959, Hellman 1953, Hinckley 1881, Thibaudeau and Altig 1988, Trauth et al. 2004. *Reproductive Biology:* Collins and Wilbur 1979, Fouquette and Littlejohn 1960, Hinckley 1880, Komoroski et al. 1998, Livezey and Wright 1947, Parmalee et al. 2002, Ritke et al. 1990, Trauth et al. 1990, A. H. Wright 1910, 1932. *Ecobehavior:* Akers et al. 2008, Altig et al. 1975, Audo et al. 1995, Beachy et al. 1999, Bragg 1947a, 1962a, Figiel and Semlitsch 1990, 1991, Gosner and Black 1957a–b, S. P. Lawler 1989, McCallum and McCallum 2005, McCollum and Leimberger 1997, McCollum and Van Buskirk 1996, Relyea 2003b, Resetarits 1998, Resetarits et al. 2004, J. L. Richardson 2006, J. M.L. Richardson 2001, Schoeppner and Relyea 2005, Schorr et al. 1990, Seale 1980, Semlitsch 1990, G. R. Smith et al. 2004, G. R. Smith and Jennings 2004, M. J. Smith et al. 2005, Steinwascher and Travis 1983, Van Buskirk 2000, Van Buskirk and McCollum 2000, B. Walters 1975, Wilbur and Alford 1985, B. K. Williams et al. 2008.

Hyla wrightorum (Mountain Treefrog)   Fig. 84; PL 9D

IDENTIFICATION LTRF 2/3; A-2 gap 20–40% of one side of LTR A-2; LTR P-1 with medial gap; LTR P-3 length about 30% of P-2 length; uniformly brown dorsally, fins speckled, side of tail muscle unicolored in younger stages, becoming blotched to crudely bicolored at advanced stages; 28.0–40.0 mm TL/stage 36.

NATURAL HISTORY Mountain Treefrogs oviposit between June and August in temporary and permanent alpine pools and ponds. Eggs are laid as small masses; see Duellman (1970) for data on the closely related *H. eximia* in Mexico.

RANGE *H. wrightorum* occurs in an east-west band through central Arizona and extreme southwestern Arizona and mountains of northwestern Mexico.

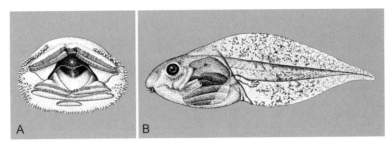

Figure 84. **Hyla wrightorum.** (A) oral apparatus and (B) tadpole (Durango, Mexico; 32.0 mm TL, stage 36) of a similar species, *H. eximia* (both modified from Duellman 1970). See also Plate 9D.

CITATIONS *Development/Morphology*: Duellman 1970 (*H. eximia*), Korky and Webb 1991 (*H. eximia*), Zweifel 1961. *Reproductive Biology:* Chapel 1939, Livezey and Wright 1947. *Ecobehavior:* Duellman 1970 (*H. eximia*).

Osteopilus septentrionalis (Cuban Treefrog)   Fig. 85

IDENTIFICATION LTRF 2/4; A-2 gap is present in contrast to figure in Conant and Collins (1998); A-2 gap about 60% of one side of LTR A-2; LTR P-1 with medial gap; midventral marginal papillae biserial; eyes lateral. Small tadpole: uniformly light tan to brown. Large tadpole: tail muscle bicolored; dorsum dark with slight darker eye line; 30.0–45.0 mm TL/stage 36.

NATURAL HISTORY Cuban Treefrogs breed in every month, but oviposition peaks in the rainy season from May to October, or opportunistically with rains in small ponds and pools. The clutches contain 130–3961 black and gray ova (OD 1.0–1.5.0 mm, ED about 2.8 mm, 1 jelly layer) with 50–200 per ovipositional bout laid as a very sticky film about 140.0 mm diameter. Eggs hatch in 1–2 days at about 10.9 mm TL, and tadpoles metamorphose in 1 month at about 15.0 mm SVL. Sexual maturity is attained in 3–8 months. This exotic inquiline is an extreme ecological generalist that breeds opportunistically in birdbaths, buckets, and rain pools.

RANGE Cuban Treefrogs are found along the Atlantic Coast of Florida from Key West to Jacksonville, perhaps north to the Savannah, Georgia area, and on the Gulf side north to at least the Tampa Bay area and perhaps to the Florida panhandle. Reports from Oahu, Hawaii have not been substantiated.

CITATIONS *General:* Duellman and Crombie 1970. *Development/Morphology*: Duellman and Schwartz 1958. *Reproductive Biology:* Babbitt and Meshaka 2000, Livezey and Wright 1947. *Ecobehavior:* H. A. Brown 1969, K. G. Smith 2005a–b, 2006.

Pseudacris brachyphona (Mountain Chorus Frog)   PL 9E

IDENTIFICATION LTRF 2/3; A-2 gap 20–30% of one side of LTR A-2; LTR P-1 without medial gap; LTR P-3 length 30–40% of P-2 length; lightly pigmented brown, sometimes with dorsal spots; tail muscle unicolored; no marks on tail

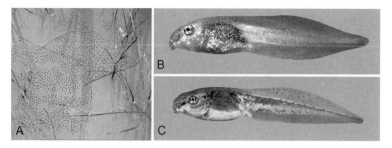

Figure 85. ***Osteopilus septentrionalis.*** (A) egg film (Hillsborough Co., FL; SAJ), (B–C) tadpoles: (B) 21.0 mm TL, stage 28, (C) 35.0 mm TL, stage 33; Dade Co., FL; RA.

Figure 86 ***Pseudacris brimleyi.*** (A) tadpole (27.0 mm TL, stage 33, SC; RA). See also Plate 9F.

fins; 22.0–30.0 mm TL/stage 36. The drawing of the mouthparts in Green (1938) has notable errors.

NATURAL HISTORY Mountain Chorus Frogs oviposit from January to June, sometimes later, and deposit their eggs in small woodland ponds, bogs, puddles, ditches, and other ephemeral wetlands, often in or near forests. Clutches number between 300 and 1500 pigmented ova (OD 1.6 mm, ED 6.0–8.5 mm, 1 jelly layer) that are laid in small masses of 10–60 eggs attached to vegetation below the surface or lying free on bottom of the pool. Eggs hatch in 3–10 days at 4.0–9.0 mm TL, and tadpoles metamorphose in 2.0–2.1 months at 8.0–13.0 mm SVL.

RANGE Mountain Chorus Frogs occur throughout the Appalachian Mountains up to 1100 m elevation and on the Cumberland Plateau at lower elevations from southwestern Pennsylvania and southeastern Ohio south to central Alabama and northeastern Mississippi.

CITATIONS *Development/Morphology*: Gosner 1959, N. B. Green 1938.

Pseudacris brimleyi (Brimley's Chorus Frog)   **Fig. 86; PL 9F**

IDENTIFICATION LTRF 2/3; A-2 gap about 50% of one side of LTR A-2; LTR P-1 without a medial gap; LTR P-3 length about 30% of P-2 length; darkly pigmented dorsally, throat and venter spotted, strikingly bicolored tail muscle, fins

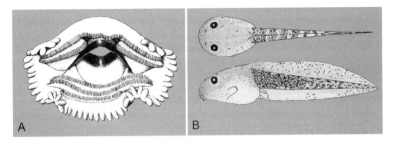

*Figure 87. **Pseudacris cadaverina**.* (A) oral apparatus (modified from Duellman 1970), (B) dorsal and lateral views of a tadpole (34.7 mm TL, stage 31; Los Angeles Co., CA; modified from Gaudin 1964). See also Plate 9G.

clear to moderately marked, pale stripe from naris through eye to tail; 22.0–30.0 mm TL/stage 36.

NATURAL HISTORY Brimley's Chorus Frogs breed from late winter to early spring (oviposit January–April) usually in swampy ponds and hardwood swamps. Clutches include 260–300 dark brown and white ova (OD 1.3–1.7 mm, ED 6.7–8.6 mm, 1 jelly layer) that are laid in small masses attached to stems and other structures in the pond. Eggs hatch in 4–7 days at about 6.0 mm TL, and larvae metamorphose in 1–2 months at 8.0–11.0 mm SVL.

RANGE *Pseudacris brimleyi* can be found on the Coastal Plain from southeastern Virginia to northeastern Georgia.

CITATIONS *Development/Morphology*: Altig and Pace 1974, Gosner and Black 1958b. *Reproductive Biology:* Livezey and Wright 1947, J. C. Mitchell 1986.

Pseudacris cadaverina (California Chorus Frog)   Fig. 87; PL 9G

IDENTIFICATION LTRF 2/3; A-2 gap about 50% of one side of LTR A-2; LTR P-1 without medial gap; LTR P-3 length 45–60% of P-2 length; tail muscle banded, especially in younger individuals; larger tadpoles boldly mottled; 29.0–40.0 mm TL/stage 36.

NATURAL HISTORY California Chorus Frogs oviposit in slow parts of rocky streams and backwaters. Eggs are laid singly and may stick together in clumps and commonly sink into the gravel. Larvae metamorphose after 1.2–2.5 months in July and August.

RANGE The California Chorus Frog occurs in the Coast Range of southwestern California and northern Baja California, Mexico.

CITATIONS *Development/Morphology*: Gaudin 1964, 1965, Grismer 2002, Storer 1925. *Reproductive Biology:* Livezey and Wright 1947. *Ecobehavior:* Cunningham 1964.

Pseudacris clarkii (Spotted Chorus Frog)   Fig. 88; PL 9H

IDENTIFICATION LTRF 2/3; A-2 gap about 60% of one side of LTR A-2; LTR P-1 without medial gap; LTR P-3 length about 45% of P-2 length; lightly

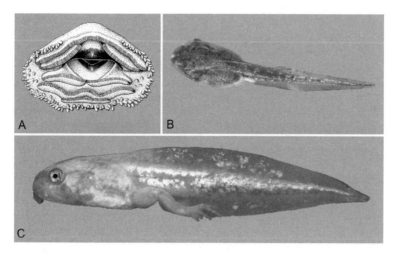

Figure 88. ***Pseudacris clarkii.*** (A) oral apparatus (modified from Duellman 1970), (B) tadpole (32.0 mm TL, stage 33, Cameron Co., TX; DLM), (C) tadpole (34.0 mm TL, stage 37, Lubbock Co., TX, turbid site; RA). See also Plate 9H.

pigmented brown, uniformly brassy in turbid water; tail muscle mottled to faintly striped; no prominent marks in tail fins; to 30.0 mm TL/stage 36.

NATURAL HISTORY The Spotted Chorus Frog oviposits from March through May, in ponds and temporary grassland pools. Occasionally individuals breed later with fall rains and in any month in the southern parts of its range. Clutches contain 150–1000 gray ova (OD 0.6–1.3 mm, ED 2.2–2.4 mm, 2 jelly layers) and eggs are deposited in masses of 3–50 and hatch in 2.5–3.0 days. Tadpoles grow to 30.0 mm TL and metamorphose in 1.0–1.5 months at 8.0–17.0 mm SVL.

RANGE *Pseudacris clarkii* is distributed in a wide band from central Kansas to southern Texas, and just into Tamaulipas, Mexico.

CITATIONS ***General:*** Pierce and Whitehurst 1990. ***Development/Morphology:*** Duellman 1970, T. H. Eaton and Imagawa 1948, Gosner 1959. ***Reproductive Biology:*** Bragg 1943, 1957c, Livezey and Wright 1947.

Pseudacris crucifer (Spring Peeper)   Fig. 89; PL 9I

IDENTIFICATION LTRF 2/3, less commonly 2/2; A-2 gap about 55% of one side of LTR A-2; LTR P-1 without medial gap; LTR P-3 length about 33–45%, often less, of P-2 length; small medial gap in the marginal papillae may or may not occur independently of LTR P-3; midventral marginal papillae biserial; although extreme variations in coloration occur, the fins of smaller tadpoles are usually clear, while older ones may have extensive contrasting marks; the dorsum of the tail muscle is sometimes banded in small tadpoles; in general, the coloration is muted browns and tans with a paler belly; to 35.0 mm TL/stage 36.

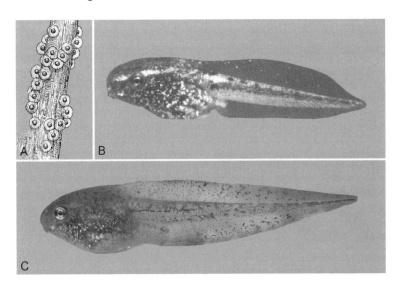

*Figure 89. **Pseudacris crucifer**.* (A) single eggs attached to vegetation (modified from A. H. Wright 1914), (B) tadpoles (28.0 mm TL, stage 32, Oktibbeha Co., MS; clear fins altered by photographic background; RA), (C) tadpole (27.0 mm TL, stage 33, Liberty Co., GA; DJS). See also Plate 9I.

NATURAL HISTORY Female Spring Peepers oviposit from November through June generally at the edges of vegetated ponds and marshes, in roadside ditches, and along slow-moving streams in forested habitats. Clutches consist of 250–1500 lightly pigmented ova (OD 0.9–1.5 mm, ED 2.0–2.8 mm, 1 or 2 jelly layers) that are laid as single eggs, small clumps of 2–4 on vegetation, or free on the bottom. Eggs hatch in 2–15 days, and tadpoles grow to 24.0–35.0 mm TL in 1.5–3.3 months. Metamorphs measure 8.0–15.0 mm SVL.

RANGE Spring Peepers occur throughout eastern North America from eastern Manitoba south to eastern Texas and eastward to the Atlantic except most of peninsular Florida.

CITATIONS *Development/Morphology*: Dodd 2004, Gosner 1959, Gosner and Black 1957a, Gosner and Rossman 1960, Hinckley 1881, Trauth et al. 2004, A. H. Wright 1910. *Reproductive Biology:* Collins and Wilbur 1979, Komoroski et al. 1998, Livezey and Wright 1947, Loraine 1984, Oplinger 1966, Parmalee et al. 2002, Travis et al. 1987, 1988. *Ecobehavior:* Brattstrom 1962, Brodie and Formanowicz 1987, Crump 1984, Fedewa 2006, Formanowicz and Brodie 1982, Gosner and Black 1957b, S. P. Lawler 1989, S. P. Lawler and Morin 1993, McCollum and Van Buskirk 1996, Morgan 1891, J. M. L. Richardson 2001, Seale 1980, Seale and Beckvar 1980, Skelly 1995a–b, 1996, Skelly et al. 2002, Van Buskirk 2000, B. K. Williams et al. 2008, Woodward and Travis 1991.

Pseudacris ocularis (Little Grass Frog)   Fig. 90; PL 9L

IDENTIFICATION LTRF 2/3; A-2 gap about 15–25% of one side of LTR A-2; LTR P-1 without medial gap; LTR P-3 length about 25–30% of P-2 length;

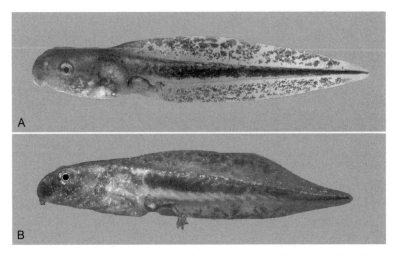

Figure 90. **Pseudacris ocularis.** (A) tadpole (28.0 mm TL, stage 25, Marion Co., FL; BM), (B) tadpole (29.0 mm TL, stage 27, SC; RA). See also Plate 9L.

dorsum dark with silvery to reddish blotches laterally, tail muscle bicolored and dorsum of tail muscle with crude bands, pale stripe from naris to eye extends as broken blotches to tail; to 30.0 mm TL/stage 36.

NATURAL HISTORY Females of *P. ocularis* deposit eggs from January to September in open, wet savannas and on vegetation along edges of pools; males can be heard calling in any month from grass stems above the water. Clutches consist of 100–1000 brown ova (OD 0.6–1.0 mm, ED 1.2–2.0 mm, 1 jelly layer) that are laid as single eggs on the bottom or on vegetation. Eggs hatch in 1.5–3.5 days. Larvae grow to 30.0 mm TL, and tadpoles metamorphose in 7–70 days at 7.0–9.0 mm SVL.

RANGE The Little Grass Frog, the smallest species in the United States, is found on the Coastal Plain from southeastern Virginia to southeastern Alabama and peninsular Florida.

CITATIONS *General:* Franz and Chantell 1978. *Development/Morphology*: Altig and Pace 1974, Gosner and Rossman 1960, A. H. Wright 1932. *Reproductive Biology:* Gosner and Rossman 1960, A. H. Wright 1932. *Ecobehavior:* Alford 1986, Gosner and Rossman 1960, Kehr 1997.

Pseudacris ornata (Ornate Chorus Frog)   Fig. 91; PL 10A

IDENTIFICATION LTRF 2/3; A-2 gap equal to or smaller than 25% of one side of LTR; LTR P-1 with medial gap; LTR P-3 length about 65–75% of P-2 length; jaw sheaths massive. Small tadpole: dark with clear fins and distinctly bicolored tail. Large tadpole: heavily marked fins and progressive loss of bicolored tail pattern; to 65.0 mm TL/stage 36.

NATURAL HISTORY Even though males may call in the late fall, oviposition occurs from late November into April, usually in ponds and pools, often

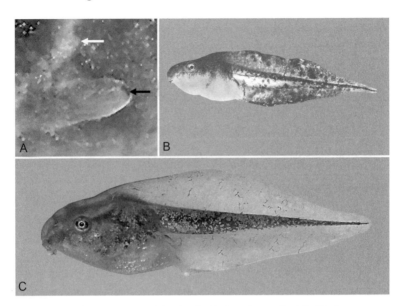

Figure 91. **Pseudacris ornata.** (A) sinistral spiracle (anterior to the left; black arrow = narrow caliber exit aperture, white arrow = front limb visible through opercular wall; RA), (B) tadpole (61.0 mm TL, stage 32, FL; BM), (C) young tadpole (36.0 mm TL, stage 28, Evans Co., GA; DJS). See also Plate 10A.

within or adjacent to swampy, forested areas (e.g., cypress ponds, Carolina bays). Their eggs contain lightly pigmented ova (OD 0.9–1.0 mm, ED 3.6–4.2 mm, 1 jelly layer) and are laid in small masses of 6–100 attached to twigs. Tadpoles grow to 42.0 mm TL or larger and metamorphose in about 3 months at 14.0–17.0 mm SVL.

RANGE   This strikingly marked chorus frog occurs on the southeastern Coastal Plain from central North Carolina south into the northern half of Florida and west through southern Georgia, Alabama to southeastern Louisiana.

CITATIONS *Development/Morphology*: Altig 1971, Altig and Pace 1974. *Reproductive Biology:* J. P. Caldwell 1987, A. H. Wright 1932. *Ecobehavior:* Alford 1986, Seyle and Trauth 1982.

Pseudacris regilla (Northern Pacific Treefrog), *P. hypochondriaca* (Baja California Treefrog), and *P. sierra* (Sierran Treefrog)   Fig. 92; PL 9J

IDENTIFICATION   LTRF 2/3; A-2 gap about 40–50% of one side of LTR A-2; LTR P-1 with medial gap; LTR P-3 length 50–55% of P-2 length; jaw sheaths massive; highly variable but usually uniformly dark dorsally, fins with or without considerable black marks, venter varies from coppery or bronze to silvery white. Variations in fin heights (Fig. 92B, E) and color among sites can be extreme over short distances, but the causes and the potential ecological significance of such differences are not known; to 55.0 mm TL/stage 36.

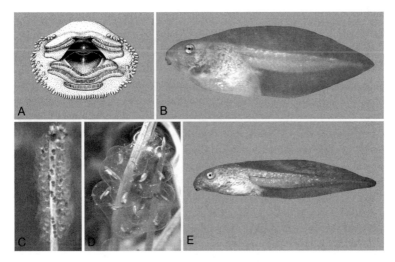

Figure 92. **Pseudacris regilla and relatives.** (A) oral apparatus of *P.* "*regilla*" (modified from Duell-man 1970), (B) tadpole of *P. regilla* (48.0 mm TL, stage 33, Lane Co., OR; RA), (C–D) egg masses: (C) *P. regilla* (South Pender Island, British Columbia; KO), (D) *P. hypochondriaca* (Fresno Co., CA; RLG), (E) tadpole of *P. hypochondriaca* with low fins (31.0 mm TL, stage 33, Santa Barbara Co., CA; RA). See also Plate 9J.

NATURAL HISTORY Pacific treefrogs oviposit from November to July typically in ponds, pools, and slow parts of streams. Clutches include 100–1000 gray to brown and cream to white ova (OD 1.2–1.4 mm, ED 4.6–6.7 mm, 2 jelly layers) that are laid as several masses of 9–118, most commonly 25-68. They hatch in 3–5 days at 6.0–8.0 mm TL and tadpoles grow to 40.0–50.0 mm TL. Metamorphosis occurs 1.6–2.7 months at 11.0–17.0 mm SVL. Fin heights vary greatly even among adjacent sites. This wide-ranging species varies in breeding phenology across its range depending on elevation and habitat (e.g., alpine versus temperate lowlands versus desert areas).

RANGE As currently understood, *P. hypochondriaca* is found in southern California, Nevada and adjacent Arizona south into Baja California. *P. regilla* occurs in the Pacific Northwest from northern California, adjacent Oregon, Washington, and British Columbia to southern Alaska. *P. sierra* provisionally is found in central California, eastern Oregon and Idaho, and western Montana.

CITATIONS *Development/Morphology*: Eakin 1947, Gaudin 1965, Wassersug 1976a–b. *Reproductive Biology:* Gardner 1995, Gaudin 1965, Livezey and Wright 1947, Perrill and Daniel 1983, Schaub and Larsen 1978. *Ecobehavior:* Brattstrom 1962, Brattstrom and Warren 1955, H. A. Brown 1969, 1975c, Claussen 1973, J. E. Johnson et al. 2007, Kiesecker and Blaustein 1997, Kupferberg 1997b, Kupferberg et al. 1994, O'Hara and Blaustein 1988, Pearl et al. 2003, J. A. Peterson and Blaustein 1992, Schechtman and Olson 1941, Wagner 1986, T. B. Watkins 1997, 2000, 2001.

Figure 93. Pseudacris streckeri. (A) tadpole (68.0 mm TL, stage 35, Yell Co., AR; SET). See also Plate 10B.

Pseudacris streckeri (Strecker's Chorus Frog) and *P. illinoensis* (Illinois Chorus Frog)   Fig. 93; PL 10B

IDENTIFICATION LTRF 2/3; A-2 gap about 20% of one side of LTR A-2; LTR P-1 with medial gap; LTR P-3 length about 40–55% of P-2 length; varies from uniform pale in turbid sites to rather uniformly brownish bronze with black dots on the dorsum; fins suffused with slightly dark color and scattered vermiform marks; to 65.0 mm TL/stage 36.

NATURAL HISTORY Females oviposit from December to May in ponds and pools in open areas and along streams, usually near grasslands. Between 300 and 700 brownish gray ova (OD 1.2–1.8 mm, ED 3.0–5.0 mm, 1 sticky jelly layer that accumulates silt) laid as small masses of 10–100 eggs attached to weeds near the surface. The eggs hatch at 4.7–6.2 mm TL and tadpoles grow to 65.0 mm TL. Metamorphosis is in 5–8.6 weeks at 11.0–16.0 mm SVL.

RANGE The Illinois Chorus Frog occurs in disjunct areas of sandy soil in west-central Illinois to southeastern Missouri and northeastern Arkansas. Strecker's Chorus Frog ranges from central Kansas south through eastern Oklahoma and west-central Arkansas to the coast of eastern Texas and into northwestern Louisiana.

CITATIONS *General:* P. W. Smith 1966a. *Development/Morphology*: Bragg 1942, 1947a, 1957c, Gosner 1959, Tucker 1997, Trauth et al. 2004. *Reproductive Biology:* Livezey and Wright 1947, Trauth et al. 1990. *Ecobehavior:* Black 1971, 1974a, 1975, Bragg 1947a, 1962a, Butterfield et al. 1989, Fouquette and Littlejohn 1960, Strecker 1926a, Tucker 1995.

Pseudacris triseriata group: *P. feriarum* (Upland Chorus Frog), *P. fouquettei* (Cajun Chorus Frog), *P. kalmi* (New Jersey Chorus Frog), *P. maculata* (Boreal Chorus Frog), *P. nigrita* (Southern Chorus Frog), and *P. triseriata* (Western Chorus Frog)   Fig. 94; PL 9K, 10C

IDENTIFICATION LTRF 2/3, sometimes 2/2, sometimes a narrow ventral gap in marginal papillae; A-2 gap about 20–45% of one side of LTR A-2; LTR P-1

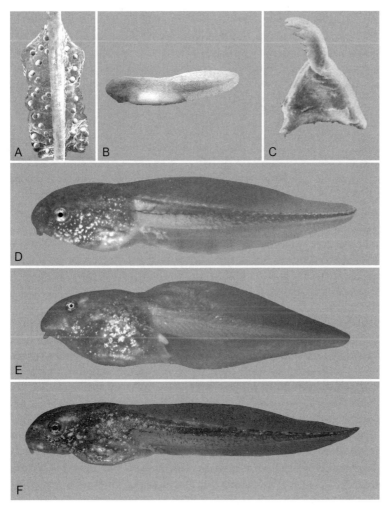

Figure 94. **Pseudacris triseriata and relatives.** (A) egg mass with ova distributed uniformly throughout the volume as is typical of hylids (Oktibbeha Co., MS; RA), (B) hatchling (6.0 mm TL, stage 20, Oktibbeha Co., MS; RA), (C) isolated tooth from (D) tadpole of *P. feriarum* (27.0 mm TL, stage 35, Oktibbeha Co., MS; RA), (E) tadpole of *P. maculata* (30.0 mm TL, stage 33, Alberta, Canada; RA), (F) tadpole of *P. nigrita* (30.0 mm TL, stage 33; Escambia Co., AL; all RA). See also Plates 9K, 10C.

with or without medial gap; LTR P-3 length about 30–45% of P-2 length; mid-ventral marginal papillae uniserial; quite variable, but usually unicolored dor-sally, sometimes with dark dots, tail muscle distinctly bicolored or graded top-to-bottom, fins range from clear to moderately speckled; to 52.0 mm TL/stage 36.

NATURAL HISTORY Females of the Western Chorus Frog oviposit from October to July in temporary ditches and pools, usually near grasslands. Up to 1500 brown to black and cream to white ova (OD 0.5–1.3 mm, ED 3.2–7.8 mm, 1–2 jelly layers) are deposited in several cylindrical egg masses 10.0–40.0 mm diameter and containing 20–190 ova. The masses are placed well below the surface and hatch in 3–18 days at 4.0–8.0 mm TL, depending on temperature. Tadpoles grow to 26.0–43.0 mm TL and metamorphose in 40–90 days at 7.0–14.0 mm SVL. The metamorphs mature in less than a year.

RANGE *P. feriarum* occurs from northern New Jersey and northwestern Kentucky south to the Gulf Coastal Plain. *P. fouquettei* ranges from the eastern third of Texas north to northern Oklahoma, east to southern Missouri, and south to coastal Mississippi. *P. kalmi* is known from Staten Island, New York, south through the Delmarva Peninsula. *P. maculata*, the Western Chorus Frog, occurs from east-central Arizona and west-central Idaho to the Northern Territories, and east to central Ontario and south to southeastern Missouri, disjunct in southeastern Ontario and adjacent Québec. *P. nigrita* is found on the Coastal Plain from southeastern Mississippi to central North Carolina and the Florida Peninsula. *P. triseriata* is found from southern Illinois northeast to northern Michigan and east to western New York.

CITATIONS **Development/Morphology**: Altig and Pace 1974, Bresler and Bragg 1954, Dodd 2004, Gates 1988, Gosner and Black 1957a, Nichols 1937, Trauth et al. 2004, Youngstrom and Smith 1936. **Reproductive Biology:** Bragg 1948b, 1957c, Collins and Wilbur 1979, Hecnar and Hecnar 1999, Livezey and Wright 1947, Meisler 2005, Morgan 1891, Pack 1920, Parmalee et al. 2002, Trauth et al. 1990, A. H. Wright 1910, 1932, A. H. Wright and Allen 1908. **Ecobehavior:** Britson and Kissell 1996, Dupré and Petranka 1985, Gee and Waldick 1995, Gosner and Black 1957b, Hay 1889b, Hecnar and Hecnar 1999, Livezey 1952, C. H. Pope 1919a, Skelly 1995a, 1996, Sours and Petranka 2007, Sredl and Collins 1991, Travis 1981b, Van Buskirk 2000, Van Buskirk et al. 1997, Voris and Bacon 1966, B. Walters 1975, Wassersug and Seibert 1975, Wassersug and Sperry 1977, Whitaker 1971.

Smilisca baudinii (Mexican Treefrog)   Fig. 95; 10D

IDENTIFICATION Nonflowing water, nektonic; LTRF 2/3; A-2 gap about 30–36% of one side of LTR A-2; LTR P-1 without medial gap; LTR P-3 length about 95% of P-2 length; dorsal gap in marginal papillae about 30% of oral disc width; marginal papillae biserial midventrally, multiserial laterally; jaw sheaths medium; dorsum uniformly pale to dark brown; sometimes a pale (= nonpigmented) stripe extends from near tail-body junction anteriorly; fins faintly marbled; to 50.0 mm TL/stage 36.

NATURAL HISTORY Females oviposit from May to August in temporary ponds, pools, and slow streams following summer rains. Clutches of 480–3400 pigmented ova (OD about 1.3 mm, ED about 1.5 mm, 1 jelly layer) are laid as a film.

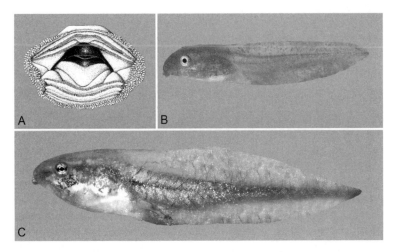

Figure 95. **Smilisca baudinii.** (A) oral apparatus (modified from Duellman 1970), (B) tadpole (39.0 mm TL, stage 32, Mexico; RWV), (C) tadpole (43.0 mm TL, stage 35, Cameron Co., TX; DLM). See also Plate 10D.

These fast-growing tadpoles hatch and metamorphose at about 13.0 mm SVL in about 3 weeks.

RANGE The Mexican Treefrog occurs from extreme southern Texas south through Mexico to Costa Rica.

CITATIONS *General:* Duellman 1968. *Development/Morphology*: Duellman 1970, Duellman and Trueb 1966, Maslin 1963b, Savage 1980, Schmidt 1995. *Reproductive Biology:* Livezey and Wright 1947, Webb 1971. *Ecobehavior:* Duellman 1970, L. C. Stuart 1948.

Smilisca fodiens (Lowland Burrowing Treefrog)   PL 10E

IDENTIFICATION Nonflowing water, nektonic; LTRF 2/3; A-2 gap about 50% of one side of LTR A-2; LTR P-1 without a medial gap; LTR P-3 length about 80% of P-2 length; small, biserial marginal papillae with medium dorsal gap. Small tadpole: distinctly bicolored tail muscle; pale line extending from the tail-body junction anteriorly at least to eye. Large tadpole: especially in turbid water, stripe less prominent; fins faintly to prominently marbled or blotched. The description in Webb (1963) is correct; that by Duellman (1970) is based on a ranid tadpole; to 50.0 mm TL/stage 36.

NATURAL HISTORY Female Burrowing Treefrogs are explosive breeders and oviposit between July and August in temporary pools formed by seasonal rains in desert areas. The pigmented ova presumably are oviposited as a film, and perhaps as scattered singles on the bottom. Eggs hatch at about 10.0 mm TL and tadpoles grow to 45.0–50.0 mm TL, undergoing metamorphosis at 18.0–24.0 mm TL. The metamorphs are green and lack the casque head of adults.

RANGE *Smilisca fodiens* is found in extreme south-central Arizona and south to near Guadalajara, Mexico.

CITATIONS **General:** Trueb 1969. **Development/Morphology:** Webb 1963. **Reproductive Biology:** L. M. Hardy and McDiarmid 1969, Livezey and Wright 1947.

Leiopelmatidae (Tailed Frogs)

Tailed Frogs occur in mountain streams of the Pacific Northwest; suctorial morphotype; LTRF 2–3/9–12, rows A-1 and last lower row multiserial, rows A-2–3 and P-1 biserial, rows A-1 and A-2 about equal in length; large, suctorial oral disc, not emarginate, small marginal papillae on the lower labium and a fleshy surface along the upper labium; submarginal papillae numerous posterior to last lower tooth row; upper jaw sheath immense, platelike with straight, finely serrate edge, oriented parallel with the substrate, partially divided medially; lower sheath small, U-shaped, usually hidden beneath upper sheath; vent medial with vent flap beneath; eyes dorsal; dorsal fin low with rounded tip; body depressed; nostrils with tubular extensions; dorsal fin low, originates near tail-body junction, with broadly rounded tip; body depressed; spiracle midventral on chest, without dorsal wall; skin glands obvious on belly; tail of males and large outer toe visible in large tadpoles, toe pads absent; black to pale yellow, often with rocklike mottling; tail tip often with contrasting ocellus; metamorphs resemble adult; 40.0–64.0 mm TL/stage 36; 1 genus, 2 species. *Ascaphus:* as for family.

Ascaphus montanus (Rocky Mountain Tailed Frog) and *A. truei* (Coastal Tailed Frog)   Fig. 96; PL 10F

IDENTIFICATION As for family.

NATURAL HISTORY Female Tailed Frogs oviposit from May to September, probably more likely in the fall and in alternate years at some sites. Eggs are placed under rocks and among cobble in cool mountain steams. Clutch size is 25–98 (fewer in coastal populations) nonpigmented ova (OD 3.7–7.0 mm, ED 6.0–8.0 mm, 1–3 jelly layers) laid in a rosary that may be grouped or attached as a monolayer to rocks. Multiple females may oviposit at a single microsite. Temperatures of 5.0–18.5 °C are required for hatching, which occurs in 3.0–6.5 weeks (4 weeks at 11 °C). Hatchlings are 10.0–15.0 mm TL and probably do not disperse from the nest site for some time. Yolk absorption is completed by 20.0–21.0 mm TL and larvae grow to 29.0–41.0 mm TL the first year and 40.0–60.0 mm TL in subsequent years. In some places, larvae may grow for up to 4 years. Metamorphosis occurs between July and September and usually takes about a month. Metamorphs measure between 14.0 and 20.0 mm TL and reach sexual maturity 4–9 years after metamorphosis. Front legs emerge through the midventral spiracle, and compared with pond dwelling tadpoles, metamorphic reduction of mouthparts is delayed relative to limb development.

These suctorial tadpoles hide during the day under stones and among cobble (usually 10–30 cm diameter) in clear, cool mountain streams (less than 22 °C) and forage mostly at night on the thin algal film on tops of rocks. The idea that the tadpoles ingest food through the nostrils (Noble 1927) is surely erroneous. Feeding activities can profoundly influence periphyton communities, and they commonly

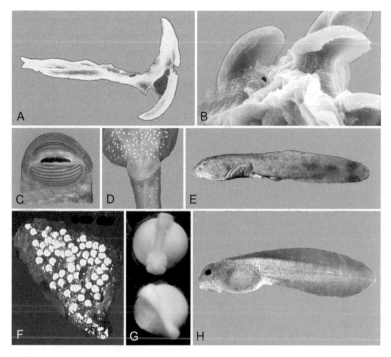

Figure 96. **Ascaphus montanus** and **Ascaphus truei**. (A) lateral view of a labial tooth from row P-1 (working surface at upper right; modified from Altig and Pace 1974), (B) teeth of last posterior row in situ, viewed from mouth (GFJ), (C) oral apparatus adhered to glass (Shoshone Co., ID; RA), (D) vent flap covering vent tube and limb buds (RA), (E) tadpole (38.0 mm TL, stage 35, Benton Co., OR; RA), (F) egg rosary (Humboldt Co., CA; LVD), (G) embryo (stage 18, Benton Co., OR; RMS), (H) hatchling (11.0 mm TL, stage 25, Jefferson Co., WA; CCC). See also Plate 10F.

move onto wet rocky surfaces out of water. They use the large oral disc to move and feed while attached to rocky substrates, and the size and structure of the substrate must be large enough to provide daytime refugia. Tadpoles seem more sensitive to vibration than to light and are commonly positioned in contact with each other.

RANGE *A. montanus* is known from the northern Rocky Mountains and the Blue, Seven Devil, and Wallowa mountains of Idaho and Oregon. *A. truei* is known from the Cascade Mountains and Pacific lowlands from southwestern British Columbia to northern California.

CITATIONS *General:* Metter 1968. *Development/Morphology:* Altig 1973, Altig and Pace 1974, H. A. Brown 1989a, Bury and Adams, 1999, Gosner 1959, Metter 1964, 1967, Mittleman and Myers 1949. *Reproductive Biology:* Adams 1993, Bury and Adams 1999, Bury et al. 2001, L. R. Franz 1970a, Goldsworthy 2007, Karraker and Beyersdorf 1997, Karraker et al. 2006, Noble and Putnam 1931, Salthe 1963, Stephenson and Verrell 2003, Wernz 1969, Wernz and Storm 1969. *Ecobehavior:* Altig and Brodie 1972, H. A. Brown 1975b, 1977, 1990, Claussen 1973, Daugherty and Sheldon 1982, DeVlaming and Bury 1970,

Feminella and Hawkins 1994, L. R. Franz 1970b, Gaige 1920, Gradwell 1971, 1973, Hawkins et al. 1988, L. L. C. Jones and Raphael 1998, Kiffney and Richardson 2001, Lamberti et al. 1992, Mallory and Richardson 2005, Metcalf 1928, Metter 1964, 1966, Mittleman and Myers 1949, G. S. Myers 1931, 1943, C. L. Taylor and Altig 1995, Van Denburgh 1912, Wahbe and Bunnell 2003, R. L. Wallace and Diller 1998, Welsh 1990, Welsh and Lind 2002.

Leptodactylidae (White-lipped Frogs)

One species of leptodactylid occurs in southern Texas; nonflowing water, benthic in ephemeral sites; LTRF 2/3, all rows uniserial; A-2 gap about 5% of one side of LTR A-2; LTR P-1 with medial gap; rows A-1 and A-2 and rows P-2 and P-3 about equal in length; row of marginal papillae with wide dorsal gap, composed of small, uniserial papillae; submarginal papillae sparse to absent; oral disc not emarginate; jaw sheaths narrow; vent medial; eyes dorsal; spiracle sinistral; dorsal fin low with rounded tip, originates near tail-body junction; body depressed; nostrils small, closer to snout than eye; digital pads absent; varies from uniformly black with unicolored, partially clear fins to mottled with considerable contrasting marks in fins; pale area just posterior to oral disc typical; to 33.0 mm TL/ stage 36; 1 genus, 1 species. *Leptodactylus*: as for family.

Leptodactylus fragilis (Mexican White-lipped Frog)   Fig. 97

IDENTIFICATION As for family; this tadpole is the only leptodactylid in the United States; it grossly resembles sympatric bufonid and small ranid tadpoles,

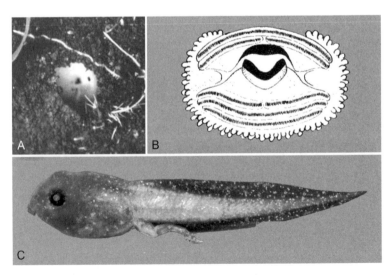

Figure 97. ***Leptodactylus fragilis.*** (A) partially exposed, subterranean foam nest near a temporary pool (Costa Rica; RWM), (B) oral apparatus (modified from Orton 1952), (C) tadpole (preserved, 23.0 mm TL, stage 38, Guanacaste, Costa Rica, USNM 330416; RA).

but it has smaller, round nostrils, complete papillae on the lower labium, and a nonemarginate oral disc.

NATURAL HISTORY Females of *L. fragilis* oviposit in a foam nest in a burrow or beneath a rock where an ephemeral pond will form following the first heavy summer rains (May–July). The clutch size is 25–250 yellow ova (OD about 0.6 mm) among foam. Depending on timing, larvae escape the foam nest when the pond fills and after 1.0–1.2 months of growth, undergo metamorphosis at about 16.0 mm SVL.

RANGE The Mexican White-lipped Frog is relatively common from extreme southern Texas, south through Mexico and Middle America to northern South America.

CITATIONS *General:* Heyer 1971. *Development/Morphology*: Heyer 1970, Savage 1980. *Reproductive Biology:* Livezey and Wright 1947, Maslin 1963a. *Ecobehavior:* Muliak 1937.

Microhylidae (Small-mouthed Toads and Sheep Frog)

Members of this family occur east of the Sierra Nevada-Cascade ranges to the Atlantic Coast; nonflowing water, suspension feeder; oral disc, labial tooth rows and jaw sheaths absent; hemispherical oral flaps pendent over mouth; vent medial; eyes lateral; spiracle midventral on abdomen; dorsal fin low with pointed tip; strongly depressed body round in dorsal view; nostrils absent until near metamorphosis; digital pads absent; dorsum uniformly dark or pale, venter often with contrasting marks; to 40.0 mm TL/stage 36; 2 genera, 3 species. *Gastrophryne*: margins of oral flaps uniform and divergent. *Hypopachus*: margins of oral flaps papillate and convergent

Gastrophryne carolinensis (Eastern Narrow-mouthed Toad)   Fig. 98; PL 10G

IDENTIFICATION As for genus; dorsum dark with minor iridophore flecking, venter dark with prominent pale mottling, lateral surface of tail muscle with definite white stripe; clear or with dark marks peripherally; to 40.0 mm TL/stage 36.

NATURAL HISTORY Eastern Narrow-mouthed Toads congregate around ponds, temporary pools, and ditches, usually after heavy rains between March and October to breed. Clutches include 400–1100 black ova (OD 1.0–1.2 mm, ED 2.8–4.0 mm with 1 or perhaps 2 jelly layers laid as a coherent film, often in rafts of 10–150 eggs. They hatch in 1.5–3 days, and the tadpoles grow to 40.0 mm TL, undergoing metamorphosis in 2.9–10 weeks at 8.0–12.0 mm SVL.

Tadpoles usually hide among and under bottom debris during the day and suspension-feed quiescently at or near the surface at night. They move about intermittently while feeding, apparently in response to the local density of food particles, via an unknown modulation of the spiracular water jet (i.e., no tail movements). One individual will commonly draft another by positioning the side of its body in the depression at the tail-body junction of the lead tadpole. During actual feeding, exaggerated buccal pumping can be seen—the oral flaps momentarily flex anteriorly and the tadpole may rock back and forth slightly. Tadpoles swim downward at about 45° to escape a disturbance; contact by other tadpoles usually makes a feeding tadpole stop feeding and move. Microhylid

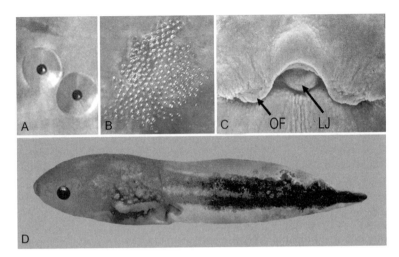

Figure 98. ***Gastrophryne carolinensis.*** (A) two freshly laid, coherent eggs (part of a coherent film; part of the upper hemisphere of outer jelly layer protruding above the water surface; Oktibbeha Co., MS), (B) an entire clutch of eggs (Oktibbeha Co., MS), (C) oral apparatus (OF = oral flap, LJ = lower jaw), (D) tadpole (28.0 mm TL, stage 32, Noxubee Co., MS; all RA). See also Plate 10G.

tadpoles may have a profound ability to withstand periods of reduced food availability.

RANGE The Eastern Narrow-mouthed Toad ranges from southern Missouri east to southern Maryland and south through eastern Oklahoma and east Texas to the Gulf Coast and the Florida Keys.

CITATIONS *General:* Nelson 1972a. ***Development/Morphology***: Bragg 1947a, Dodd 2004, Orton 1946, Ryder 1891, Strauss and Altig 1992, Thibaudeau and Altig 1988, Trauth et al. 2004. ***Reproductive Biology:*** Livezey and Wright 1947, A. H. Wright 1932. ***Ecobehavior:*** Altig et al. 1975, Altig and McDearman 1975, P. K. Anderson 1954, Bragg 1947a, 1950, Dodd 1995, Freiberg 1951, Hubbs and Armstrong 1961, Orton 1946, Ryder 1891, Schorr et al. 1990, Walls et al. 2002.

Gastrophryne olivacea (Western Narrow-mouthed Toad) PL 10H

IDENTIFICATION As for genus; dorsum olive-gray to brown to brassy caused by abundant iridophores; venter white or coppery, sometimes with small contrasting marks; lateral tail stripe faint to absent; to 35.0 mm TL/stage 36.

NATURAL HISTORY Western Narrow-mouthed Toads oviposit between March and September in ponds and temporary pools, often in arid regions. About 600 pigmented ova with 1 jelly layer are laid as a coherent film. Larval development is not well studied and presumably is similar to that of the Eastern Narrow-mouthed Toad.

RANGE *Gastrophryne olivacea* is found from southeastern Nebraska and western Missouri south to the Texas Gulf Coast and west through most of Texas and much of northern Mexico. A separate population in a small area in south-central Arizona extends south through much of northwestern Mexico.

CITATIONS *General:* Nelson 1972b. *Development/Morphology*: Bragg 1947a, Nelson and Cuellar 1968, Trauth et al. 2004. *Reproductive Biology:* Livezey and Wright 1947, J. N. Stuart and Painter 1996. *Ecobehavior:* Bragg 1947a, 1950, H. S. Fitch 1956.

Hypopachus variolosus (Northern Sheep Frog)   Fig. 99

IDENTIFICATION As for genus; uniformly brownish, often with middorsal stripe, venter faintly marked; to 35.0 mm TL/stage 36.

NATURAL HISTORY Northern Sheep Frogs oviposit from March to September in ponds and temporary pools filled by seasonal rains. Clutches consist of about 700 pigmented ova (OD 1.0 mm, ED 1.5–2.0 mm, 1 jelly layer) laid as a film of coherent eggs that hatch in 12–24 h. Tadpoles grow to 27.0–35.0 mm TL in about a month and metamorphose at 10.0–16.0 mm SVL.

RANGE In the United States, the Northern Sheep Frog occurs only in extreme south Texas. Other populations are found in northern Mexico, from Tamaulipas and Nuevo León on the east and Sinaloa on the west, south through Mexico and Central America to northwestern Costa Rica.

CITATIONS *Development/Morphology*: Nelson and Cuellar 1968, E. H. Taylor 1942. *Reproductive Biology:* Livezey and Wright 1947.

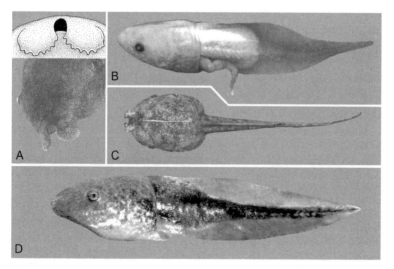

Figure 99. **Hypopachus variolosus.** (A) schematic drawing (top; PCU) and posterodorsal view of the mouthparts of a live tadpole (bottom; Willacy Co., TX; DLM), (B) tadpole (preserved, 30.0 mm TL, stage 33, Oaxaca, Mexico, USNM 304943; RA), living tadpole: (C) dorsal view, (D) lateral view (28.0 mm TL, stage 31, Willacy Co., TX; DLM).

Pipidae (Tongueless Frogs)

This African exotic occurs in southern California and southern Arizona; non-flowing water, suspension feeder; oral disc, jaw sheaths and labial teeth absent; mouth slitlike with single mobile barbel at each corner; vent medial; eyes lateral; spiracles dual and lateral, medial wall absent; dorsal fin low with acute tip, originates near dorsal tail-body junction or posteriorly on tail muscle, filiform, mobile flagellum beats continually, ventral fin higher than dorsal fin and with prominent anterior lobe where fin extends onto abdomen; body and especially snout depressed; nostrils transversely oval, nearer snout than eye; toe of large tadpoles with extensive webbing, tips keratinized, digital pads absent; usually uniformly pale with darker pigmentation around nasal capsules and brain, large individuals often have a dark suffusion on posterior third of tail; to 70.0 mm TL/ stage 36; 1 genus, 1 species. *Xenopus*: as for family.

Xenopus laevis (African Clawed Frog)   Fig. 100

IDENTIFICATION As for family.

NATURAL HISTORY Females of *X. laevis* oviposit in ponds and slow streams. Between 500 and 27,000 pale brown ova (OD 1.1–1.2 mm, ED about 1.6 mm, 2 jelly layers) are laid as singles and small groups stuck to vegetation. Tadpoles metamorphose in 1.0–4.2 months at 17.0–25.0 mm SVL and remain fully aquatic after metamorphosis.

These exotic, suspension-feeding tadpoles aggregate in midwater where they maintain a static, head-down posture by constant undulations of the tail flagellum. The exaggerated feeding pumps of the throat apparatus make the quiescent tadpole rock back and forth.

RANGE Introduced populations of this African exotic occur in southwestern California and adjacent Mexico and in south-central Arizona.

CITATIONS *Development/Morphology*: Dreyer 1915, Nokhbatolfoghahai and Downie 2007, Rot-Nikcevic and Wassersug 2004, Weiss 1945. *Reproductive Biology:* Bles 1906. *Ecobehavior:* Moriya et al. 1996, Rot-Nikcevic et al. 2005,

*Figure 100. **Xenopus laevis.** (A) tadpole (62.0 mm TL, stage 33, culture; RA).*

Schoonbee et al. 1992, Seale 1982, Seale and Wassersug 1979, Seale et al. 1982, Sherman and Levitis 2003, Viertel 1999, Wassersug 1989, 1992, 1996, Wassersug and Feder 1983, Wassersug and Hessler 1971, Wassersug and Murphy 1987, Wassersug et al. 1981, R. S. Wilson et al. 2000.

Ranidae (True Frogs)

True frogs are introduced throughout Hawaii and occur naturally on the bottom of nonflowing water sites over the entire mainland or uncommonly in flowing water; LTRF 2/3–6/7, all rows uniserial, rows A-1 and A-2 about equal in length; large marginal papillae with wide dorsal gap; submarginal papillae sparse to absent; oral disc emarginate although weakly so in stream forms, remains approximately oval at rest; jaw sheaths narrow to wide; vent dextral; eyes dorsal; spiracle sinistral; dorsal fin low to medium with broadly pointed tip; body shape slightly depressed; nostrils small to medium, without ornamentation; digital pads absent; coloration variable but seldom contrasty; to more than 100.0 mm TL, often smaller; 3 genera, 30 species. Specific generic characters not known for *Glandirana* (1 species), *Lithobates* (21 species), and *Rana* (8 species).

Glandirana rugosa (Japanese Wrinkled Frog)   Fig. 101

IDENTIFICATION LTRF 1/3; P-1 with medial gap; P-3 length about 75% of P-2 length; almost uniformly medium brown to slightly russet with small, scattered black marks; belly pale; to 80.0 mm TL/stage 36.

NATURAL HISTORY Female Japanese Wrinkled Frogs deposit eggs from February to August in ponds and slow streams. Clutches consist of 400–1360 pale brown ova (OD 0.8–1.0 mm, ED 6.0–9.0 mm, 2 flimsy jelly layers) laid as small groups of 10–60 eggs attached to vegetation and hatch in 5 days. Tadpoles grow to 72.0–76.0 mm TL and metamorphose in a year at 19.0–35.0 mm SVL.

RANGE *Glandirana rugosa* is exotic on the major islands of Hawaii.

CITATIONS **Development/Morphology**: Maeda and Matsui 1989, Okada 1931. **Reproductive Biology:** Okada 1931, Svihla 1936. **Ecobehavior:** Gilbertson and

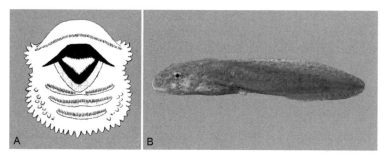

Figure 101. **Glandirana rugosa.** (A) mouthparts (modified from Okada 1931), (B) tadpole (48.0 mm TL, stage 33, Kauai, HI; RA).

Watermolen 1998, Maeda and Matsui 1989, Okada 1931, J. A. Oliver and Shaw 1953.

Lithobates areolatus (Crawfish Frog), *L. capito* (Gopher Frog), and *L. sevosus* (Dusky Gopher Frog)   Fig. 102; PL 10I

IDENTIFICATION LTRF 2/3; A-2 gap 60–100% of one side of LTR A-2; LTR P-1 with medial gap; LTR P-3 length about 66–83% of P-2 length; origin of dorsal fin near the plane of the spiracle; tail length about 1.7 times body length; coloration greatly influenced by conditions of habitat—uniformly pale with greenish-golden tint in turbid sites, often uniformly gray in clear water and occasional marks in the fins not bold with fuzzy borders; to 80.0 mm TL/stage 36.

NATURAL HISTORY Female Crawfish and Gopher frogs breed in temporary and permanent ponds from December to August, especially following episodic rains at southern localities. Clutches of 2200–7000 large, widely spaced pigmented ova (OD 1.8–2.5 mm, ED 4.4–6.0 mm, 2 jelly layers) are laid in clumps (120–150 mm diameter). Eggs hatch in 2.0–4.5 days, and tadpoles grow to 74.0–81.0 mm TL in 2.0–5.2 months and metamorphose at 27.0–35.0 mm SVL. The egg clumps are usually more spherical, compact, and less often oviposited in groups compared with those of sympatric members of the *R. pipiens* group. The tadpoles are often found in deeper water within a site than sympatric tadpoles of leopard frogs.

RANGE *L. areolatus* occurs from southeastern Iowa, central Illinois, and southwestern Indiana south through western Kentucky, Tennessee, northern Mississippi and west to the Gulf Coast in east Texas, and in eastern Oklahoma and Kansas. Large gaps exist in the range in the Ozark and Ouachita regions. *L. capito* occurs on the Coastal Plain from northern North Carolina and most of peninsular

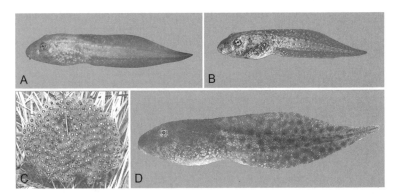

Figure 102. **Lithobates areolatus and relatives.** (A) Tadpole of *L. capito* (62.0 mm TL, stage 35, Covington Co., AL; RA), (B) Tadpole of *L. sevosus* (52.0 mm TL, stage 32, Jackson Co., MS; RA), (C) egg clump of *L. sevosus* (Jackson Co., MS; GNJ), (D) tadpole of *L. capito* (72.0 mm TL; RWV). See also Plate 10I.

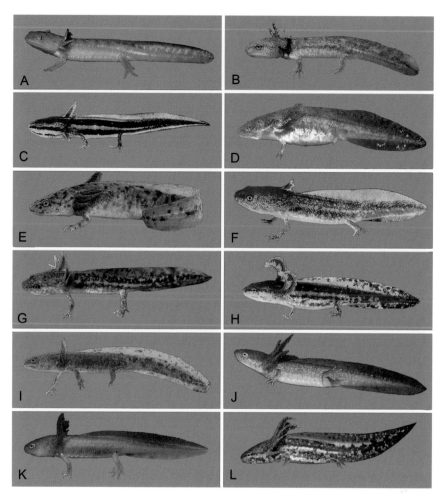

Plate 1. (A) *Ambystoma annulatum* (48.0 mm TL, Stone Co., MO; RA; Fig. 7B), (B) *A. barbouri* (51.0 mm TL, Fayette Co., KY; RA; Fig. 8C–E), (C) *A. bishopi* (66.0 mm TL, Santa Rosa Co., FL; BM; Fig. 9A–B), (D) *A. californiense* (75.0 mm TL, Madera Co., CA; RWH; Fig. 9A), (E) *A. gracile* (190.0 mm TL, Pierce Co., WA; WPL; Fig. 11B, D), (F) *A. jeffersonianum* (52.0 mm TL, Hocking Co., OH; RWV), (G) *A. laterale* (57.0 mm TL, Washtenaw Co., MI; RWV; Fig. 12B), (H) *A. mabeei* (61.0 mm TL, Scotland Co., NC; RWV; Fig. 13A–B), (I) *A. macrodactylum* (56.0 mm TL, WA; WPL; Fig. 14D–E), (J) *A. maculatum* (68.0 mm TL, Noxubee Co., MS; RA; Fig. 15E), (K) *A. opacum* (60.0 mm TL, Oktibbeha Co., MS; RA; Fig. 16D), (L) *A. talpoideum* (71.0 mm TL, Noxubee Co., MS; RA; Fig. 17D, F)

Plate 2. (A) *Ambystoma texanum* (58.0 mm TL, Union Co., IL; RA; Fig. 18B–D), (B) *A. tigrinum* (78.0 mm TL, Covington Co., AL; RA), (C) *Dicamptodon copei* (172.0 mm TL, Multnomah Co., OR; RWV), (D) *D. ensatus* (162.0 mm TL, Santa Cruz Co., CA; RA), (E) *D. tenebrosus* (92.0 mm TL, Wahkiakum Co., WA; WPL; Fig. 20B), (F) *Amphiuma tridactylum* (60.0 mm TL, Oktibbeha Co., MS; RA; Fig. 21B–C), (G) *Cryptobranchus alleganiensis* (87.0 mm TL, Watauga Co., NC; RWV; Fig. 22C), (H) *Desmognathus apalachicolae* (26.0 mm TL, Liberty Co., FL; DBM), (I) *D. auriculatus* (48.0 mm TL, Hanover Co., NC; RWV; Fig. 23C–D), (J) *D. brimleyorum* (45.0 mm TL, MO; DMD), (K) *D. carolinensis* (21.0 mm TL, Yancy Co., NC; RA), (L) *D. conanti* (22 mm TL, Tishimingo Co., MS; RA)

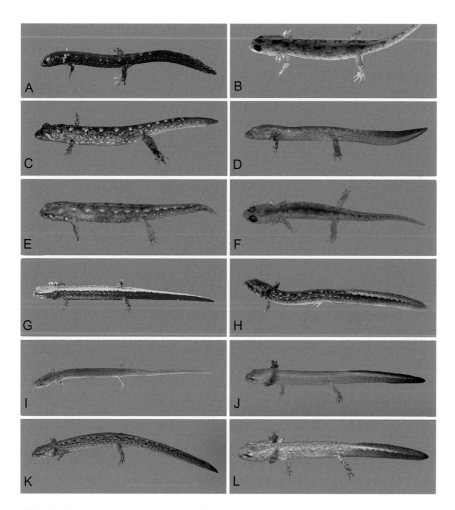

Plate 3. (A) *Desmognathus marmoratus* (64 mm TL, CO., NC; RA), (B) *D. monticola* (38.0 mm TL; RWV), (C) *D. orestes* (24.0 mm TL, Burke Co., NC; RA), (D) *D. quadramaculatus* (68.0 mm TL, Clay Co., NC; RA; Fig. 26D), (E) *D. santeetlah* (17.0 mm TL, Graham Co., NC; RA; Fig. 23F), (F) *D. welteri* (38.0 mm TL, Harlan Co., KY; DMD), (G) *Eurycea aquatica* (44.0 mm TL, Shelby Co., AL; RWV), (H) *E. bislineata* (58.0 mm TL, MA; JM), (I) *E. chisholmensis* (57.0 mm TL, Bell Co., TX; DMH), (J) *E. cirrigera* (56.0 mm TL, Webster Co., MS; RA), (K) *E. junaluska* (62.0 mm TL, TN; RWV), (L) *E. longicauda* (58.0 mm TL, Union Co., IL; RA; Fig. 30A)

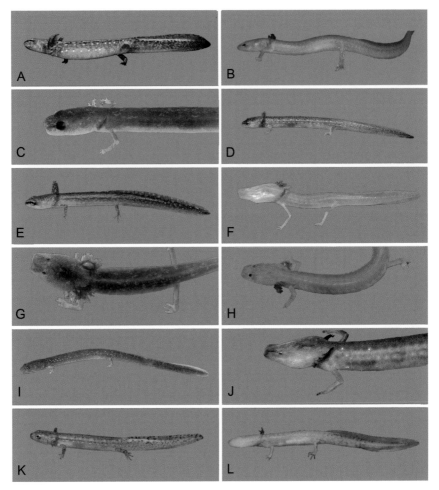

Plate 4. (A) *Eurycea lucifuga* (52.0 mm TL, GA; JBJ; Fig. 31A), (B) *E. multiplicata* (48.0 mm TL, Montgomery Co., AR; RA), (C) *E. nana* (52.0 mm TL, Hays Co., TX; RWV; Fig. 32A–B), (D) *E. pterophila* (46.0 mm TL, Helotes Co., TX; RA), (E) *E. quadridigitata* (36.0 mm TL, Okaloosa Co., FL; RA; Fig. 33A–B), (F) *E. rathbuni* (67.0 mm TL, Hays Co., TX; RWV), (G) *E. sosorum* (50.0 mm TL, Travis Co., TX; WM; Fig. 35C), (H) *E. spelaea* (61.0 mm TL, Phelps Co., MO; TRJ; Fig. 36B–C), (I) *E. tynerensis* (66.0 mm TL; RWV), (J) *E. waterlooensis* (80.0 mm TL, Travis Co., TX; DMH), (K) *E. wilderae* (52.0 mm TL, Watauga Co., NC; RA), (L) *E. wallacei* (64.0 mm TL, Jackson Co., FL; RA; Fig. 37A–B)

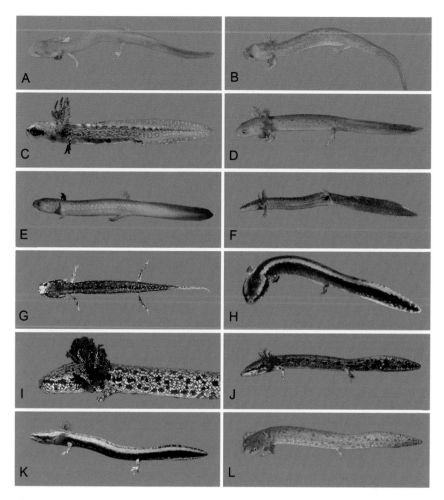

Plate 5. (A) *Gyrinophilus palleucus* (177.0 mm TL, Franklin Co., TN; RA; Fig. 38B), (B) *G. por-phyriticus* (162.0 mm TL; DMD; Fig. 39A–B), (C) *Hemidactylium scutatum* (22.0 mm TL, KY; DMD; Fig. 40A, inset), (D) *Pseudotriton montanus* (58.0 mm TL; DMD; Fig. 41B–C), (E) *P. ruber* (48.0 mm TL, Winston Co., AL; RA; Fig. 42B, D), (F) *Stereochilus marginatus* (68.0 mm TL, Bladen Co., NC; RA; Fig. 43F), (G) *Urspelerpes brucei* (38.0 mm TL, Stephens Co, GA; JB; Fig. 44A–B), (H) *Necturus alabamensis* (44.0 mm TL, Winston Co., AL; BM; Fig. 45A), (I) *N. beyeri* (182.0 mm TL; RWV; Fig. 46A–C), (J) *N. lewisi* (124.0 mm TL; RWV; Fig. 47A–D), (K) *N. macu-losus* (51.0 mm TL; DMD; Fig. 48B–D), (L) *N. punctatus* (152.0 mm TL; RWV; Fig. 49A–B)

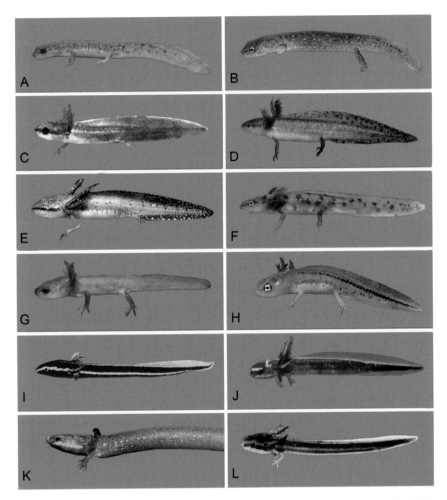

Plate 6. (A) *Rhyacotriton cascadae* (53.0 mm TL, Skamania Co., WA; WPL; Fig. 50B–D), (B) *R. variegatus* (50.0 mm TL, Humboldt Co., CA; WPL), (C) *Notophthalmus meridionalis* (21.0 mm TL, culture from Cameron Co., TX; DLM; Fig. 51C–D), (D) *N. perstriatus* (51.0 mm TL, Bryan Co., GA; DJS; Fig. 52A–B), (E) *N. viridescens* (34.0 mm TL, Noxubee Co., MS; RA; Fig. 53A–C), (F) *Taricha granulosa* (46.0 mm TL, Klickitat Co., WA; WPL; Fig. 54D), (G) *T. rivularis* (43.0 mm TL, Humboldt Co., CA; RA; Fig. 55B–C), (H) *T. torosa* (34.0 mm TL, Santa Clara Co., CA; RA; Fig. 56E), (I) *Pseudobranchus striatus* (18.0 mm TL, Levy Co., FL; BM; Fig. 57B–C), (J) *Siren intermedia* (hatchling, 20.0 mm TL, Mobile Co., AL; RA; Fig. 58A–C), (K) *S. intermedia* (larva, 50.0 mm TL, Oktibbeha Co., MS; RA), (L) *S. lacertina* (22.0 mm TL, Alachua Co., FL; BM; Fig. 59B–C)

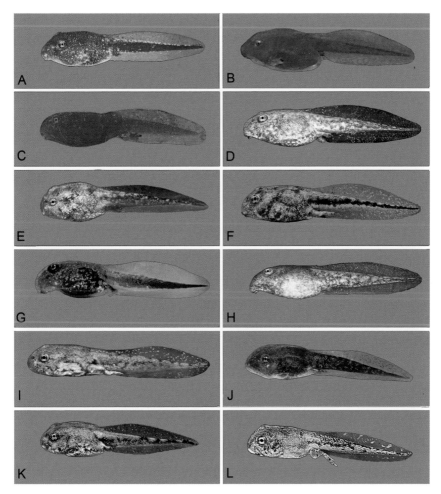

Plate 7. (A) *Anaxyrus americanus* (23.0 mm TL, stage 33, Webster Co., MO; RA; Fig. 62B–C), (B) *A. boreas* (39.0 mm TL, stage 31, Hood River Co., OR; RA; Fig. 63D), (C) *A. canorus* (29.0 mm TL, stage 35, Alpine Co., CA; JET; Fig. 64B), (D) *A. cognatus* (34.0 mm TL, stage 34, Yavapai Co., AZ; RA), (E) *A. debilis* (18.0 mm TL, stage 28, Eddy Co., NM; RA), (F) *A. fowleri* (28.0 mm TL, stage 35, Oktibbeha Co., MS; RA; Fig. 65B, D), (G) *A. hemiophrys* (19.0 mm TL, stage 32, Grand Forks Co., ND; RA; Fig. 66B), (H) *A. microscaphus* (31.0 mm TL, stage 34, Gila Co., AZ; RA), (I) *A. californicus* (28.0 mm TL, stage 34, Santa Barbara Co., CA; RA), (J) *A. punctatus* (22.0 mm TL, stage 28, Clark Co., NV; RA; Fig. 67A–C), (K) *A. quercicus* (23.0 mm TL, stage 34, Hancock Co., MS; RA), (L) *A. speciosus* (26.0 mm TL, stage 37, Brewster Co., TX; RA; Fig. 68A–B)

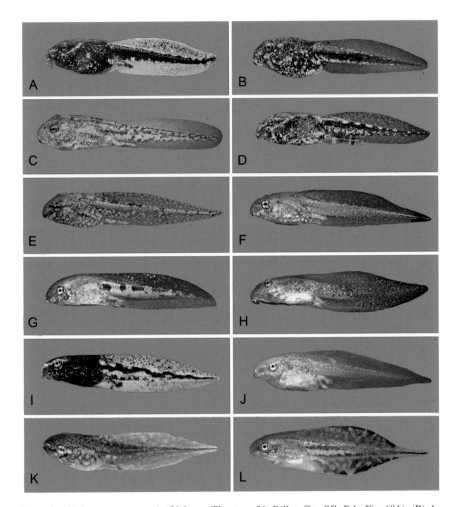

Plate 8. (A) *Anaxyrus terrestris* (29.0 mm TL, stage 36, Dillon Co., SC; RA; Fig. 69A), (B) *A. woodhousii* (31.0 mm TL, stage 33, Yavapai Co., AZ; RA; Fig. 70B–C), (C) *Incilius alvaria* (30.0 mm TL, stage 36, Pima Co., AZ; RA; Fig. 71A–B), (D) *I. nebulifer* (33.0 mm TL, stage 30, Orange Co., TX; RA), (E) *Acris crepitans* (38.0 mm TL, stage 36, Jackson Co., IL; RA; Fig. 74A–B), (F) *A. gryllus* (36.0 mm TL, stage 35, Seminole Co., AL; RA; Fig. 75C), (G) *Hyla andersonii* (27.0 mm TL, stage 36, Okaloosa Co., FL; RA; Fig. 76A–B), (H) *H. arenicolor* (41.0 mm TL, stage 35, Gila Co., AZ; RA; Fig. 77A–C), (I) *H. avivoca* (27.0 mm TL, stage 35, Oktibbeha Co., MS; RA; Fig. 78A–B), (J) *H. chrysoscelis* (28.0 mm TL, stage 33, Oktibbeha Co., MS; RA; Fig. 83K), (K) *H. cinerea* (43.0 mm TL, stage 38, Berkeley Co., SC; RWV; Fig. 79A–B, D–E), (L) *H. femoralis* (35.0 mm TL, stage 35, Okaloosa Co., FL; RA; Fig. 80B)

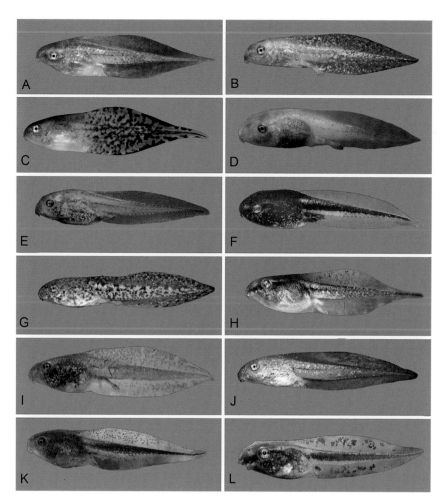

Plate 9. (A) *Hyla gratiosa* (59.0 mm TL, stage 35, Okaloosa Co., FL; RA; Fig. 81C–D), (B) *H. squirella* (28.0 mm TL, stage 37, Hillsborough Co., FL; RA; Fig. 82B, D), (C) *H. versicolor* (38.0 mm TL, stage 37, Warren Co., MO; TRJ; Fig. 83L), (D) *H. wrightorum* (25.0 mm TL, stage 36, Coconino Co., AZ; RA; Fig. 84A–B), (E) *Pseudacris brachyphona* (31.0 mm TL, stage 36, Wayne Co., WV; RA), (F) *P. brimleyi* (20.0 mm TL, stage 35, SC; RA; Fig. 86A), (G) *P. cadaverina* (32.0 mm TL, stage 37, Santa Barbara Co., CA; RA; Fig. 87A–B), (H) *P. clarkii* (32.0 mm TL, stage 33, Cameron Co., TX; DLM; Fig. 88A–C), (I) *P. crucifer* (36.0 mm TL, stage 37, MA; JM; Fig. 89B–C), (J) *P. hypochondriaca* (37.0 mm TL, stage 34, Santa Barbara Co., CA; RA; Fig. 92A–B, E), (K) *P. nigrita* (30.0 mm TL, stage 34, Liberty Co., GA; DJS; Fig. 94F), (L) *P. ocularis* (29.0 mm TL, stage 28, Bryan Co., GA; DJS; Fig. 90A–B)

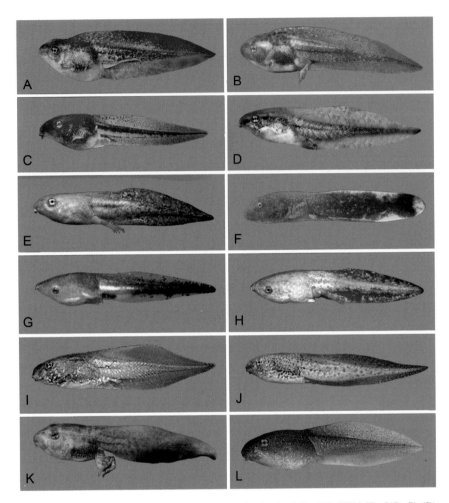

Plate 10. (A) *Pseudacris ornata* (52.0 mm TL, stage 37, Scotland Co., NC; RWV; Fig. 91B–C), (B) *P. streckeri* (64.0 mm TL, stage 38, Mississippi Co., AR; RA; Fig. 93A), (C) *P. triseriata* (36.0 mm TL, stage 36, Cass Co., IL; RA; Fig. 94D–E), (D) *Smilisca baudinii* (43.0 mm TL, stage 35, Cameron Co., TX; DLM; Fig. 95A–C), (E) *S. fodiens* (46.0 mm TL, stage 36, Pima Co., AZ; RA), (F) *Ascaphus truei* (37.0 mm TL, stage 33, Benton Co., OR; RA; Fig. 96C, E), (G) *Gastrophryne carolinensis* (23.0 mm TL, stage 34, Okaloosa Co., FL; RA; Fig. 98C–D), (H) *G. olivacea* (30.0 mm TL, stage 37, Lubbock Co., TX; RA), (I) *Lithobates areolatus* (58.0 mm TL, stage 37, Jackson Co., IL; RA), (J) *L. berlandieri* (81.0 mm TL, stage 34, Bastrop Co., TX; RA; Fig. 103A–B), (K) *L. blairi* (57.0 mm TL, stage 32, Webster Co., MO; RA; Fig. 104C–E), (L) *L. catesbeianus* (72.0 mm TL, stage 33, Evans Co., GA; DJS; Fig. 105B, F)

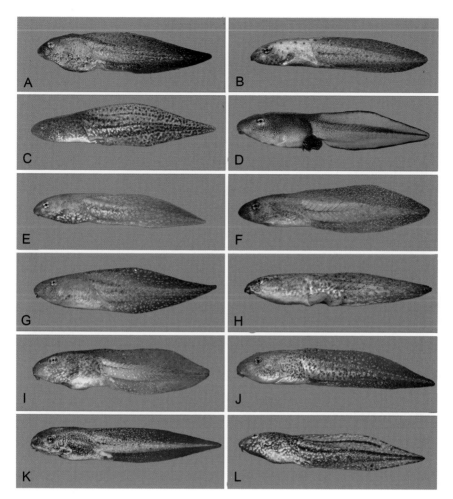

Plate 11. (A) *Lithobates chiricahuensis* (51.0 mm TL, stage 33, Catron Co., NM; RA; Fig. 106B–C), (B) *L. clamitans* (66.0 mm TL, stage 35, Berkeley Co., SC; RA; Fig. 107B, D–E), (C) *L. grylio* (78.0 mm TL, stage 36, Long Co., GA; DJS; Fig. 108B–C), (D) *L. heckscheri* (101.0 mm TL, stage 40, Columbia Co., FL; RWV; Fig. 109B, D), (E) *L. okaloosae* (47.0 mm TL, stage 34, Okaloosa Co., FL; RA; Fig. 110A, C), (F) *L. onca* (52.0 mm TL, stage 33, Clark Co., NV; RA; Fig. 111B–C), (G) *L. palustris* (55.0 mm TL, stage 25, Webster Co., MO; RA; Fig. 112A–B), (H) *L. pipiens* (71.0 mm TL, stage 38, Randolph Co., WV; RA; Fig. 113D, G), (I) *L. septentrionalis* (67.0 mm TL, stage 33, Clearwater Co., MN; RA), (J) *L. sphenocephalus* (82.0 mm TL, stage 35, Oktibbeha Co., MS; RA; Fig. 113H–J), (K) *L. sylvaticus* (41.0 mm TL, stage 35, Polk Co., TN; RA; Fig. 114B), (L) *L. virgatipes* (72.0 mm TL, stage 34, Bladen Co., NC; RA; Fig. 116A–B)

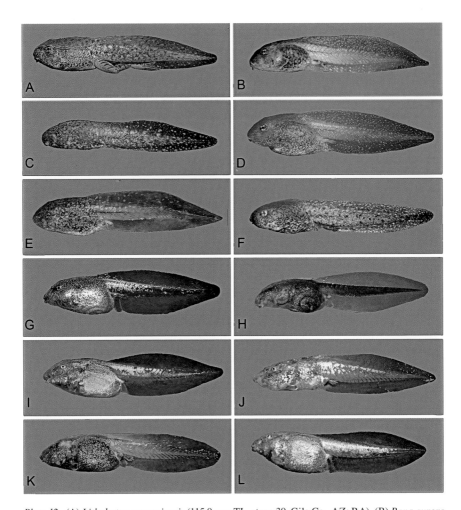

Plate 12. (A) *Lithobates yavapaiensis* (115.0 mm TL, stage 39, Gila Co., AZ; RA), (B) *Rana aurora* (63.0 mm TL, stage 34, Lincoln Co., OR; RA; Fig. 117A–B), (C) *R. boylii* (45.0 mm TL, stage 32, Sonoma Co., CA; RA; Fig. 118A), (D) *R. cascadae* (60.0 mm TL, stage 34, Clackamas Co., OR; RA; Fig. 119B), (E) *R. pretiosa* (54.0 mm TL, stage 33, Lane Co., OR; RA), (F) *R. sierrae* (50.0 mm TL, stage 36, Mono Co., CA; RA; Fig 120B–C), (G) *Scaphiopus couchii* (32.0 mm TL, stage 33, Eddy Co., NM; RA; Fig. 123C), (H) *S. holbrookii* (28.0 mm TL, stage 27, Alexander Co., IL; RA; Fig. 124F–G), (I) *Spea bombifrons* (omnivore, 32.0 mm TL, stage 32, Eddy Co., NM; RA; Fig. 125B–C), (J) *S. hammondii* (71.0 mm TL, stage 38, Santa Barbara Co., CA; RA; Fig. 126B), (K) *S. intermontana* (omnivore, 31.0 mm TL, stage 35, Mono Co., CA; RA; Fig. 127A, D), (L) *S. multiplicata* (omnivore, 47.0 mm TL, stage 33, Eddy Co., NM; RA; Fig. 128A, C)

Florida to southeastern Alabama. *L. sevosus* is now found only in a small area in southern Mississippi.

CITATIONS *General:* Altig and Lohoefener 1983. ***Development/Morphology***: Altig and Pace 1974, Bragg 1955b, Gosner 1959, H. M. Smith et al. 1947, 1948, Trauth et al. 2004. ***Reproductive Biology:*** Livezey and Wright 1947, Parmalee et al. 2002, Redmer 2000, Volpe 1957b, A. H. Wright 1932. ***Ecobehavior:*** Aresco and Reed 1998, Bragg 1953, 1964b, Cronin and Travis 1986, Parris and Semlitsch 1998, Semlitsch et al. 1995.

Lithobates berlandieri (Rio Grande Leopard Frog)   Fig. 103; PL 10J

IDENTIFICATION LTRF commonly 2/3, short A-3 often present; A-2 gap 80 to greater than 100% of one side of LTR A-2; LTR P-1 with medial gap; LTR P-3 length about 75% of P-2 length; ventral marginal papillae large; body rather uniform gray to brown, sometimes greenish; tail muscle, and to a lesser extend the fins, often with bold marbling; iris with almost no melanophores visible except at four major compass points; to 90.0 mm TL/stage 36.

NATURAL HISTORY The Rio Grande Leopard Frog breeds in ponds and pools of clear, slow streams, and females oviposit from February to September, perhaps with a spring peak. Egg masses are 7–9 mm across and attached to emergent vegetation. Tadpoles may overwinter in some areas, and metamorphs measure 30–32 mm SVL. Other life history data are few.

RANGE Rio Grande Leopard Frogs are known from southeastern New Mexico east through central Texas and south to the Gulf of Mexico and adjacent areas of northeastern Mexico. Introduced populations in the drainages of the Lower Colorado and Gila rivers of southwestern Arizona have spread into Imperial Co., southeastern California.

CITATIONS **General**: Platz 1991. ***Development/Morphology***: D. M. Hillis 1982, N. J. Scott and Jennings 1985. ***Reproductive Biology:*** Bragg 1944a,

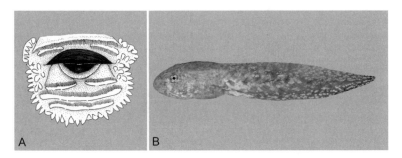

Figure 103. **Lithobates berlandieri.** (A) oral apparatus (PCU), (B) tadpole (80.0 mm TL, stage 34, TX; RWV). See also Plate 10J.

Livezey and Wright 1947. *Ecobehavior:* Bragg 1962a, Wassersug and Feder 1983, Wassersug et al. 1981.

Lithobates blairi (Plains Leopard Frog)   Fig. 104; PL 10K

IDENTIFICATION LTRF 2/3; A-2 gap about 66–70% of one side of LTR A-2; LTR P-1 without medial gap; LTR P-3 length about 90–100% of P-2 length; ventral marginal papillae large; considerable variation throughout the range of this poorly described species. Western parts of range: iris sometimes lacks iridophores; general visual impression uniformly honey brown to gray. Eastern parts of range: iris with iridophores, body and tail with dark mottling. To 65.0 mm TL/stage 36.

NATURAL HISTORY Females oviposit from February to October and often in shallow, turbid prairie ponds and streams. Females produce clutches of 4000–6500

Figure 104. **Lithobates blairi.** (A) egg clump (Lyon Co., KS; GS), (B) embryos at stages 21 and 23 (Lyon Co., KS, GS), (C) mouthparts (Lyon Co., KS, GS), (D) tadpole (Webster Co., MO; RA), (E) tadpole (72.0 mm TL, stage 35, McLean Co., IL; MR). See also Plate 10K.

eggs and deposit them in small masses attached to plant stems and branches just below the water surface. Eggs hatch in 2–3 weeks, and larvae are often in deeper portions of a pond and may spend more time in the water column than other ranids. Metamorphosis usually occurs in midsummer, but tadpoles derived from late breedings may overwinter. Metamorphs measure between 27 and 30 mm SVL.

RANGE The Plains Leopard Frog is known from southeastern South Dakota southwest to eastern New Mexico east to Indiana and southwest to central Texas.

CITATIONS *General:* L. E. Brown 1992. *Development/Morphology*: Korky 1978, N. J. Scott and Jennings 1985, Wassersug 1976c. *Reproductive Biology:* Livezey and Wright 1947, Parmalee et al. 2002. *Ecobehavior:* Bragg 1961, Seale 1980, Wassersug and Seibert 1975.

Lithobates catesbeianus (American Bullfrog)   Fig. 105; PL 10L

IDENTIFICATION LTRF 2/3, a short LTR A-3 is sometimes present; A-2 gap about 30–50% of one side of LTR A-2; LTR P-1 with medial gap; LTR P-3 length about 60–70% of P-2 length. Tadpoles with large hind legs: toe webbing curved downward between separated toes and does not extend to toe tips. Small tadpole less than 28.0 mm TL: appear grossly black but have abundant small blotches of golden iridophore at various layers of the skin and subintegumentary layers, fins clear. Large tadpole greater than 25.0 mm TL: usually with a muted green dorsum with black dots with distinct margins over the dorsum and dorsal part of tail. Tadpoles from some parts of Florida and adjacent Georgia and Alabama are uniformly bright green and lack black dots; to 178.0 mm TL/stage 36, occasionally larger.

NATURAL HISTORY Bullfrogs occur in ponds, lakes, and slow rivers and females oviposit December–October. Eggs are in an adherent film (0.5–1.5 m², 1000–47,800, ova black and whitish, OD 6.4–0.4 mm, ED 1.2–1.8 mm, 1 jelly layer) that is sometimes multilayered. The eggs commonly sink before hatching in 3–5 days, and tadpoles grow to 70.0–170.0 mm TL. Metamorphosis is in 5–24 months or more at 30.0–60.0 mm SVL. Females lay 2 clutches per year in some places; tadpoles often overwinter once or twice at northern latitudes.

This large tadpole commonly enters shallow water apparently in search of warmer temperatures. It often escapes to deeper water by swimming away rapidly near the bottom for a considerable distance and then turning about 90° to hide in bottom debris. Tadpoles in deep, warm water commonly come to the surface to gulp air.

RANGE The natural and exotic ranges of the American Bullfrog include most of North America east of the Mississippi River, much of the Great Plains south of southern South Dakota, most of the western tier of states west of the mountain ranges, several adjacent areas in northern Mexico, and all the major islands of Hawaii.

CITATIONS *Development/Morphology*: Altig 1972, Altig and Pace 1974, Bolek and Janovy 2004, Dodd 2004, Gosner 1959, Hinckley 1881, Provenzano and Boone 2009, Trauth et al. 2004. *Reproductive Biology:* Collins and Wilbur

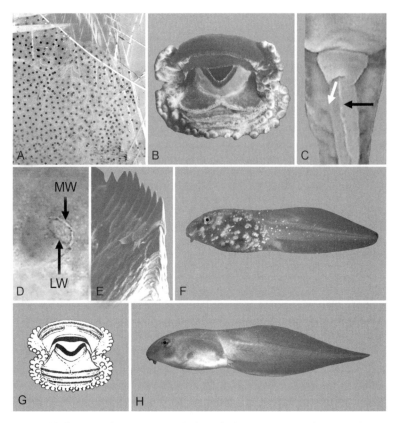

Figure 105. **Lithobates catesbeianus.** (A) partial film of adherent eggs (Lynn Co., MO; TRJ), (B) oral apparatus (GFJ), (C) dextral vent tube (black arrow = plane of ventral fin, white arrow = exit direction of vent; RA), (D) sinistral spiracle (anterior to the left; LW = lateral wall, MW = medial wall; RA), (E) jaw sheath (viewed obliquely at a cut edge in foreground and rear surface of the sheath; GFJ), (F) tadpole (28.0 mm TL, stage 26, Benton Co., OR; RA), (G) schematic of the oral apparatus (modified from Orton 1952), (H) tadpole (68.0 mm TL, stage 34, Hillsborough Co., FL; RA). See also Plate 10L.

1979, Livezey and Wright 1947, J. A. Moore 1940, Parmalee et al. 2002, Trauth et al. 1990, Viparina and Just 1975, A. H. Wright 1910. *Ecobehavior:* Altig et al. 1975, Altig and McDearman 1975, Anholt et al. 2000, Auburn and Taylor 1979, Brattstrom 1962, Bruneau and Magnin 1980, Cecil and Just 1979, Collins 1979, Crawshaw et al. 1992, Dudley et al. 1991, Eklöv and Werner 2000, Garcia et al. 2012, Gosner and Black 1957b, Hedeen 1975, Hoff and Wassersug 1985, Hutchison and Hill 1978, Justis and Taylor 1976, Kruse and Francis 1977, Kupferberg 1995, 1997b, Licht 1969a, Menke and Claussen 1982, J. A. Moore 1939, 1942, Nie et al. 1999, Pearl et al. 2003, Petranka 1985, Pryor 2003, Pryor and Bjorndal 2005a–b, Punzo 1992, Regula Meyer et al. 2007, Relyea and Werner 1999, 2000, J. M. L. Richardson 2001, Rogers 1996, S. M. Rose 1960, M. J. Ryan

1978, Schiesari et al. 2009, Schorr et al. 1990, Seale 1980, Seale and Beckvar 1980, Seixas et al. 1998, G. R. Smith 1999a–b, G. R. Smith et al. 2004, 2007, G. R. Smith and Doupnik 2005, Steinwascher 1978, Ultsch et al. 2004, Van Buskirk 2000, D. Walker and Busack 2000, Walston and Mullin 2007, B. Walters 1975, Wassersug 1989, Wassersug and Hoff 1985, Wassersug and Yamashita 2001, Werner 1991, 1994, Werner and Arnholt 1996, Werner and McPeek 1994, Wilbur and Semlitsch 1990, Wollmuth and Crawshaw 1988, Woodward 1982a.

Lithobates chiricahuensis (Chiricahua Leopard Frog)   Fig. 106; PL 11A

IDENTIFICATION LTRF usually 2/3; A-2 gap greater than 200% of one side of LTR A-2; LTR P-1 with medial gap; LTR P-3 length 60–70% of P-2 length; ventral marginal papillae medium; dorsum dark with subtle paler mottling, mostly from iridophores; tail boldly marked throughout; iris fairly densely flecked throughout; to 87.0 mm TL/stage 38.

NATURAL HISTORY Females oviposit from early February through early September in relatively permanent ponds, streams, and stock tanks, often with breeding peaks in the spring and late summer months. In warm springs they may breed all year. Ova (300–1485/clutch) have a sharp demarcation between the black animal and white vegetal poles and are laid as clumps attached to vegetation in shallow water. Larvae metamorphose in 3–9 months at 35–40 mm SVL, and at higher or colder sites, they often overwinter.

RANGE Chiricahua Leopard Frogs are known from central Arizona southeast into southeastern New Mexico and adjacent Mexico.

CITATIONS *General:* Platz and Mecham 1984. *Development/Morphology*: R. D. Jennings and Scott 1993, N. J. Scott and Jennings 1985. *Reproductive*

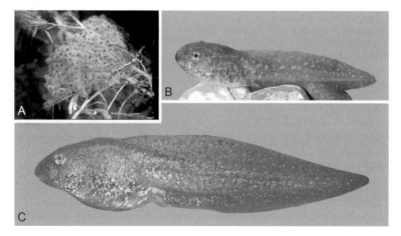

Figure 106. **Lithobates chiricahuensis.** (A) egg clump (Catron Co., NM; RDJ), (B) tadpole (36.0 mm TL, stage 27; RWV), (C) tadpole (86.0 mm TL, stage 31; Cochise Co., AZ; RA). See also Plate 11A.

Biology: Field and Groebner 2005, Livezey and Wright 1947. *Ecobehavior:* Zweifel 1968d, 1977.

Lithobates clamitans (Green Frog)   Fig. 107; PL 11B

IDENTIFICATION LTRF 2/3; A-2 gap greater than 300% of one side of LTR A-2; LTR P-1 with medial gap; LTR P-3 length 50–55% of P-2 length. Tadpoles with large hind legs, first and third toes about same length and webbing stops well short of toe tips; quite variable throughout large range, but usually brownish to gray dorsum with scattered dark spots with diffuse borders; fin pigmented throughout and posterior third of tail often with irregular dark marks; some individuals from the Northeast have a row of dots on the dorsal fin similar to those of *R. grylio* and *R. virgatipes*; to 65.0 mm TL/stage 36.

NATURAL HISTORY Green Frogs occur around permanent ponds, lakes, and reservoirs and along swampy river and stream margins. Females oviposit March–October, laying 1000–7000 black ova (OD 1.0–1.8 mm, 2 jelly layers) as a film usually less than 500 mm diameter. The eggs hatch in 3–6 days at 10.0–12.0 mm TL and tadpoles grow to 85.0–100.0 mm TL. Metamorphosis occurs in 70 days to 13 months at 21.0–40.0 mm SVL. Females sometimes lay 2 clutches a year, and tadpoles commonly overwinter, especially those from second clutches.

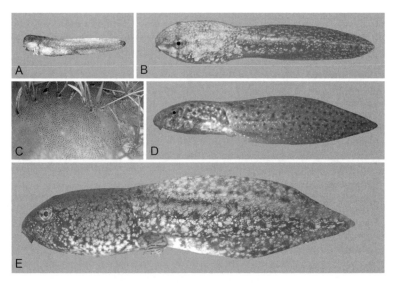

*Figure 107. **Lithobates clamitans.*** (A) embryo (10.0 mm TL, stage 21, NY; AMR), (B) tadpole (35.0 mm TL, stage 26, Okaloosa Co., FL; RA), (C) egg film (Fairfax Co., VA; KDW), (D) tadpole (72.0 mm TL, stage 32, Oktibbeha Co., MS; RA), (E) tadpole (70.0 mm TL, Burlington Co., NJ; JFB) with extreme numbers of iridophores and an indistinct row of dots in dorsal fin. See also Plate 11B.

This tadpole frequents swampy sites and is seldom seen while walking along the banks; it remains hidden among dense vegetation and within bottom debris.

RANGE Green Frog populations are found throughout eastern North America south of central Quebec and east of central Texas except in the prairie areas in Illinois and southeastern Indiana and the southern two-thirds of peninsular Florida. Exotic populations of the Green Frog are on Vancouver Island, adjacent mainland British Columbia, and several other sites in the Pacific Northwest.

CITATIONS *Development/Morphology*: Altig and Pace 1974, Dodd 2004, Gosner 1959, Korky and Smallwood 2011, Trauth et al. 2004. *Reproductive Biology:* Collins and Wilbur 1979, Livezey and Wright 1947, Morgan 1891, Parmalee et al. 2002, Ting 1951, Trauth et al. 1990, K. D. Wells 1977, A. H. Wright 1910, 1932. *Ecobehavior:* Anholt et al. 2000, Eklöv and Werner 2000, Formanowicz and Brodie 1982, Gosner and Black 1957b, Holomuzski 1998, D. H. Ireland et al. 2007, Jenssen 1967, Licht 1969a, Martof 1952, 1956, J. A. Moore 1939, 1940, M. K. Moore and Townsend 1998, Petranka and Kennedy 1999, Regula Meyer et al. 2007, Relyea and Werner 1999, 2000, J. M. L. Richardson 2001, Schiesari et al. 2009, Schorr et al. 1990, G. R. Smith et al. 2004, 2010, G. R. Smith and Jennings 2004, Smith-Gill et al. 1979, Steinwascher and Travis 1983, Tarr and Babbitt 2002, Thiemann and Wassersug 2000, Van Buskirk 2000, B. Walters 1975, Warkentin 1992a–b, Warny et al. 2012, Wassersug and Hoff 1985, B. W. Watkins and McPeek 2006, Werner 1991, 1994, Werner and Anholt 1996, Werner and McPeek 1994, Wilbur and Fauth 1990.

Lithobates grylio (Pig Frog)   **Fig. 108; PL 11C**

IDENTIFICATION LTRF 2/3; A-2 gap greater than 200% of one side of LTR A-2; LTR P-1 with medial gap; LTR P-3 length about 65% of P-2 length. Tadpoles

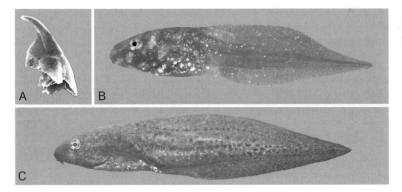

Figure 108. **Lithobates grylio.** (A) individual labial tooth (viewed from side; modified from Altig and Pace 1974), (B) tadpole (22.0 mm TL, stage 26, Baldwin Co., AL; RA), (C) tadpole (84.0 mm TL, stage 36, Baldwin Co., AL; RA). See also Plate 11C.

with large hind legs: toe webbing nearly straight between separated toes. Small tadpole: slightly opaque fins; low-contrast pale band at midbody; becomes paler at night. Large tadpole: brown to slightly greenish or maroon dorsum with many darker specks and small spots; pigment around neuromasts forms a longitudinal row of dots in dorsal fin; to 100.0 mm TL/stage 36.

NATURAL HISTORY Females of *L. grylio* oviposit from February to September in permanent water bodies, including lakes, ponds, sloughs, and swamps. Eggs number 5000–15,000 and have black ova (OD 1.4–2.0 mm, ED 4.0–6.0 mm, 2 jelly layers) that are laid as a film (305.0–610.0 mm diameter). They hatch in 2.0–3.5 days, and tadpoles grow to 110.0 mm TL. They metamorphose in 10–26 months, sometimes less, and metamorphs measure 30.0–52.0 mm SVL.

This tadpole inhabits larger bodies of water and is seldom seen while walking along the shore. It seems solitary and spends most of its time hidden within bottom debris, often quite close to shore.

RANGE Pig Frogs live on the Coastal Plain from central South Carolina to southeastern Texas including peninsular Florida.

CITATIONS *Development/Morphology*: Altig 1972, Altig and Pace 1974. *Reproductive Biology:* Livezey and Wright 1947, A. H. Wright 1932.

Lithobates heckscheri (River Frog)   Fig. 109; PL 11D

IDENTIFICATION LTRF 2/3, short A-3 sometimes present; A-2 gap about 200% of one side of LTR A-2; LTR P-1 with medial gap; LTR P-3 length about

*Figure 109. **Lithobates heckscheri.*** (A) tadpole school moving along a bank (area: about 1.5 × 8.0 m), (B) tadpole (25.0 mm TL, stage 25; RA), (C) individual tooth (viewed from the side; modified from Altig and Pace 1974) and several labial teeth (in situ, viewed from the back; modified from Altig 1973), (D) tadpole (108.0 mm TL, stage 32; arrow = site of *Aeromonas* infection noted multiple years; all from Okaloosa Co., FL; RA). See also Plate 11D.

85% of P-2 length; tail length 1.6 times body length; neuromasts visible on larger tadpoles. Tadpoles less than 25.0 mm TL: dark black with prominent gold band at midbody, fins clear. Tadpoles greater than 26.0: body dark gray to black with abundant gold specks; gold body band becomes less prominent with size; dorsal half of tail muscle black; fins clear to hyaline with prominent black rim; large tadpoles and metamorphs have distinctly red eyes; to 100.0 mm TL/stage 36.

NATURAL HISTORY Females of *L. heckscheri* oviposit between April and August in swampy habitats with abundant aquatic vegetation in and near ponds, oxbows, slow streams, and rivers. Between 5000 and 8000 eggs (OD 1.5–2.9 mm) are laid as a film. Hatching begins after 3–15 days and tadpoles grow to 95.0–160.0 mm TL; they metamorphose in 1.3–2 years at 30.0–52.0 mm SVL. Larvae sometimes overwinter, and metamorphosis may occur en masse.

The behavior of these social tadpoles is striking but poorly known—sometimes they form immense, traveling schools. Individual younger tadpoles move by dart-and-coast swimming within polarized, midwater schools, while older tadpoles move about in enormous nonpolarized schools, at least during the day. The schools are surely made up of multiple clutches, based on numbers and size distributions of their members. The numbers of individuals coming to the surface to breathe sometimes produce a crackling noise that is audible at some distance.

RANGE River Frogs can be found on the Coastal Plain from central South Carolina to southern Alabama except for southern two-thirds of peninsular Florida.

CITATIONS *General:* Sanders 1984. *Development/Morphology*: Altig 1972, 1973, Altig and Pace 1974, A. H. Wright 1924. *Reproductive Biology:* Livezey and Wright 1947, A. H. Wright 1932. *Ecobehavior:* Altig and Christensen 1981, Altig and McDearman 1975, Goin and Ogren 1956, Punzo 1991, 1992, Simmons and Hardy 1959.

Lithobates okaloosae (Florida Bog Frog)   Fig. 110; PL 11E

IDENTIFICATION LTRF 2/3; A-2 gap greater than 400% of one side of LTR A-2; LTR P-1 without medial gap; LTR P-3 length about 20% of P-2 length; tail length about 2.2 times body length; fins low; dorsum russet brown; diagnostic large silvery patches of iridophores ventrolaterally; to 56.0 mm TL/stage 37.

NATURAL HISTORY The Florida Bog Frog breeds in boggy seepage areas associated with slow-flowing small streams that are highly acidic. From April to September females deposit a few hundred pigmented ova as an adherent film, often in pockets isolated from the main stream. Larvae spend the winter feeding, and metamorphs emerge the following spring at 21.0–23.0 mm SVL.

RANGE *L. okaloosae* has a small distribution at several sites in the Yellow River drainage of Okaloosa, Santa Rosa, and Walton cos., Florida.

CITATIONS *General:* Moler 1993. *Development/Morphology*: Moler 1985, 1993. *Ecobehavior:* Moler 1985.

Figure 110. ***Lithobates okaloosae.*** (A) mouthparts (modified from Moler 1985), (B) adherent egg film (Okaloosa Co., FL; RA), (C) tadpole (43.0 mm TL, stage 36; Okaloosa Co., FL; RWV). See also Plate 11E.

Lithobates onca (**Relict Leopard Frog**) and *L. fisheri* (**Vegas Valley Leopard Frog**)   **Fig. 111; PL 11F**

IDENTIFICATION LTRF 2/3; data on other regional species of the *L. pipiens* group presumably apply. *L. fisheri*: 69.0–85.0 mm TL/stage 36 (CAS 4034); body and fins uniform. *L. onca*: general coloration varies from uniformly medium brown to boldly mottled with pale flecks on the tail; white lip line absent or weak; to 52.0 mm TL/stage 33.

NATURAL HISTORY *L. fisheri* has not been seen in the Vegas Valley since 1942 and little is known of its biology. It probably bred in the spring, and metamorphs were 28.0–33.0 mm SVL. *L. onca* is poorly known and females probably oviposit in November–February.

RANGE *L. fisheri* was known only from the Vegas Valley, Clarke Co., Nevada but is now probably extinct in that area. Recently, Hekkala et al. (2011) provided ample evidence that populations previously thought to be *L. chiricahuensis* from the northwestern rim of the Mogollon Rim in Coconino Co., Arizona actually represent *L. fisheri*. *L. onca* occurs in the Virgin River drainage in Mohave Co., Arizona and downstream into the Black Canyon of the Colorado River below Hoover Dam in Clark Co., Nevada and adjacent Washington Co., Utah.

Figure 111. **Lithobates fisheri and L. onca.** (A) tadpole of *L. fisheri* (preserved, 84.0 mm TL, stage 40, Clark Co., NV; CAS 4034; RA), (B) tadpole of *L. onca* (45.0 mm TL, stage 36, Clark Co., NV; DLD), (C) oral apparatus of *L. onca* (Clark Co., NV; DLD), (D) egg clump (Clark Co., NV; DLD). See also Plate 11F.

CITATIONS *L. onca: General:* M. R. Jennings 1988. *Development/Morphology:* A. H. Wright and Wright 1949 (as *L. fisheri*). *Reproductive Biology:* Livezey and Wright 1947.

Lithobates palustris (Pickerel Frog)   Fig. 112; PL 11G

IDENTIFICATION LTRF 2/3; A-2 gap 85–100% of one side of LTR A-2; LTR P-1 with medial gap; LTR P-3 length 60–70% of P-2 length; marginal papillae larger and tail fin often taller and extends further forward than other sympatric

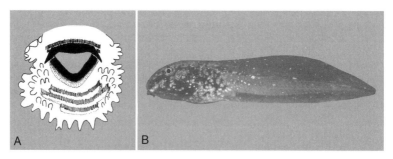

Figure 112. **Lithobates palustris.** (A) oral apparatus (PCU), (B) tadpole (28.0 mm TL, stage 25, Bladen Co., NC; RWV). See also Plate 11G.

members of the *R. pipiens* group; usually rather uniform brownish, sometimes with purplish tint; no prominent contrasting marks; to 77.0 mm TL/stage 36.

NATURAL HISTORY Pickerel Frogs occur in a variety of aquatic habitats, including floodplain swamps, trout streams, ponds and slow streams, bogs, and grassy meadows. Females oviposit from December to June and lay 1000–3000 brown and yellow ova (OD 1.6–2.0 mm, ED 3.6–5.0 mm, 2 jelly layers, inner one distinct) as a clump (75.0–100.0 mm diameter) attached to vegetation, often near the surface. The eggs hatch in 1.4–3.4 weeks, and tadpoles grow to 76.0 mm TL. Larvae metamorphose in 2.0–3.3 months at 19.0–27.0 mm SVL. This tadpole seems to be a more active swimmer and spends more time in deeper water than do tadpoles of leopard frogs.

RANGE The Pickerel Frog is known from populations in Wisconsin south through southeastern Texas to the Gulf Coast, east through the Maritime Provinces of southeastern Canada to the Atlantic Ocean, south to northern Georgia, and west across the southern states to Texas; it does not occur on the southeastern Coastal Plain or in the midwestern prairie regions.

CITATIONS *General:* Schaaf and Smith 1971. ***Development/Morphology***: Dodd 2004, Hinckley 1881, Trauth et al. 2004. ***Reproductive Biology:*** Collins and Wilbur 1979, Livezey and Wright 1947, Parmalee et al. 2002, Trauth et al. 1990, A. H. Wright 1910. ***Ecobehavior:*** Alford 1989, Formanowicz and Brodie 1982, Gosner and Black 1957b, Mills and Barnhart 1999, J. A. Moore 1939, 1940, Wilbur and Fauth 1990.

Lithobates pipiens (**Northern Leopard Frog**), *L. sphenocephalus* (**Southern Leopard Frog**), and *Lithobates* sp. nov.   **Fig. 113; PL 11H,J**

IDENTIFICATION LTRF 2/3; A-2 gap 50–70% of one side of LTR A-2; LTR P-1 with or without medial gap; LTR P-3 length about 90% of P-2 length; ventral marginal papillae medium; highly variable in all aspects; dorsum ranges from very dark to almost white, tail coloration ranges from uniform to extensively blotched; dorsum usually rather uniform or with subtle marbling; to 100.0 mm TL/stage 36.

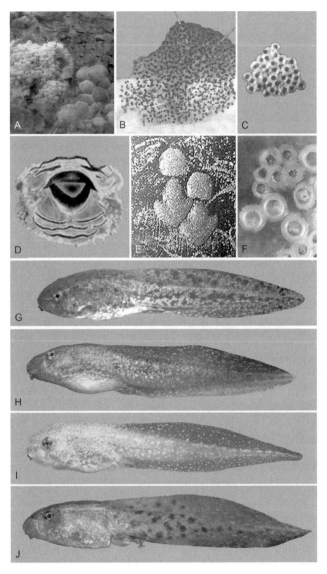

Figure 113. **Lithobates pipiens and L. sphenocephalus.** (A) In situ comparison of (left) the lobate egg clumps of *L. pipiens* and (right) the egg masses of *Ambystoma maculatum* with smooth surfaces (Scioto Co., OH; DMD), (B) egg clump of *L. sphenocephalus* (FL; BM), (C) partial clump of *L. sphenocephalus* showing details of construction (Oktibbeha Co., MS; RA), (D) oral apparatus (uncommon third upper tooth row present on left side) of *L. sphenocephalus* (RA), (E) cast of 2 successive feeding marks of *L. sphenocephalus* feeding on agar-based food (upper jaw entry at top, lower jaw entry at bottom, vertical streaks made by labial teeth; RA), (F) size comparisons of eggs of (upper, smaller) *L. sphenocephalus* and (lower, larger) *L. capito* (Okaloosa Co., FL; RA), (G) tadpole of *L. pipiens* (42.0 mm TL, stage 32, DuPage Co., IL; MR), (H–J) tadpoles of *L. sphenocephalus*: (H) 73.0 mm TL, stage 35, Alexander Co., IL; RA, (I) turbid water site; 52.0 mm TL, stage 32, Oktibbeha Co., MS; RA, (J) 73.0 mm TL, stage 37, Alachua Co., FL; RWV. See also Plate 11H, J.

NATURAL HISTORY Leopard Frogs oviposit every month of the year throughout their ranges but often in early spring. They deposit eggs in all kinds of nonflowing water and in slow streams, with communal ovipositions at some sites or in some seasons. Clutches include 300–6500 black and white ova (fall and spring clutches differ in clutch size and mass/egg; OD 1.0–2.2 mm, ED 3.4–7.0 mm, 2 jelly layers) laid as a clump (75.0–160.0 mm diameter), often 13.0–18.0 cm below water surface, usually attached to vegetation. Eggs are smaller and more closely spaced than those of members of *L. areolatus* group, and many females may oviposit in the same general area of a pond. Eggs hatch in 3–20 days at 6.2–6.8 mm TL. Tadpoles grow to 65.0–83.0 mm TL and metamorphose after 2.2–5.7 months at 15.0–45.0 mm SVL. There are no reports of overwintering. The tadpoles commonly enter shallow water, apparently to bask, and in pools choked with submerged vegetation, tadpoles often lie on the vegetation with their backs exposed.

RANGE *R. pipiens* occurs from eastern Alberta to Labrador southwest to Ohio and westward to northern Arizona. *R. sphenocephalus* is found from eastern Texas north and east to central Illinois and east to the Atlantic Ocean except for the central Appalachians. A recently recognized species that has not been named occurs in northern New Jersey, southeastern mainland of New York, and Staten Island.

CITATIONS *Development/Morphology*: Altig and Pace 1974, Dodd 2004, D. M. Hillis 1982, Korky 1978, Nichols 1937, N. J. Scott and Jennings 1985, A. C. Taylor and Kollros 1946, Thibaudeau and Altig 1988, Trauth et al. 2004. *Reproductive Biology:* Collins and Wilbur 1979, Corn and Livo 1989, M. Gilbert et al. 1994, Livezey and Wright 1947, McAlister 1962, McCallum et al. 2004, Merrell 1977, Parmalee et al. 2002, Petzing and Phillips 1999, Saenz et al. 2003, Steinke and Benson 1970, Trauth 1989, A. H. Wright 1910, 1932. *Ecobehavior:* Alford 1986, Alford and Crump 1982, Altig et al. 1975, Anholt et al. 2000, Awan and Smith 2007, Babbitt 2001, Babbitt and Tanner 1998, Borland and Rugh 1943, Bragg 1944a, 1948a, Casterlin and Reynolds 1979, Corn 1981, Cross and Hoyt 1972, Dayton et al. 2005, DeBenedictis 1974, Dupré and Petranka 1985, Eichle and Gray 1976, Faragher and Jaeger 1998, Gee and Waldick 1995, Golden et al. 2001, Gosner and Black 1957b, Hassinger 1970, F. S. Hendricks 1973, Lefcort 1996, McLaren 1965, Mills and Barnhart 1999, J. A. Moore 1939, 1940, 1949, Relyea and Werner 2000, Richards 1958, 1962, J. M. L. Richardson 2001, S. M. Rose 1958, 1960, Schiesari 2006, Schiesari et al. 2006, 2009, Schorr et al. 1990, D. E. Scott et al. 2007, Seale 1980, 1982, Smith-Gill et al. 1979, C. L. Taylor et al. 1995, Venesky et al. 2012, Volpe 1957c, B. Walters 1975, Ward and Sexton 1981, Werner 1992, Werner and Glennemeier 1999, West 1960, Wilbur 1987, Wilbur and Alford 1985, Wilbur et al. 1983, B. K. Williams et al. 2008, M. L. Wright et al. 1988.

Lithobates septentrionalis (Mink Frog)   **PL11I**

IDENTIFICATION LTRF 2/3; A-2 gap about 250% of one side of LTR A-2; LTR P-1 with medial gap; LTR P-3 length about 65% of P-2 length; tail length 1.6

times body length; dorsum greenish, tail sometimes with pinkish spots; tadpoles with large hind legs: third toe much shorter than first and webbing extends to toe tips; complex mottling of brown to black, yellow, and green without discrete contrasting dots on dorsum or contrasting black marks on posterior third of tail; to 99.0 mm TL/stage 36.

NATURAL HISTORY Mink Frogs breed along the margins of ponds and lakes, and where cold streams empty into them. Females oviposit from May to August when 500–4000 brown to black ova (OD 1.3–1.6 mm, ED 6.0–9.0 mm; 2 jelly layers) are laid as a clump (75.0–125.0 mm diameter) attached to vegetation and often in deep water. Eggs hatch in 5–13 days, and tadpoles grow to 99.0 mm TL in 12–15 months. Metamorphs measure 25.0–42.0 mm SVL, or 72.0 mm SVL if the tadpoles overwinter. These tadpoles also occur in slow-flowing water and often hide among bottom debris as those of *R. clamitans* do.

RANGE *L. septentrionalis* is a northern species that ranges throughout northeastern North America from northern Minnesota and Wisconsin north to James Bay and east to northern Maine.

CITATIONS *General:* Hedeen 1977. *Development/Morphology:* Gosner 1959. *Reproductive Biology:* Livezey and Wright 1947, Parmalee et al. 2002, A. H. Wright 1932. *Ecobehavior:* Camara and Buttrey 1961, Hedeen 1971, Hoff and Wassersug 1985, LeClair and Laurin 1996, McAlpine et al. 2001, Rondeau and Gee 2005, Wassersug 1989, Wassersug and Hoff 1985.

Lithobates sylvaticus (Wood Frog)   Fig. 114; PL 11K

IDENTIFICATION LTRF 3(2–3)/4; A-2 gap 60–100% of one side of LTR A-2; LTR P-1 with medial gap; Preston (1982) reported some specimens with 4 upper tooth rows; grossly uniform dark brown to black, sometimes with dark specks; abundant iridophore frosting visible at slight magnification; definite pale lip line; flanks sometimes bronzy or pinkish; to 66.0 mm TL/stage 36, usually smaller.

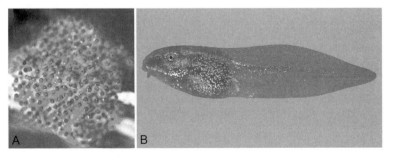

Figure 114. **Lithobates sylvaticus.** (A) egg clump (northwestern British Columbia; CCC), (B) tadpole (25.0 mm TL, stage 26, Polk Co., TN; RA). See also Plate 11K.

NATURAL HISTORY Wood Frogs are explosive breeders in ponds, pools, and slow streams in deciduous forests. Clutches consist of 350–5000 darkly pigmented ova (OD 1.7–3.1 mm, ED 5.0–17.3 mm, 2 jelly layers, inner one indistinct) laid early in the year (January–April) near the surface. Clumps vary between 60.0 and 100.0 mm, and many females frequently oviposit in restricted area of the pond. Eggs hatch in 15–35 days at 9.8–11.0 mm TL, and tadpoles grow to 38.0–48.0 mm TL. Metamorphosis is in 1.2–3.8 months and metamorphs measure 15.0–28.0 mm SVL. This darkly-colored tadpole swims actively in midwater more than most *Lithobates* tadpoles.

RANGE *L. sylvaticus* occurs throughout most of Canada and Alaska southeastward through the northeastern United States and at higher elevations in the Appalachian Mountains into southeastern Tennessee, the Carolinas, northern Georgia and Alabama. Isolated populations are known from northern Idaho, North Dakota, Wyoming, northern Colorado and Arkansas, and Missouri.

CITATIONS **General:** Martof 1970. **Development/Morphology:** Altig and Pace 1974, Dodd 2004, Gosner 1959, Hinckley 1881, Pollister and Moore 1937, Trauth et al. 2004. **Reproductive Biology:** Camp et al. 1990, Collins and Wilbur 1979, Corn and Livo 1989, M. S. Davis and Folkerts 1986, Livezey and Wright 1947, Meeks and Nagel 1973, Morgan 1891, Parmalee et al. 2002, Redmer 2002, Sutherland et al. 2009, Touchon et al. 2006, Trauth et al. 1990, Van Buskirk and Relyea 1998a, Waldman 1982a, A. H. Wright 1910. **Ecobehavior:** A. R. Anderson and Petranka 2003, Anholt et al. 2000, K. F. Baldwin and Calhoun 2002, Berven 1987, 1990, Berven and Boltz 2001, Berven and Chadra 1988, Bleakney 1958, Brodie and Formanowicz 1987, Cory and Manion 1953, Cupp 1980, DeBenedictis 1974, B. R. Eaton and Pazakowski 1999, Eidietis 2005, Forester and Lykens 1988, Formanowicz and Brodie 1982, Gascon and Planas 1986, Gee and Waldick 1995, Gervasi and Foufopoulos 2008, Gomez-Mestre et al. 2006, Hassinger 1970, Herreid and Kinney 1966, Hinckley 1882a, Holbrook and Petranka 2004, K. E. Johnson and Eidietis 2005, Lynn and Edelman 1936, Mathis et al. 2008, McClure et al. 2009, J. A. Moore 1939, 1940, 1949, Petranka 1985, Petranka et al. 1998, Petranka and Hayes 1998, Petranka and Kennedy 1999, Pinder and Friet 1994, Relyea 2001, 2003, Relyea and Auld 2004, 2005, Relyea and Werner 2000, J. M. L. Richardson 2001, Rondeau and Gee 2005, Rot-Nikcevic et al. 2005, Rowe and Dunson 1995, Schiesari 2006, Schiesari et al. 2006, 2009, Seale and Beckvar 1980, Seale and Wassersug 1979, Seigel 1983, Seymour 1995, Skelly 1994, 2004, Skelly et al. 2002, Smith-Gill et al. 1979, Sours and Petranka 2007, C. N. Taylor et al. 2004, Thiemann and Wassersug 2000, Trauth et al. 1989a, 1995a, Van Buskirk 2000, Van Buskirk and Relyea 1998a–b, Waldman 1984, 1989, Waldman and Ryan 1983, Walls and Williams 2001, Walston and Mullin 2007, B. Walters 1975, Wassersug 1989, T. B. Watkins and Vraspir 2006, Werner 1992, Werner and Glennemeier 1999, Wilbur 1972, 1976, 1977c.

Lithobates tarahumarae (Tarahumara Frog)   Fig. 115

IDENTIFICATION LTRF 4–6/3; A-2 gap about 73% of one side of LTR A-2; LTR P-1 with medial gap; dorsal fin nearly parallel with tail muscle; jaw sheaths

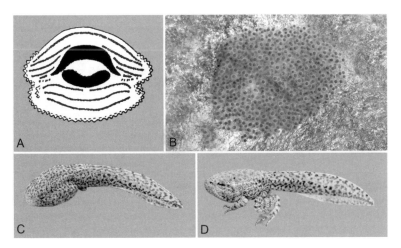

Figure 115. **Lithobates tarahumarae.** (A) mouthparts (modified from Zweifel 1955), (B) egg clump (Cochise Co., AZ; MS), (C) tadpole (87.0 mm TL, stage 36, culture originating from Sonora, Mexico; JCR), (D) metamorph (91.0 mm TL, stage 43, culture, Sonora, Mexico; JCR).

wide and robust; tail length about 1.9 times body length; dark brown to greenish brown with small spots on body; prominent and profuse spotting on both fins; belly white; to 101.0 mm TL/stage 36.

NATURAL HISTORY Females oviposit in the slow, deeper parts of streams and stream pools from April to August. Egg clumps measure 62.0 x 75.0 mm and contain 527–2200 black and white eggs (OD 2.0–2.2 mm, ED 3.7–5.0 mm, 2 jelly layers). Tadpoles are in the streams for 2.9–10 months or more and metamorphose at 21.0–40.0 mm SVL.

RANGE The Tarahumara Frog is restricted to stream habitats in a small area in the Pajarito, Tumucacori, and Santa Rita mountains of south-central Arizona and south into the Sierra Madre Occidental to central Sinaloa, Mexico.

CITATIONS *General:* Zweifel 1968b. *Development/Morphology:* Webb and Korky 1977. *Reproductive Biology:* Livezey and Wright 1947. *Ecobehavior:* Zweifel 1955.

Lithobates virgatipes (Carpenter Frog)   Fig. 116; PL 11L

IDENTIFICATION LTRF 2/3; A-2 gap greater than 400% of one side of LTR A-2; LTR P-1 with medial gap; LTR P-3 length 70–75% of P-2 length; generally dark maroon to brown with small black spots and specks; longitudinal row of dots on dorsal fin may fuse into a solid line; to 92.0 mm TL/stage 36.

NATURAL HISTORY Carpenter Frogs live in ponds, marshes, and swamps with low pH and lots of sphagnum moss. They breed from March to August and place between 200 and 600 widely spaced, pigmented ova (OD 1.4–1.8 mm, ED 3.8–6.9 mm, 1 jelly layer) in a clump (diameter 75.0–100.0 mm). The eggs hatch in 3–5 days, and tadpoles grow to 92.0 mm TL in 42.8–52 weeks. Metamorphs are

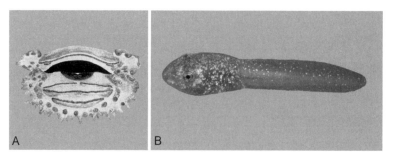

*Figure 116. **Lithobates virgatipes.*** (A) oral apparatus (modified from A. H. Wright 1932), (B) tadpole (41.0 mm TL, stage 27, Bladen Co., VA; RA). See also Plate 11L.

23.0–36.0 mm SVL. Metamorphs of *Lithobates grylio, L. okaloosae* and *L. virgatipes* are very similar.

RANGE Carpenter Frogs are known from scattered sites on the Coastal Plain from New Jersey to northern Florida.

CITATIONS *General:* Gosner and Black 1968. ***Development/Morphology***: Altig and Pace 1974, Gosner 1959. ***Reproductive Biology:*** Livezey and Wright 1947, A. H. Wright 1932. ***Ecobehavior:*** Gosner and Black 1957b.

Lithobates yavapaiensis (Lowland Leopard Frog)   PL 12A

IDENTIFICATION LTRF 2/3, short A-3 often present; A-2 gap about 70–90% of one side of LTR A-2; LTR P-1 with medial gap; LTR P-3 length about 75% of P-2 length; ventral marginal papillae medium; neuromasts obvious; fairly uniform gray dorsum; dorsolateral folds (after about stage 36) and neuromasts prominent; tail darker posteriorly, muscle with large iridophore patches; to 90.0 mm TL/stage 36.

NATURAL HISTORY The Lowland Leopard Frog is usually found in stream pools and side channels in arid foothills when and where flooding is not likely, and sometimes in permanent springs and cattle tanks in habitats ranging from desert scrub to piñon-juniper. Depending on habitat and locality, females oviposit in March–May, September–October, and sometimes December–January. Eggs can tolerate water temperatures of 11–29 °C and hatch in 14–18 days. Metamorphosis is in 3–4 months at 25.0–29.0 mm SVL. The progeny of late breeding pairs may overwinter.

RANGE Lowland Leopard Frogs occur from southeastern California across southern Arizona and adjacent Nevada to southwestern New Mexico and south into Sinaloa, Mexico, but may be extinct in southeastern California and at some sites in Arizona as a result of habitat destruction and severe floods and drought.

CITATIONS *General:* Platz 1988. ***Development/Morphology***: N. J. Scott and Jennings 1985. ***Reproductive Biology:*** Collins Lewis 1979, Livezey and Wright 1947, Sartorius and Rosen 2000. ***Ecobehavior:*** Collins and Lewis 1979, Rosen 2007.

Rana aurora (Northern Red-legged Frog) and *R. draytonii* (California Red-legged Frog)   Fig. 117; PL 12B

IDENTIFICATION LTRF either 3(2–3)/4 (common for *aurora*) or 2/3 (common for *draytonii*); A-2 gap greater than 100% of one side of LTR; P-1 with medial gap; if LTRF 2/3, LTR P-3 length about 80% of P-2 length; tail about 1.5 times body length; coloration varies from grossly appearing dark lavender-brown with extensive numbers of golden iridophores visible at slight magnification to generally golden with iridophores; throat often dark; belly often pinkish; faint dorso-lateral pale body stripes when small; to 70.0 mm TL/stage 36.

NATURAL HISTORY Red-legged Frogs breed in ponds, marshes, and slow streams from November to April in the south and February to March in the north (6–7 °C). Clutches vary from 500 to 1300 eggs in the north to 2000–5000 in the south and contain dark brown to black ova (OD 1.8–3.6 mm, ED 8.5–14.0 mm; with 3 rather indistinct jelly layers) that are laid in clumps (75.0–250.0 mm diameter). Jelly layers are loose and viscous and have a bluish tint when first laid. The clumps are attached to vertical vegetation and well submerged (north) or near the surface (south). Eggs hatch in 2–14 days or longer at 11.0–12.0 mm TL. Tadpoles metamorphose in 2.7–14 months at 17.0–33.0 mm SVL. Embryonic temperature tolerances range from 4 to 21 °C and salinity tolerance is less than 4.5%.

RANGE *R. aurora* occurs in the Sierra Nevada and Cascade ranges and western lowlands from southwestern British Columbia and Vancouver Island south to about Mill Creek, Mendicino Co., California, usually at less than 1000 m elevation. *R. draytonii* is known from the Coast and Sierra Nevada ranges south of Mill Creek, Mendicino Co., California, overlaps *R. aurora* between Mill Creek

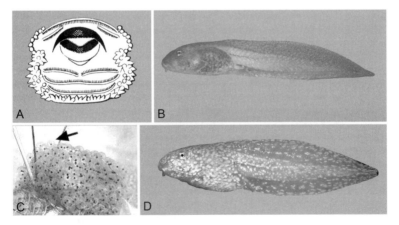

Figure 117. **Rana aurora and R. draytonii.** (A) oral apparatus (modified from Stebbins 1985), (B) tadpole of *R. aurora* (62.0 mm TL, stage 36, Lincoln Co., OR; RA), (C) egg clump of *R. aurora* (Lower Fraser Valley, British Columbia; arrow = differences in ovum diameter versus egg diameter; KO), (D) tadpole of *R. draytonii* (71.0 mm TL, stage 33, Santa Barbara Co., CA; RA). See also Plate 12B.

and Bad River, and extends south into northern Baja California. Records from Elko, Nye, and White Pine cos., Nevada are exotic.

CITATIONS *General:* Altig and Dumas 1972. *Development/Morphology:* Altig and Pace 1974, Gosner 1959. *Reproductive Biology:* Briggs 1987, Licht 1969b, 1971, 1974, Livezey and Wright 1945, 1947, Storm 1960. *Ecobehavior:* Blaustein et al. 1993, H. A. Brown 1975a, Calef 1973, Dickman 1968, D. Cook and Jennings 2001, Fellers et al. 2001, Pearl et al. 2003, Petranka 1985, Rathbun 1998, J. A. Wiens 1970, D. J. Wilson and Lecort 1993.

Rana boylii (Foothill Yellow-legged Frog)   Fig. 118; PL 12C

IDENTIFICATION LTRF 6(6)/6–7(1); A-2 gap about 15% of one side of LTR A-2; LTR P-1 with medial gap; tail length about 2.9 times body length; marginal papillae small; labial teeth arranged more densely than in tadpoles that inhabit nonflowing water; generally dark brown to olive with fine; paler mottling throughout; basal third of dorsal fin usually pigmented, remainder of fins clear with pale spotting; to 56.0 mm TL/stage 36.

NATURAL HISTORY Foothill Yellow-legged Frogs live in the slow parts of streams, usually below 1800 m where water depth is less than 50 cm and flow rate less than 21 cm/s. Females oviposit between March and June and deposit a clump (28.0–100.0 mm diameter) of 100–1400 ova that are mostly black with a small white vegetal pole (OD 1.0–2.5 mm, ED 2.9–5.4 mm, 3 jelly layers). Clumps are usually attached to downstream side of rocks in slow streams and accumulate silt. Eggs hatch in 4–36 days at 7.0–8.0 mm TL. Tadpoles metamorphose in 3–4 months at 20.0–30.0 mm SVL. This tadpole inhabits slower reaches of rivers and streams than the tadpoles of *Ascaphus.*

RANGE *R. boylii* is found in the Sierra Nevada and Coast ranges from northern Oregon south to southern California but is absent from much of the Central Valley.

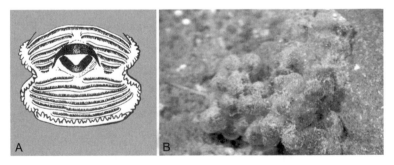

Figure 118. ***Rana boylii.*** (A) oral apparatus (modified from Stebbins 1985), (B) egg clump (Jackson Co., OR; WPL). See also Plate 12C.

CITATIONS *General:* Zweifel 1968c. *Development/Morphology*: Altig and Pace 1974, Gosner 1959. *Reproductive Biology:* Livezey and Wright 1947. *Ecobehavior:* Brattstrom 1962, Kupferberg 1995, 1997a–b, Rombough and Hayes 2005, Wiseman et al. 2005, Zweifel 1955.

Rana cascadae (Cascades Frog)   Fig. 119; PL 12D

IDENTIFICATION LTRF 3(2–3)/4; A-2 gap about 70% of one side of LTR A-2; LTR P-1 without medial gap; dorsal fin with medium arch; body slightly depressed but often appears pot-bellied; tail about 1.5 times body length; general impression is uniformly medium brown, sometimes with slight dark marks on the dorsal fin; venter white to silverish; to 60.0 mm TL/stage 36.

NATURAL HISTORY *R. cascadae* lives in lake margins, boggy sites, lakes, and along small streams through coniferous forests. They oviposit a clump of 300–800 dark brown ova (OD 2.2–2.8 mm, ED 11.8 mm, 3 jelly layers) between March and August. Females commonly oviposit communally, with 60–70 clumps/0.5 m². Eggs hatch at about 13.0 mm TL, and tadpoles grow for 2.0–3.2 months (sometimes overwinter) and metamorphose at 13.0–30.0 mm SVL. The eggs often have a green alga in the jelly layers. The tadpoles spend considerable time swimming lazily in midwater and commonly aggregate.

RANGE *R. cascadae* occurs in the Olympic Mountains of Washington and the Cascade Mountains throughout Washington, Oregon, and northern California.

CITATIONS *General:* Altig and Dumas 1971. *Development/Morphology*: Altig and Pace 1974, Slater 1939. *Reproductive Biology:* Briggs 1987, Livezey and Wright 1947. *Ecobehavior:* Blaustein and O'Hara 1987, Hews and Blaustein 1985, Kiesecker and Blaustein 1997, O'Hara and Blaustein 1981, J. A. Peterson and Blaustein 1992, Slater 1939, J. A. Wiens 1972, Wollmuth et al. 1987.

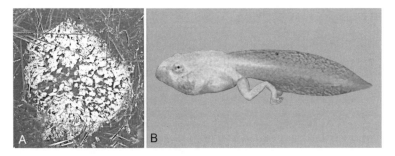

Figure 119. **Rana cascadae.** (A) egg clump floating and dispersed to resemble a film; frothy because fermentation bubbles caught in jelly and embryos are hatching; jelly starting to disintegrate (Linn Co., OR; JMR), (B) albino tadpole (53.0 mm TL, stage 37, Deschutes Co., OR; RA). See also Plate 12D.

Rana muscosa (Southern Mountain Yellow-legged Frog) and *R. sierrae* (Sierra Nevada Yellow-legged Frog)   Fig. 120; PL 12F

IDENTIFICATION LTRF 2(2)/4(1) to 4(2–4)/4(1); A-2 gap about 160% of one side of LTR A-2; LTR P-1 with medial gap; dorsal fin nearly parallel with tail muscle; tail length 1.8–2.4 times body length; dorsum dark with marbled or densely mottled pattern, few darker spots on tail muscle; to 75.0 mm TL/stage 36.

NATURAL HISTORY These frogs live in various slow, rocky streams in the south and in nonflowing, alpine water (lake edges, meadows, river banks) in the high Sierra Nevada. Females oviposit clumps (diameter 25.0–50.0 mm) in March–July that have 15–350 grayish tan (*R. sierrae*) creamy white ova (OD 1.8–2.3 mm, ED 6.4–7.9 mm, 3 jelly layers) that are attached to vegetation, sometimes under overhanging banks. Because of the outer jelly diameter, ova are more widely spaced than those of *R. boylii*. Eggs hatch in 2.6–3 weeks, and tadpoles may overwinter at least once at some sites. Metamorphose in 1–3 years at 22.0–27.0 mm SVL. The tadpoles often make daily movements from warmer, shallow areas during the day to deeper areas at night. At high elevations, the tadpoles tolerate hypoxia and low osmotic pressures for long periods during winter.

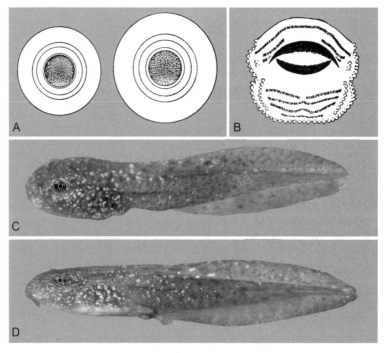

Figure 120. **Rana muscosa** and **R. sierrae.** (A) schematic comparison of individual eggs of (left) *R. boylii* and (right) *R. muscosa* (modified from Zweifel 1955), (B) mouthparts (modified from Zweifel 1955), (C–D) tadpoles of *R. muscosa*: (C) 26.0 mm TL, stage 31, (D) 62.0 mm TL, stage 38; San Bernadino Co., CA; NJS. See also Plate 12F.

RANGE   *R. muscosa* is found in the transverse ranges and adjacent mountains in southern California and in the south-central Sierra Nevada Range at 1370–3660 m. *R. sierrae* occurs in the central Sierra Nevada range of California and extreme western Nevada.

CITATIONS   *General:* Zweifel 1968a. *Development/Morphology:* Rachowicz 2002. *Reproductive Biology:* Livezey and Wright 1945, 1947. *Ecobehavior:* Kupferberg 1997a, K. L. Pope 1999, Vredenburg 2000, Zweifel 1955.

Rana pretiosa (Oregon Spotted Frog) and *R. luteiventris* (Columbia Spotted Frog)   Fig. 121; PL 12E

IDENTIFICATION   LTRF 2/3; A-2 gap about 200% of one side of LTR A-2; LTR P-1 with medial gap; LTR P-3 length about 65% of P-2 length; tail about 1.5 times

Figure 121. **Rana luteiventris and R. pretiosa.** (A) communal oviposition site of *R. pretiosa* in shallow water (WPL), (B) egg clump of *R. luteiventris* (Sublette Co., WY; DWZ), (C–E) *R. pretiosa*: (C) metamorph (32.0 mm SVL, stage 45), (D) hatchling (9.0 mm TL, stage 22), (E) metamorph (34 mm SVL, stage 43) (all from Thurston Co., WA; WPL), tadpoles of *R. luteiventris*: (F) 40.0 mm TL, stage 28, Okanogan Co., WA; WPL, (G) 64.0 mm TL, stage 34, Deer Lodge Co., MT; RA. See also Plate 12E.

body length; usually uniformly brown to grayish, usually with iridophores; fins usually uniform, dorsal fins usually not marked boldly, except in some far northern populations; to 105.0 mm TL/stage 36.

NATURAL HISTORY Spotted Frogs live around permanent water in ponds, lakes, springs, and slow-flowing streams. Females oviposit, often communally, in March–June in clumps (diameter 75.0–300.0 mm) that may contain 500–3000 black and cream to white ova (OD 1.8–3.0 mm, ED 5.0–21.0 mm, 1–2 jelly layers). Jelly often has a bluish tint when first laid but soon accumulates silt. Eggs hatch in 3.3–3.6 weeks at 7.0–10.0 mm TL, and tadpoles grow to 71.0 mm TL before they metamorphose in 1.0–6.9 months at 16.0–33.0 mm SVL. Tadpoles may overwinter at higher elevations. These tadpoles commonly hide in dense vegetation, often near the bank, and escape to deeper water when disturbed.

RANGE *R. luteiventris* is known from southeastern Alaska to northwestern Wyoming and west and south to central Nevada and Utah with a number of disjunct populations. *R. pretiosa* is found in disjunct populations in the central Cascade Mountains from southern British Columbia to northeastern California.

CITATIONS *General:* Turner and Dumas 1972. *Development/Morphology*: Altig and Pace 1974. *Reproductive Biology:* Hossack 2006, Leonard et al. 1997, Licht 1975b, Livezey and Wright 1945, 1947, Morris and Tanner 1969, Rombaugh and Hayes 2008. *Ecobehavior:* Briggs 1987, Carpenter 1953, Cuellar 1994, Licht 1969b, 1971, O'Hara and Blaustein 1988, Petranka 1985, Svihla 1935, H. B. Thompson 1913, Turner 1958, 1960.

Rhinophrynidae (Burrowing Toad)

This unusual frog in southern Texas is a suspension feeder in ephemeral sites; oral disc, labial teeth, and jaw sheaths absent; several pairs of nonmotile barbels around slitlike mouth; vent medial; eyes lateral; spiracles dual and lateral; dorsal fin low with pointed tip, originates near dorsal tail-body junction; body and especially snout depressed; nostrils transversely rounded, closer to mouth than eyes; toe pad absent, but large metatarsal spade evident in older tadpoles; different cohorts at same or different sites may be uniformly pale or uniformly black, if pale, posterior half of tail may have a dark suffusion; to 75.0 mm TL/stage 36; 1 genus, 1 species. *Rhinophrynus*: as for family.

Rhinophrynus dorsalis (Burrowing Toad)   Fig. 122

IDENTIFICATION As for family.

NATURAL HISTORY *R. dorsalis* is an explosive breeder that may oviposit at almost any time during the summer months (May–September) when rain has been sufficient to fill temporary ponds. Clutches include 2000–8000 lightly pigmented ova laid singly but in small bunches (6–12 in rapid succession) that sink and may adhere into groups on vegetation or sometimes break free and float at the surface. Tadpoles are suspension feeders that swim about lazily in midwater, often in large, disc- or column-shaped schools, and frequently break the surface

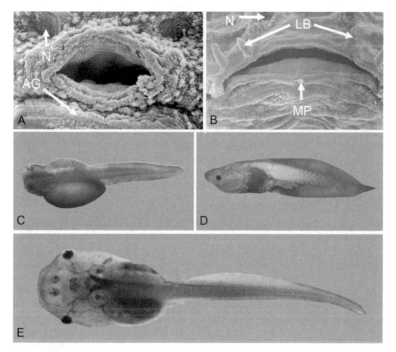

Figure 122. **Rhinophrynus dorsalis.** SEMs of oral apparatuses: (A) embryo (9.0 mm TL, stage 23; AG = transversely linear adhesive gland, N = naris), (B) tadpole (13.0 mm TL, stage 25, Costa Rica; LB = lateral barbels, MP = medial papilla, N = naris; modified from Altig and McDiarmid 1999; GT), (C) embryo (preserved, 9.0 mm TL, stage 21, Guanacaste, Costa Rica; RA), (D) living tadpole (66.0 mm TL, stage 34, Oaxaca, Mexico; RA), (E) tadpole at stage 34; PL.

to gulp air. They develop for 2–3 months and metamorphose at about 30.0 mm SVL. Mobile aggregations of metamorphs have been observed.

RANGE In the United States, the Burrowing Toad occurs only in two counties in southern Texas but is widely distributed in Mexico and Central America as far south as Costa Rica.

CITATIONS *General:* Fouquette 1969. *Development/Morphology*: Orton 1943, Starrett 1960, E. H. Taylor 1942, Thibaudeau and Altig 1988. *Reproductive Biology:* L. C. Stuart 1961. *Ecobehavior:* Altig et al. 1975, Foster and McDiarmid 1982, L. C. Stuart 1961.

Scaphiopodidae (Spadefoots)

Spadefoot benthic tadpoles are found throughout most of the continent, usually in ephemeral and often xeric sites; LTRF 4/4–6/6 with numerous rows with a medial gap, LTR A-1 much shorter than A-2; vent medial, all rows uniserial; small marginal papillae complete or with medial gap the width of 1–2 marginal papillae; submarginal papillae sparse; oral disc not emarginate, disc approximately circular at rest; jaw sheaths medium to wide, finely serrate in omnivores, larger serrations and cuspate in carnivores; vent medial; eyes dorsal; spiracle sinistral but low on side; dorsal fin low to medium with rounded tip, originates at tail-body junction; body globular to depressed; nostrils on a low hillock; digital pads absent; uniformly pale to uniformly dark, often with abundant iridophore pigment; to 100.0 mm TL; 2 genera, 7 species. *Scaphiopus*: only omnivore morphotypes; keratinized knob in roof of buccal cavity absent; tarsal spade of advanced tadpole sickle-shaped; to 40.0 mm TL; 3 species. *Spea*: omnivore and carnivore (Fig. 128A) morphotypes; keratinized knob in roof of buccal cavity present; tarsal spade of advanced tadpole wedge-shaped; to 100.0 mm TL; 4 species.

Scaphiopus couchii (Couch's Spadefoot)   Fig. 123; PL 12G

IDENTIFICATION LTRF 2/4–5/5, commonly 4/4; P-1 with medial gap; marginal papillae uniserial; uniform and bright brassy coloration with a freckled appearance; irregular dark marks common on posterior third of tail muscle; to 40.0 mm TL/stage 36.

NATURAL HISTORY Females of Couch's Spadefoot oviposit from April to September, usually explosively in temporary ponds formed following first heavy

Figure 123. **Scaphiopus couchii.** (A) wrapped rosary of eggs (Brewster Co., TX; less than 8, about 12, and about 18 h after oviposition; hatched embryos attached to jelly; arrows = hatching ports in turgid jelly; RA), (B) group of tadpoles feeding on bare soil within a bottom algal mat (rough areas to left and upper right; Lynn Co., TX; RA), (C) preserved tadpole (18.0 mm TL, stage 34, Lynn Co., TX; RA). See also Plate 12G.

summer rains in creosote desert, mesquite grassland, thorn forest, and short-grass prairie habitats. Clutches consist of 350–3000 pigmented ova (OD 1.4–1.6 mm, ED 5.0–7.0 mm) laid as wrapped rosaries; the single jelly layer permits wrapping the rosary around vegetation so that the clutch is sometimes mistaken for a mass. Eggs hatch in 1.0–1.5 days and tadpoles grow quickly, metamorphosing in 7–16 and up to 40 days at 7.0–13.0 mm SVL. These benthic tadpoles, like other scaphiopodids, often swim off the bottom with exaggerated beats of their tails, which are noticeably shorter than in most *Spea* tadpoles. They make only feeble attempts to escape.

RANGE *S. couchii* occurs in extreme southeastern California east to central Oklahoma and eastern Texas, south throughout much of northern Mexico including Baja California.

CITATIONS *General:* Wasserman 1970. *Development/Morphology*: Altig and Pace 1974, Bragg and Hayes 1963. *Reproductive Biology:* Livezey and Wright 1947, Strecker 1908b, Woodward 1987a, Zweifel 1968d. *Ecobehavior:* Bragg 1944c, 1945a, 1947a, 1957a, 1962c, 1964a, H. A. Brown 1969, Buchholz and Hayes 2000, Dayton and Jung 1999, Dayton and Wapo 2002, Hubbs and Armstrong 1961, Morey and Janes 1994, Morey and Reznick 2000, R. A. Newman 1987, 1988, 1994, 1998, Woodward 1982a–b, Woodward and Johnson 1985.

Scaphiopus holbrookii (Eastern Spadefoot) and *S. hurterii* (Hurter's Spadefoot)   Fig. 124; PL 12H

IDENTIFICATION LTRF 6/6 in *S. holbrookii*, commonly 4/4–5/5 in *S. hurterii*; grossly appears uniformly medium brown but abundant brassy iridophores are visible throughout at slight magnification; fins translucent and unmarked, paler at night; hourglass pattern on back of adults apparent on tadpoles by stage 36; 25.0–30.0 mm TL/stage 36.

NATURAL HISTORY These frogs breed most often in temporary pools; in southern parts of range may breed any time of year with heavy rains; oviposit February–September; 200–2500 brown and creamy white ova (OD 1.4–2.3 mm, ED 3.8–6.7 mm, 1 jelly layer) laid as wrapped rosaries with the tube about 25.0 mm in diameter and the entire structure up to about 130.0 mm long, wrapped around vegetation and thus may appear like a mass because of flimsy outer layer; hatch in 1.5–2.0 up to 15 days; grow to 28.0 mm TL; metamorphose in 1.7–9 weeks at 8.0–15.0 mm SVL. Egg morphology of *S. hurterii* may differ considerably; eggs sometimes laid as singles with 1 jelly layer (ED 6.7 mm). The benthic tadpoles spend considerable time in midwater but are rather weak swimmers. They sometimes metamorphose en masse and exit the pool in a narrow span of shoreline.

RANGE *S. holbrookii* is scattered across the landscape in sandy habitats from southern New England and southeastern Missouri south to the Gulf Coast except in the Appalachian Mountains. *S. hurterii* is found in the eastern half of Oklahoma and southeastern Missouri south to southern Texas and east through western Arkansas and northeast Louisiana, with a disjunct population in north-central Arkansas.

CITATIONS *General:* Wasserman 1968. *Development/Morphology*: Altig 1973, Altig and Pace 1974, Bragg 1944c, 1947a, 1964a, Bragg et al. 1964, Bresler

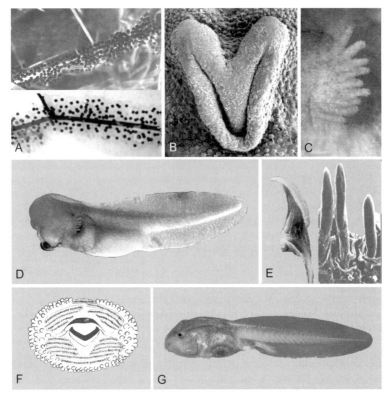

Figure 124. **Scaphiopus holbrookii** and **S. hurterii.** (A–G) *S. holbrookii*: (A) wrapped rosary of eggs: (top) in the field (MA; AMR) and (bottom) on a white background (Noxubee Co., MS; RA), (B) SEM of the adhesive gland (stage 19; RA), (C) gills (stage 19; RA), (D) hatchling (9.0 mm TL, stage 19; RA), (E) SEM of labial teeth (left: single tooth viewed from the side, modified from Altig and Pace 1974; right: teeth in situ viewed from the back, modified from Altig 1973), (F) oral apparatus (modified from Orton 1952), (G) tadpole (28.0 mm TL, stage 36, Alexander Co., IL; RA). See also Plate 12H.

and Bragg 1954, Dodd 2004, Feinberg and Hoffman 2004, Gosner 1959, Gosner and Black 1954, S. H. Hampton and Volpe 1963, Trauth et al. 2004. *Reproductive Biology:* Bragg 1944b, 1945b, 1951a, 1957c, 1964a, Komoroski et al. 1998, Livezey and Wright 1947, Richmond 1947, Trauth and Holt 1993, Trauth et al. 1990. *Ecobehavior:* Bragg 1945c, 1947a, 1948a, c, 1951b, 1954, 1957b, 1958a, 1959, 1961, 1962b, Dayton et al. 2005, Neill 1957, Semlitsch and Caldwell 1982, Wilbur 1987, Wilbur et al. 1983.

Spea bombifrons (Plains Spadefoot) Fig. 125; PL 12I

IDENTIFICATION LTRF variable but commonly 4/4; omnivorous and carnivorous morphotypes occur; uniformly pale or dark, often bronzy or silvery; to 80.0 mm TL/stage 36.

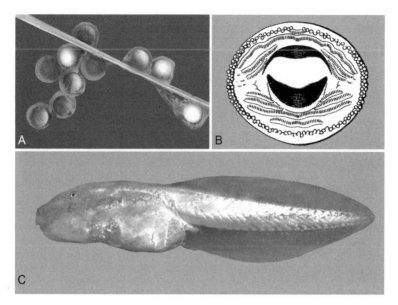

Figure 125. **Spea bombifrons.** (A) independent eggs (Lea Co., NM; RDJ), (B) oral apparatus (modified from Stebbins 1985), (C) carnivorous tadpole (44.0 mm TL, stage 34, Eddy Co., NM; RA). See also Plate 12I.

NATURAL HISTORY Females of the Plains Spadefoot oviposit from April to August, usually in temporary pools after seasonal rains in open grassland habitats. Clutches consist of dark brown and pale yellow ova (OD 1.5 mm, ED 3.0 mm, 3 jelly layers) that are laid as a wrapped rosary around vegetation or in groups of 10–200 or more (clutch up to 3800). Eggs hatch in 20 h to 2 days, and tadpoles develop quickly, often at temperatures of 39–40 °C, and metamorphose in 13–20 days up to 1.3 months.

RANGE *S. bombifrons* occurs in much of the central Great Plains from southern Alberta and southwestern Manitoba south to eastern Arizona and northwestern Texas and into Chihuahua, Mexico. Its range also extends east along the Missouri River through central Missouri and most of western Oklahoma. Disjunct populations occur in northwestern Arkansas, extreme south Texas, and south-central Colorado.

CITATIONS **Development/Morphology**: Bragg 1941a, 1956, Bragg and Bragg 1959, Bresler and Bragg 1954, Orton 1954, Potthoff and Lynch 1986, H. M. Smith 1934, 1956, Trauth et al. 2004, A. H. Trowbridge and Trowbridge 1937, M. S. Trowbridge 1941, 1942, Turner 1952, W. J. Voss 1961. **Reproductive Biology:** Klassen 1998, Livezey and Wright 1947, Mabry and Christiansen 1991, Parmalee et al. 2002, A. P. Russell and Bauer 1993, Trauth et al. 1990. **Ecobehavior:** Arendt 2008, Black 1970, 1974b, Bragg 1944c, 1945a, 1946a–b, 1947a, 1948a, 1960,

1965a–b, 1966, 1967, 1968, Bragg and Bragg 1957, 1958, Bragg and King 1961, H. A. Brown 1967, Hoyt 1960, Pfennig 1999, Pfennig and Murphy 2002, Storz 2004, Wassersug and Seibert 1975.

Spea hammondii (Western Spadefoot)   Fig. 126; PL 12J

IDENTIFICATION LTRF commonly 4/4; carnivorous morphotypes seemingly do not exist in the eastern part of the range; in the western portion, perhaps all individuals progress toward a carnivore-like morphotype with age; varies from uniformly pallid to silvery gray in turbid water to uniformly finely mottled in clear water; to 100.0 mm TL/stage 36.

NATURAL HISTORY Western Spadefoots breed in temporary rain pools, shallow lakes, and small streams in sandy soils in valley and foothill grassland, open chaparral, and pine-oak woodland habitats. Females oviposit from January to May and lay 300–500 eggs as a wrapped rosary. Eggs hatch in 3–4 days and larvae metamorphose in 8–10 weeks.

RANGE The Western Spadefoot is widely distributed in the Central Valley and southern coastal areas of California and south into northwestern Baja California.

CITATIONS *Development/Morphology*: Arendt and Hoang 2005, Turner 1952. *Reproductive Biology:* Livezey and Wright 1947. *Ecobehavior:* Arendt 2003, Ervin et al. 2005, Morey 1996, Morey and Reznick 2001, 2004.

Figure 126. **Spea hammondii.** (A) large, dense array of eggs from multiple females (San Diego Co., CA; EE), (B) tadpole (68.0 mm TL, stage 36, Madera Co., CA; RWH). See also Plate 12J.

Spea intermontana (Great Basin Spadefoot)   Fig. 127; PL 12K

IDENTIFICATION   LTRF 2/4–4/4, commonly 4/4; omnivore and carnivore morphotypes occur; to 70.0 mm TL/stage 36.

NATURAL HISTORY   Great Basin Spadefoots use temporary pools and other nonflowing water for breeding; rain is not always needed to stimulate reproduction. Oviposition occurs from April to July and the lightly pigmented ova (clutches equal 300–1000) are laid as wrapped rosaries around vegetation or in small groups (15.0–20.0 mm in diameter of 20–40 eggs). Eggs hatch in 2–4 days at 5.0–7.0 mm TL, and tadpoles grow to 30.0–70.0 mm TL. Metamorphosis occurs in 1–2 months at 14.0–25.0 mm SVL.

RANGE   *S. intermontana* occurs east of the Sierra Nevada-Cascade Range from southern British Columbia south to southern Nevada and east to central Wyoming and northeastern Colorado.

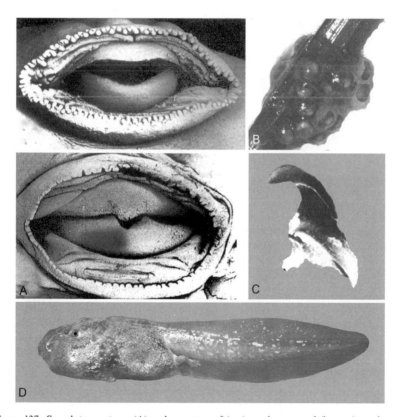

Figure 127. **Spea intermontana.** (A) oral apparatus of (top) omnivorous and (bottom) carnivorous tadpoles (modified from J. A. Hall 1998), (B) egg clump (WPL), (C) individual labial tooth viewed from the side (modified from Altig and Pace 1974), (D) carnivorous tadpole (46.0 mm TL, stage 34, Mono Co., CA; RA). See also Plate 12K.

CITATIONS *General:* J. A. Hall 1998. *Development/Morphology:* Altig and Pace 1974, H. A. Brown 1989b, J. A. Hall 1998, J. A. Hall et al. 1997, 2002, Orton 1954, V. M. Tanner 1939, Turner 1952. *Reproductive Biology:* Hovingh et al. 1985, Livezey and Wright 1947. *Ecobehavior:* Fox 2008.

Spea multiplicata (Mexican Spadefoot)   Fig. 128; PL 12L

IDENTIFICATION LTRF typically 4/4 but variable; omnivore and carnivore morphotypes occur; uniformly pale or dark, often bronzy or silvery; to 80.0 mm TL/stage 36.

NATURAL HISTORY The Mexican Spadefoot breeds in temporary sites and ponds often following summer rains in desert grassland, creosote and sagebrush desert, and piñon juniper and pine-oak woodlands. Eggs (300–500; OD 1.0–1.6 mm; ED 3.2–4.4 mm; 2–3 jelly layers, sometimes stalked) are laid in July and August in groups of 2–26, in water 18.0–25.0 cm deep, 2.0–6.0 cm below the surface, attached to vegetation. They hatch in 24 h to 6 days and metamorphose in 10–45 days as governed by temperature and food supplies. Carnivorous morphotypes occur.

RANGE *S. multiplicata* occurs in southeastern Utah and southern Colorado south through eastern Arizona, New Mexico, western Oklahoma and west Texas and south into north-central Mexico.

CITATIONS *Development/Morphology:* Bragg 1941a, Orton 1954, E. H. Taylor 1942, Turner 1952. *Reproductive Biology:* Livezey and Wright 1947. *Ecobehavior:* Arendt 2008, Bragg 1944c, 1947a, 1948a, H. A. Brown 1967, 1969, Buchholz and Hayes 2000, Pfennig 1990, 1999, Pfennig and Murphy 2002, Pfennig et al. 1991, Storz 2004, Storz, and Travis 2007, Woodward 1982a–b.

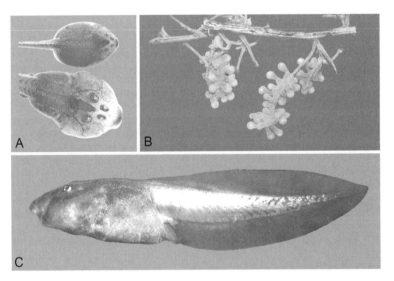

Figure 128. **Spea multiplicata.** (A) morphotypes of (upper) normal or omnivorous and (lower) carnivorous tadpoles (DWP), (B) two groups of single eggs with elongate outer jelly layers; opaqueness caused by silt accumulations (Grant Co., NM; RA), (C) carnivorous tadpole (62.0 mm TL, stage 33, Eddy Co., NM; RA). See also Plate 12L.

GLOSSARY

Terms are defined relative to the subjects, taxa, and geography of this book.

adherent, adhesive Firm association of one object to another by stickiness or some form of active attachment, not easily pushed apart; see coherent/cohesive.

adhesive gland Transitory, sticky gland behind the presumptive mouth of a frog embryo and hatchling; stabilizes the individual before full swimming capabilities develop.

adpressed Describes the positioning of the forelimbs backward and the hind limbs forward along the body of a salamander so that the digit tips of the hands and feet face each other; the resulting distance between toe tips (or amount of overlap, both often presented as the number of costal grooves) is a relative measure of leg-body proportions (e.g., short body + long limbs results in toe tips meeting or overlapping).

adult An individual that has attained sexual maturity regardless of metamorphic status; see immature, larva, metamorph.

aestivation Condition of reduced physiological and locomotor activity in response to unfavorably hot and dry conditions; see hibernation.

***Anaxyrus americanus* group** A subset of the genus *Anaxyrus* that includes *A. americanus, A. baxteri, A. fowleri, A. hemiophrys, A. houstonensis, A. terrestris,* and *A. woodhousii.*

***Anaxyrus boreas* group** A subset of the genus *Anaxyrus* that includes *A. boreas* (distinct lineages known), *A. canorus, A. exsul,* and *A. nelsoni.*

animal pole Dorsal, commonly pigmented hemisphere of an ovum with a lower concentration of yolk relative to the vegetal pole.

anterior Anatomical/morphological descriptor, refers to the front, front end, or toward the front end of an animal or structure relative to a stated landmark; see dorsal, lateral, medial, posterior, ventral.

array Specific ovipositional mode of suspended eggs each with an independent attachment point, pendent or not; see clump, cluster, film, mass, rosary, single, string.

atrophy Reduction of a structure by cellular reorganization and breakdown relative to a comparative norm, usually defined ontogenetically (e.g., a tadpole's tail atrophies at metamorphosis).

balancer Ephemeral, fleshy, filiform projection from the side of the head of hatchling ambystomatids and salamandrids.

band, banded Contrasting, transverse color markings; see bicolored, color, coloration, stripe.

barbel Fleshy papilla(e) of various lengths around the margins of the mouths of pipid and rhinophrynid tadpoles; does not refer to marginal papillae.

benthic Living on the bottom or characteristic of a larva modified for inhabiting the bottom of a water body; salamander: robust body, tall tail fins that extend well onto body, large gills (e.g., *Ambystoma*); tadpole: body depressed, eyes dorsal, fins low and often with rounded tip (e.g., *Ascaphus*, bufonids, some hylids, ranids, scaphiopodids); see morphotype, nektonic, rheophilous, suctorial, suspension feeder.

bicolored Usually refers to longitudinal pattern on side of tail muscle—dark dorsal part and light ventral part; see band, color, coloration, stripe.

biserial Structures, such as labial teeth and eggs in a tube, that occur in 2 rows or files; see multiserial, uniserial.

body That part of a larva minus the tail; in salamander larvae, the body ends at posterior end of the vent; in tadpoles, the body ends at the body terminus; see body length, tail.

body length Salamander larva: tip of snout to posterior extent of vent; see snout-vent length (SVL). Tadpole: tip of snout to the body terminus.

body terminus Preferred definition of the posterior terminus of the body of a tadpole because it is the most accurate and consistent landmark among stages and species of different shapes—junction of the posterior body wall with the axis of the tail myotomes; see body, body length, tail length.

buccal cavity Poorly delimited cavity immediately inside the mouth and bounded at the front by the upper and lower jaws, often referred to inaccurately as the mouth.

capsular chamber A chamber immediately outside the vitelline membrane caused by the dissolution of the first jelly layer soon after oviposition of salamander eggs; thus each ovum rests slightly off center at the bottom of the next jelly layer, and rotates with animal pole upward if the group is inverted.

carnivore, carnivorous Describes an organism that eats animal tissue.

cavernicolous Living in and adapted (e.g., pigmentation reduced or absent, eyes small or absent, limbs spindly, head depressed) for existence in caves; see epigean; this category (see hypogean) does not include epigean salamanders that seek refuge (i.e., suitable temperature and moisture) in caves.

chromatophore Pigment-containing cell; see color, coloration, iridophore, melanophore, pattern.

clump Specific ovipositional mode involving a group of adherent or coherent eggs without a common, surrounding surface; at least, a freshly oviposited open clump has interstices among the eggs like a stack of marbles; a melded clump lacks interstices, and the jellies meld together completely at points of contact. The jellies of older clumps may sag to fill the interstices but a lobate surface remains evident; see array, cluster, film, mass, rosary, single, string.

cluster Specific ovipositional mode involving a small group of eggs that may be suspended and pendent or not, with the attachments of all eggs at the same or adjacent points; each egg is oviposited singly, terrestrial; see array, clump, film, mass, rosary, single, string.

clutch Total egg number (= complement) ovulated by a female at one ovipositional event whether all eggs are laid in one or multiple bouts; may be more than one ovipositional event per breeding season and usually more than one per lifetime.

coherent, cohesive Loose association of one object to another, usually between 2 wet surfaces, can be easily pushed apart, not sticky or attached; see adherent.

color Visual spectral impression based on selective reflectance of wavelengths of light; see coloration, pattern.

coloration Total visual impression gained from the interaction of color and pattern.

compressed Object or structure that is higher than wide (i.e., cross-section is an oval with the long axis oriented vertically); usually describes cross-sectional shapes of bodies of caudate and anuran larvae; see depressed.

costal fold Slight bulge of body musculature between 2 successive costal grooves of a salamander.

costal groove Vertical, parallel depressions along the lateral body surface of a salamander larva at the position of a rib and myoseptum; separates successive costal folds.

cross-section (transverse section) Imaginary or real view of an organism or structure cut perpendicular to its long dimension and viewed on end (e.g., a cylinder has a circular cross-section).

cusps Projections on the working surface of a labial tooth.

cuspate Describes the cutting edges of jaw sheaths that have prominent convexities in the upper sheath that are sometimes matched by a concavity in the lower sheath (e.g., carnivorous *Spea* tadpoles) such that the cutting edges do not follow a typical uniform arc; see cusps, jaw sheaths, labial tooth.

depressed Object or structure that is wider than high (i.e., cross-section is an oval with the long axis oriented horizontally); usually describes cross-sectional shapes of bodies of caudate and anuran larvae; see compressed.

dextral To the right, refers to the aperture of the vent tube of a tadpole that opens to the right of the plane of the ventral fin; see medial.

disjunct Describes a part of a geographical range that is separated and isolated, perhaps relictual, from the main part of the range; see endemic, relict.

dorsal, dorsum Anatomical/morphological descriptor, refers to the top or toward the top or back of an animal or structure relative to a stated landmark (e.g., the eyes are dorsal to the jaws); relative to eye position, describes eyes positioned so they are not included in the dorsal silhouette; can be combined with other descriptors: dorsolateral; see anterior, lateral, medial, posterior, ventral.

ED Abbreviation for egg diameter—diameter of outermost layer of egg jelly; see OD, ovum.

eft Metamorphosed, immature, terrestrial stage between larva and sexually mature adult of *Notophthalmus*.

egg Collective term for that which is oviposited, including the ovum/ova (female gamete, from ovary) and jelly layers (from oviduct); see ovum.

emarginate Describes actual lateral indentations (i.e., not folds or tucks) in the margin of the oral disc.

embryo An individual in about stages 1–24, after fertilization and prior to hatching; full tadpole or larval morphology not yet present; see hatchling, ovum.

endemic Describes the geographical distribution of a taxon restricted to a specified region (e.g., sirenids are endemic to the southeastern United States and adjacent Mexico); see disjunct, relict.

endotroph Amphibian developmental mode in which all energy used by the embryo to attain a free-living stage is obtained from vitellogenic yolk; see exotroph.

ephemeral Of short duration or temporary, usually refers to small bodies of water.

epigean Living above ground or describes the morphology of a larva adapted (i.e., normal body proportions and pigmentation) for living above ground (i.e., not in a cave or aquifer) even if beneath surface objects or in burrows in the ground; having characteristics suited to occupy these habitats; see hypogean.

exotic Not native, introduced from some extralimital area.

exotroph Amphibian developmental mode that includes a free-living larva that consumes various kinds of foodstuffs not associated with vitellogenesis or with either parent; see endotroph.

fertilization Joining of egg and sperm; occurs externally in cryptobranchid salamanders and internally via a spermatophore in all others; occurs internally in *Ascaphus* frogs by copulation via the male's tail and externally in most other frogs.

filiform A structure (e.g., gill fimbriae) that has a long axis and an approximately circular cross-section (i.e., a long, small-diameter cylinder); see lamellar.

film Specific ovipositional mode involving eggs of some summer breeding frogs arranged in a single layer that floats at the surface; ovipositing female positions herself so that the extruded eggs fall onto the surface (all other ovipositional modes involve submerged extrusion); see array, clump, cluster, mass, rosary, single, string.

fimbria, fimbriae Well-vascularized parts of gills, usually filiform or lamellar, that provide large surface areas for gas exchange; see ramus.

fin Fleshy, unsupported (i.e., no bony rays as in fishes) vanes on the top and bottom midline of the tail of larval or larviform amphibians, provide locomotor force when tail is undulated.

flagellum A distal part of the tail tip of tadpoles that is greatly reduced in height, is often nonpigmented or pigmented differently from the remainder of the tail, and can sometimes be undulated independently of the remainder of the tail.

froglet See metamorph.

gill External (hatchling frogs and all salamanders) or internal (tadpoles) respiratory structure that provides large surface areas for gas exchange; external gills in salamanders usually composed of a branched (rarely) or unbranched (commonly) ramus with fimbriae of various shapes.

gill arch Curved pharyngeal elements that support the gills of salamander larvae, have food-gathering projections (gill rakers) on medial surface.

gill raker Linear series of knobby or pyramidal projections on the medial side of the gill arches of salamander larvae; rows are angled toward the neighboring arch and positioned such that the rakers on adjacent arches interdigitate; used for separating food items from water entering the mouth during suction feeding.

gill slit The space or opening between successive gill arches through which respiratory and feeding water is expelled in salamander larvae; all slits are covered by the gular fold.

group General term for an assemblage of eggs when the identity or ovipositional mode is undetermined; see array, clump, film, single, string.

gular fold Flap of skin across the throat of larval salamanders that covers the gill arches.

hatching port The hole through which an embryo hatches from the egg jelly; stands open after the embryos exit if the jelly is turgid enough.

hatchling Individual that has hatched (= escaped from the egg jellies) but not developed to a free-swimming, foraging larva; usually stages 21–24 in anurans; hatchling salamanders have external gills, partially developed legs, and sometimes a balancer; see adult, immature, metamorph, tadpole.

herbivore Refers to an animal that eats plant material; tadpoles typically considered to be herbivores but not true in the strictest sense; see carnivore.

hibernation Condition of reduced physiological and locomotor activity in response to unfavorably cold conditions; see aestivation.

hypogean Below ground or describes the morphology of a larva adapted (e.g., reduced pigmentation, eyes small or absent, snout depressed, limbs spindly) for living below ground (i.e., in a cave or aquifer); does not refer to being beneath objects or in burrows; see cavernicolous, epigean.

immature (juvenile) Postmetamorphic individual prior to attainment of sexual maturity; also a paedomorphic salamander prior to attainment of sexual maturity; see adult, hatchling, larva, metamorph, tadpole.

inquiline Describes an organism living in or associated with the abode of another species, used here mostly in reference to human abodes.

iridophore Chromatophore containing reflective platelets that produce a white, silver, or golden color; see melanophore.

jaw sheath Salamander: a keratinized sheath that fits over the lower (ambystomatids) or both bony jaws (sirenids) and provides a cutting edge; tadpole: keratinized sheath that covers the front and back surfaces of the jaw cartilages and provides cutting surfaces; see cuspate, labial tooth, labium, marginal papillae, mouthparts, oral apparatus, oral disc, submarginal papillae, tooth row.

jelly layer Mucopolysaccharides (complex carbohydrates) and a small amount of protein of oviductal origin that surround an ovum in various layers; does not include the vitelline membrane of ovarian origin; see egg, ovum.

juvenile See immature.

keratinized Formed of keratin (e.g., fingernails and hair), a complex protein, as in jaw sheaths and labial teeth; provides strength to working surfaces, not ossified.

labial fold Fleshy pleated folds between the upper and lower jaws of larval larviform salamanders that modify the mouth size and shape for suction feeding; see mandibular groove.

labial tooth Keratinized tooth of a tadpole in densely arranged transverse rows on labial tooth ridges on the face of the upper (anterior) and lower (posterior) labia; distal end (i.e., working surface) often with cusps; see labial tooth row formula, mouthparts, oral disc, tooth row, tooth series.

labial tooth row formula (LTRF) A fractional construction that denotes the number of tooth rows on the upper (numerator) and lower (denominator) labium of a tadpole; numbers in parentheses denote rows with medial gap; 2(2)/4(1) indicates a tadpole with 2 upper rows with a gap in the second one and 4 rows on the lower labium with a gap in the first one; rows are counted from the periphery toward the mouth on the upper labium and from the mouth distally on the lower labium.

labium, labia Upper (anterior) and lower (posterior) areas of the oral disc of most tadpoles; see jaw sheath, labial tooth, marginal papillae, mouthparts, oral apparatus, submarginal papillae.

lamellar An elongate structure (e.g., gill fimbria) that has a width greater than its thickness (e.g., a blade of grass); see filiform.

larva An amphibian between posthatchling and metamorphic stages, usually sexually immature; see tadpole; larval frogs are called tadpoles; see larviform.

larviform A sexually mature amphibian with a larval morphology; see paedomorph, pedotype.

lateral Anatomical/morphological descriptor, refers to the side or toward the side of an animal or structure relative to a stated landmark; relative to eye position, describes eyes that are positioned laterally so that some part is included in the dorsal silhouette; can be combined with other terms: dorsolateral, ventrolateral; see anterior, medial, posterior, ventral.

layer See jelly layer.

lateral line organ/system See neuromast.

limb bud Developing limb anlage up to the point of full morphological (not size) differentiation (i.e., from a small finger-shaped projection to a fully differentiation limb); the pattern of developmental progression of limbs in tadpoles is used in staging.

***Lithobates areolatus* group** A subset of the *L. pipiens* complex that includes *L. areolatus, L. capito*, and *L. sevosus*.

***Lithobates catesbeianus* group** A subset of the genus *Lithobates* that includes *L. catesbeianus, L. clamitans, L. grylio, L. heckscheri, L. okaloosae, L. septentrionalis*, and *L. virgatipes*.

***Lithobates pipiens* complex** A subset of the genus *Lithobates* that includes all species in the *L. pipiens* and *L. areolatus* groups.

***Lithobates pipiens* group** A subset of the *L. pipiens* complex that includes *L. berlandieri, L. blairi, L. chiricahuensis, L. fisheri, L. onca, L. palustris, L. pipiens, L. sphenocephalus, L. yavapaiensis*, and *Lithobates* sp. nov. (recognized but not named; C. E. Newman et al. 2012).

longitudinal, longitudinal axis Imaginary axis extending from the tip of the snout to the plane of the body terminus; used in describing locations of other structures (e.g., spiracle).

LTR Abbreviation for labial tooth row; see LTRF.

LTRF Abbreviation for labial tooth row formula.

mandibular groove Groove on the medial side of the dentary (= lower jaw) of a larval salamander into which the labial folds of larviform salamanders appear to insert.

marginal papillae Fleshy protuberances along the margin of the oral disc of a tadpole; see barbel, papilla, submarginal papillae.

mass Specific ovipositional mode involving a volume of eggs formed by the indistinguishable melding of adjacent outer jelly layers; mass appears to be ova with individual jelly layers embedded in a matrix (e.g., marbles embedded in jelly) with no interstices remaining between neighboring eggs; see array, clump, cluster, film, rosary, single, string.

matrix Used in reference to egg clutch terminology, a voluminous jelly in which ova, usually with at least one layer encompassing each, appear to be embedded; actually formed by imperceptible fusion of outer egg jellies; in ambystomatid masses, the outer jelly is wide so that peripheral ova are distant from the mass edge (i.e., an ova-free peripheral zone); in hylids the outer jelly is less voluminous and ova are positioned near the mass margin; see mass.

mature See adult.

medial Anatomical/morphological descriptor referring to the midline or toward the midline of an animal or structure relative to a stated landmark; opposite of lateral; see anterior, lateral, posterior, ventral.

medial gap A natural break in a tooth row centered on the midline of the oral disc.

melanin, melanic Brown or black pigment derived from tyrosine deposited in a melanophore; see chromatophore, iridophore.

melanophore Chromatophore that contains melanin; see iridophore.

melded clump See clump.

metamorph The product of metamorphosis, a froglet or immature salamander; usually short developmental period at the end of larval development; see adult, immature, juvenile, larva, tadpole.

metamorphosis The complicated process by which a larva is transformed into a juvenile; see embryo, hatchling, immature, adult.

midventral The centerline of a body as viewed ventrally; used to describe the position of the spiracle in microhylids; see sinistral.

morphotype Morphological type, a generalized descriptor of the conglomerate suite of morphological characters of a given individual; see benthic, nektonic, rheophilous, suctorial, suspension feeder.

mouth Refers only to the aperture from the outside into the buccal cavity; incorrectly applied to the oral apparatus of tadpoles, which is external to the mouth; see mouthparts.

mouthparts Collective term for all soft and keratinized oral structures of tadpoles; see jaw sheath, labial teeth, marginal papillae, mouth, oral apparatus, oral disc, submarginal papillae.

multiserial Structures, such as labial teeth or eggs in a tube, that occur in multiple rows or files; see biserial, uniserial.

myotome Strictly speaking, the portion of undifferentiated tissue in each body segment that mostly gives rise to muscle tissue, bounded by a myosepta (connective tissue sheets) anteriorly and posteriorly; by extension, often refers to the muscle tissue per se; see costal fold, costal groove.

nektonic Inhabiting the midwater regions of a water body or describes the morphology of a tadpole adapted for spending most time in midwater—compressed body, lateral eyes, medium to tall tail fins, usually with pointed tip (e.g., most hylids); see benthic, morphotype, rheophilous, suctorial, suspension-feeder.

neuromast (lateral line organ or pore) Integumentary sensory organ for detecting direction and force of changes in water currents; except in pipid frogs, these occur only in larvae and larviform adults; collectively termed the lateral line system.

neustonic See suspension feeder.

nostril External opening of narial or olfactory canal somewhere on the snout of an organism.

OD Abbreviation for ovum diameter; see ED, egg, ovum.

ontogeny, ontogenetic One lifetime, that period from fertilization to death of one individual and all the processes that occur during that time.

open clump See clump.

operculum Fleshy fold that grows backward over the gills of hatchlings and fuses with the surrounding epidermis to form the spiracle in tadpoles; see gular fold.

oral apparatus (mouthparts) Collective term for oral disc, jaw sheaths, and labial tooth rows or whichever of these are present in tadpoles; see labium, marginal papillae, submarginal papillae, tooth ridge.

oral disc Collective term for combined upper (anterior) and lower (posterior) labia of the oral apparatus of tadpoles; see jaw sheath, labial teeth, marginal papillae, mouth, mouthparts, oral apparatus, papillae, submarginal papillae, tooth ridge, tooth row.

oral flap Fleshy hemispherical flaps pendent over the mouth of microhylid tadpoles.

oviduct, oviductal The female reproductive canal through which ova move from the ovary to the outside; parts of the oviduct secrete the jelly layers around the ova and the foam in *Leptodactylus*).

oviposit, oviposition Process of depositing eggs, usually, except in *Ascaphus*, fertilized as they exit the female; a male is usually present during oviposition in frogs and usually not present during oviposition in salamanders; see fertilization, ovulation.

ovipositional bout One of several periods of oviposition during the deposition of an entire clutch in hylids and microhylids; see oviposit, ovipositional mode.

ovipositional mode The various structural arrangements of groups of oviposited eggs; see array, clump, film, mass, rosary, single, string.

ovulate, ovulation Process of releasing ova from the ovary, in contrast to oviposition.

ovum, ova Female gamete (from ovary) only, not including the jelly layers (from oviduct) that surround it; not synonymous with egg; see vitelline membrane.

paedomorph, paedomorphic Sexually mature but permanently larviform salamanders (i.e., *Amphiuma, Cryptobranchus,* some *Eurycea, Gyrinophilus* and *Dicamptodon, Necturus, Pseudobranchus,* and *Siren*) resulting from incomplete metamorphosis that varies in degree among taxa; most will not respond fully or at all to extraneous hormone treatments; see pedotype.

papilla, papillae Fleshy projection of various heights, most often with a circular cross-section; see barbel, filiform, lamellar, marginal papillae, submarginal papillae.

pattern The design or motif (e.g., banded, spotted, or striped) of coloration regardless of color; see band, bicolored, coloration, stripe.

pedicel The suspensory structure formed from one or more layers of the jelly coats around an egg; see pendent, suspended.

pedotype, pedotypy, pedotypic A sexually mature but larviform salamander (e.g., several *Ambystoma* spp.) caused by allometry between body and gonadal growth; will eventually metamorphose if conditions that promote a larviform existence diminish; will respond to extraneous hormone treatments; see paedomorph.

pendent Hanging down, not simply suspended; in reference to jellies of suspended eggs that droop so that the ovum is not centered within the outer jelly layers; see array, clump, film, mass, rosary, single, string.

Phytotelmon, phytotelmons A small water catchment in a plant, leaf axils of bromeliads, or a tree hole.

plasticity Having sufficient latitude in developmental genetics to produce variable phenotypes under different environmental conditions.

posterior Anatomical/morphological descriptor that refers to the rear, rear end, or toward the rear of an animal or structure relative to a stated landmark; see anterior, dorsal, lateral, medial, ventral.

postmetamorphic An organism that has completed metamorphosis, usually refers to juveniles.

***Pseudacris triseriata* group** A subset of the genus *Pseudacris* that includes *P. brachyphona, P. brimleyi, P. clarkii, P. feriarum, P. fouquettei, P. kalmi, P. maculata, P. nigrita*, and *P. triseriata*. A further subset of these taxa are often are grouped together (*P. feriarum, P. fouquettei, P. kalmi, P. maculata*, and *P. triseriata*) because their tadpoles usually cannot be distinguished.

raft A group of eggs laid as a film that results from a single ovipositional bout; a number of rafts make up the entire clutch; occurs in microhylids and some hylids; ranids that oviposit as a film most often do so in one bout and thus do not form rafts.

ramus, rami Central supporting strut, in this case the central shaft of a gill; see fimbria, gill raker.

relict Describes a relatively small geographical range that presumably is a remnant of a larger former range (e.g., *Anaxyrus exsul* restricted to a couple of springs in Inyo Co., California by desert formation); see disjunct, endemic.

replacement tooth Any labial tooth in a tooth series other than the presently erupted tooth; see tooth ridge, tooth row.

rheophilous Adapted for and living in flowing water; see suctorial.

rosary Specific ovipositional mode involving a linear series of widely-spaced eggs laid in a tube with the inter-egg part of the tube constricted and usually twisted, a modified string; those of some scaphiopodids, twined around vegetation, are termed "wrapped rosaries"; see array, clump, cluster, film, mass, single, string.

ruderal Describes the habits of a "weedy" species, or one that invades or inhabits disturbed sites; usually disappears as the habitat undergoes further succession; see inquiline.

SEM Abbreviation for scanning electron microscopy.

sexual maturity See adult.

single Specific ovipositional mode involving eggs oviposited singly although may be randomly grouped if female does not move or moves rather little between egg depositions; see array, clump, cluster, film, mass, rosary, string.

sinistral On the left, usually refers to the spiracle of a tadpole on the left side somewhere near the body axis; see midventral.

snout-vent length(SVL) The body measurement from the tip of the snout to the posterior extent of the vent in salamander larvae; not used in tadpoles (see body length) before about stage 42 because a consistently identifiable landmark at the vent tube is not available.

spermatophore Gelatinous, roughly mushroom-shaped body produced by cloacal glands of a male salamander and containing sperm; deposited during courtship whether on land or in water; female picks up part or all of the spermatophore and stores the sperm in a spermatotheca for internal fertilization at a later time.

spiracle Opening(s) where respiratory and feeding water pumped into the mouth exits the tadpole; position and number (1 or 2) vary among taxa; see midventral, sinistral.

stage, staging Means of describing the continuum of development based on attainment of specific conditions of morphology; time and size are not considered; allows developmental comparisons of larvae of the same or different species at different sizes or ages. Gosner (1960) is the most commonly used staging system.

stratify Behavioral feature of ambystomatid larvae that move at night from the pond bottom to float quietly in midwater to feed.

string Specific ovipositional mode common to bufonids; ova arranged uni- or biserially within a small-diameter, usually rather sturdy, uni- or bilayered tube; contrast with rosary; see array, clump, cluster, film, mass, single.

stripe, striped Longitudinal (e.g., sides or dorsal midline of body, lateral side of tail) or oblique (e.g., between eye and naris) contrasting color pattern; see band, bicolored, color, coloration.

submarginal papillae Papillae on the face of the oral disc of tadpoles exclusive of the margin; see marginal papillae.

suction feeding Primary method of feeding by larval salamanders, floor of buccal cavity is rapidly depressed just prior to jaws opening widely and rapidly; food items are carried into the buccal cavity with the inrush of water; mouth closure captures the food item, and the water is expelled through the gill slits; see carnivore, gill raker.

suctorial Describes a tadpole with a large oral disc that rheophilous tadpoles use to feed, move, and maintain position in fast-flowing water; a tadpole that has a suctorial oral disc (e.g., *Ascaphus*). No salamander larva has such a morphology; see benthic, jaw sheath, morphotype, nektonic, oral apparatus, rheophilous, suspension feeder.

suspended Describes aquatic or terrestrial eggs attached to the ceiling of a chamber or the lower surface of an object; may or may not be pendent.

suspension feeder Tadpole adapted for spending most of the time in midwater and feeding by filtering naturally suspended particles; body depressed, eyes lateral, fins low, usually with pointed tip; mouthparts greatly reduced and modified (e.g., microhylids, pipids, and rhinophrynids); see benthic, morphotype, nektonic, rheophilous, suctorial.

SVL See snout-vent length.

sympatric Describes ranges of species that overlap; the species in question occur at least in part in the same area.

tadpole Nonreproductive larva of a frog in stages 25–41, does not refer to a salamander larva; see adult, froglet, immature, metamorph.

tail That portion of an amphibian larva posterior to the body, with dorsal and ventral, unsupported (i.e., bony rays as in fishes are lacking) fins; vertebrae present in the permanent tail of salamander larvae are lacking in the temporary tails of tadpoles; see body axis, body length, body terminus, tail length.

tail-body junction Literally the junction between the tail and body; salamander larva: that plane at the posterior terminus of the vent; tadpole: because of the way the tail and body join, the body terminus is the most consistently accurate landmark; see body, tail.

tail length Salamander larva: linear distance from posterior extent of the vent to the tail tip; tadpole: linear distance from the body terminus to the tail tip; total length minus body length in either case; see body terminus.

tooth ridge Fleshy transverse ridge on the face of the oral disc of a tadpole, includes connective tissue and the mitotic area where new cells are formed; see tooth row, tooth series.

tooth row A uni-, bi-, or multiserial, transverse row of labial teeth on a tooth ridge on the oral face of the oral disc of a tadpole; see labial tooth, tooth series.

tooth series Presently erupted labial tooth of a tadpole and all its successive replacement teeth embedded within the tooth ridge; see tooth row.

transverse Describes a structure that extends to the side relative to the longitudinal axis of the body; see cross-section.

uniserial Structures, such as labial teeth and eggs in a tube, that occur in a single row or file; see biserial, multiserial.

vegetal pole Ventral, yolk-laden, and more lightly pigmented part of an ovum; see animal pole.

vent Exterior opening to the outside of the cloaca of an amphibian; not equal to anus; see vent tube.

ventral, venter Anatomical/morphological descriptor referring to the bottom or toward the bottom or belly of an animal or structure relative to a stated landmark; can be combined with other terms: ventrolateral; see anterior, lateral, medial, posterior.

vent tube Exit tube of the cloaca of a tadpole; not an anus; has 2 basic configurations— dextral: aperture lies to the right of the plane of the ventral fin; medial: aperture lies parallel with the plane of the ventral fin.

vitelline membrane First membrane external to the surface of an ovum, of ovarian origin; see jelly layers, oviduct.

wrapped rosary See rosary.

zooplanktors Plankton-sized animals (in contrast to planktonic algae); protozoans, crustaceans, adults and nauplii of cladocerans, copepods, and ostracods, and small insects and their larvae.

LITERATURE CITED

Abbott, C. C. 1882. Notes on the habits of the Savannah Cricket Frog. American Naturalist 16:707–711.

Adams, M. J. 1993. Summer nests of the Tailed Frog (*Ascaphus truei*) from the Oregon Coast Range. Northwestern Naturalist 74:15–18.

Ahlgren, M. O., and S. H. Bowen. 1991. Growth and survival of tadpoles *Bufo americanus* fed amorphous detritus derived from natural waters. Hydrobiologia 218:49–51.

Akers, E. C., C. M. Taylor, and R. Altig. 2008. Effects of clay-associated organic material on the growth of *Hyla chrysoscelis* tadpoles. Journal of Herpetology 42:408–410.

Albers P. H., and R. M. Prouty. 1987. Survival of Spotted Salamander eggs in temporary woodland ponds of coastal Maryland. Environmental Pollution 46:45–61.

Alexander, W. P. 1927. The Allegheny Hellbender and its habitat. Buffalo Society of Natural Science 7:13–18.

Alford, R. A. 1986. Habitat use and positional behavior of anuran larvae in a northern Florida temporary pond. Copeia 1986:408–423.

——. 1989. Competition between larval *Rana palustris* and *Bufo americanus* is not affected by variation in reproductive phenology. Copeia 1989:993–1000.

Alford, R. A., and M. L. Crump. 1982. Habitat partitioning among size classes of larval Southern Leopard Frogs, *Rana utricularia*. Copeia 1982:367–373.

Alford, R. A., and R. N. Harris. 1988. Effects of larval growth history on anuran metamorphosis. American Naturalist 131:91–106.

Altig, R. 1967. Food of *Siren intermedia nettingi* in a spring-fed swamp in southern Illinois. American Midland Naturalist 77:239–241.

——. 1970. A key to the tadpoles of the continental United States and Canada. Herpetologica 26:180–207.

——. 1971. Descriptive notes on the tadpoles of *Pseudacris ornata* and *Bufo alvarius*. Texas Journal of Science 23:301–303.

——. 1972. Notes on the larvae and premetamorphic tadpoles of four *Hyla* and three *Rana* with notes on tadpole color patterns. Journal of the Elisha Mitchell Scientific Society 88:113–119.

——. 1973. Preliminary scanning electron observations of keratinized structures of amphibian larvae. Herpetological Information Search Service News Journal 1:129–132.

——. 2007. A primer for the morphology of anuran tadpoles. Herpetology Conservation and Biology 2:73–76.

Altig, R., and E. D. Brodie Jr. 1972. Laboratory behavior of *Ascaphus truei* tadpoles. Journal of Herpetology 6:21–24.

Altig, R., and A. Channing. 1993. Hypothesis: Functional significance of colour and pattern of anuran tadpoles. Herpetological Journal 3:73–75.

Altig, R., and M. T. Christensen. 1981. Behavioral characteristics of the tadpole of *Rana heckscheri*. Journal of Herpetology 15:151–154.

Altig, R., and P. C. Dumas. 1971. *Rana cascadae*. Catalogue of American Amphibians and Reptiles 105.1–105.2.

——. 1972. *Rana aurora*. Catalogue of American Amphibians and Reptiles 160.1–160.4.

Altig, R., and P. H. Ireland. 1984. Key to salamander larvae and larviform adults of the United States and Canada. Herpetologica 40:212–218.

Altig, R., and G. F. Johnston. 1989. Guilds of anuran larvae: Relationships among developmental modes, morphologies, and habitats. Herpetological Monographs 3:81–109.

Altig, R., J. P. Kelly, M. Wells, and J. Phillips. 1975. Digestive enzymes of seven species of anuran tadpoles. Herpetologica 31:104–108.

Altig, R., and R. Lohoefener. 1983. *Rana areolata*. Catalogue of American Amphibians and Reptiles 324.1–324.3.

Altig, R., and W. McDearman. 1975. Percent assimilation and clearance times of five anuran tadpoles. Herpetologica 31:67–69.

Altig, R., and R. W. McDiarmid. 1999. Body plan: Development and morphology. *In* R. W. McDiarmid and R. Altig, eds., Tadpoles: The biology of anuran larvae, pp. 24–51. Chicago: University of Chicago Press.

——. 2007. Diversity, morphology, and evolution of egg and clutch structure in amphibians. Herpetological Monographs 21:1–32.

Altig, R., R. W. McDiarmid, K. A. Nichols, and P. C. Ustach. 1998. A key to the anuran tadpoles of the United States and Canada. Contemporary Herpetology, http://alpha.selu.edu/ch/; also http://www.pwrc.usga.gov/tadpole/.

Altig, R., and W. L. Pace. 1974. Scanning electron photomicrographs of tadpole labial teeth. Journal of Herpetology 8:247–251.

Alvarez, J. A. 2004. Overwintering California Tiger Salamander (*Ambystoma californiense*) larvae. Herpetological Review 35:344.

Anderson, A. R., and J. W. Petranka. 2003. Odonate predator does not affect hatching time or morphology of embryos of two amphibians. Journal of Herpetology 37:65–71.

Anderson, J. D. 1965. *Ambystoma annulatum*. Catalogue of American Amphibians and Reptiles 19.1–19.2.

——. 1967a. *Ambystoma texanum*. Catalogue of American Amphibians and Reptiles 37.1–37.2.

——. 1967b. *Ambystoma opacum*. Catalogue of American Amphibians and Reptiles 46.1–46.2.

——. 1967c. *Ambystoma maculatum*. Catalogue of American Amphibians and Reptiles 51.1–51.4.

——. 1967d. A comparison of the life histories of coastal and montane populations of *Ambystoma macrodactylum* in California. American Midland Naturalist 77:323–355.

——. 1968a. Thermal histories of two populations of *Ambystoma macrodactylum*. Herpetologica 24:29–35.

——. 1968b. *Rhyacotriton, R. olympicus*. Catalogue of American Amphibians and Reptiles 68.1–68.2.

——. 1968c. A comparison of the food habits of *Ambystoma macrodactylum sigillatum, Ambystoma macrodactylum croceum*, and *Ambystoma tigrinum californiense*. Herpetologica 24:273–284.

——. 1969. Dicamptodontidae, *D. ensatus*. Catalogue of American Amphibians and Reptiles 76.1–76.2.

——. 1972a. Embryonic temperature tolerance and rate of development in some salamanders of the genus *Ambystoma*. Herpetologica 28:126–130.

——. 1972b. Phototactic behavior of larvae and adults of two subspecies of *Ambystoma macrodactylum*. Herpetologica 28:222–226.

Anderson, J. D., and R. E. Graham. 1967. Vertical migration and stratification of larval *Ambystoma*. Copeia 1967:371–374.

Anderson, J. D., D. D. Hassinger, and G. H. Dalrymple. 1971a. The egg-alga relationship in *Ambystoma t. tigrinum*. Herpetological Review 3:76.

——. 1971b. Natural mortality of eggs and larvae of *Ambystoma t. tigrinum*. Ecology 52:1107–1112.

Anderson, J. D., and P. J. Martino. 1966. The life history of *Eurycea l. longicauda* associated with ponds. American Midland Naturalist 75:257–279.

Anderson, J. D., and G. K. Williamson. 1974. Nocturnal stratification in larvae of the Mole Salamander, *Ambystoma talpoideum*. Herpetologica 30:28–29.

——. 1976. Terrestrial mode of reproduction in *Ambystoma cingulatum*. Herpetologica 32:214–221.

Anderson, P. K. 1954. Studies in the ecology of the Narrow-mouthed Toad *Microhyla carolinensis carolinensis*). Tulane Studies in Zoology 2:13–38.

Anholt, B. R., E. Werner, and D. K. Skelly. 2000. Effect of food and predators on the activity of four larval ranid frogs. Ecology 81:3509–3521.

Antonelli, A. L., R. A. Nussbaum, and S. D. Smith. 1972. Comparative food habits of four species of stream-dwelling vertebrates (*Dicamptodon ensatus, D. copei, Cottus tenuis, Salmo gairdneri*). Northwestern Science 46:277–289.

Arendt, J. D. 2003. Reduced burst speed is a cost of rapid growth in anuran tadpoles: Problems of autocorrelation and inferences about growth rates. Functional Ecology 17:328–334.

——. 2008. Influence of sprint speed and body size on predator avoidance in New Mexican spadefoot toads (*Spea multiplicata*). Oecologia 159:455–461.

Arendt, J., and L. Hoang. 2005. Effect of food level and rearing temperature on burst speed and muscle composition of Western Spadefoot Toad (*Spea hammondii*). Functional Ecology 19:982–987.

Aresco, M. J. 2001. *Siren lacertina* (Greater Siren). Aestivation chamber. Herpetological Review 32:32–33.

Aresco, M. J., and R. N. Reed. 1998. *Rana capito sevosa* (Dusky Gopher Frog). Predation. Herpetological Review 29:40.

Arndt, R. G. 1989. Notes on the natural history and status of the Tiger Salamander, *Ambystoma tigrinum*, in Delaware. Bulletin of the Maryland Herpetological Society 25:1–21.

Ashton, R. E., Jr. 1985. Field and laboratory observations on microhabitat selection, movements, and home range of *Necturus lewisi* (Brimley). Brimleyana 10:83–106.

——. 1990. *Necturus lewisi*. Catalogue of American Amphibians and Reptiles 456.1–456.2.

Ashton, R. E., Jr., and P. S. Ashton. 1988. Handbooks of reptiles and amphibians of Florida. Part Three. The Amphibians. Miami: Windward Publishing.

Ashton, R. E., Jr., and A. L. Braswell. 1979. Nest and larvae of the Neuse River Waterdog, *Necturus lewisi* (Brimley) (Amphibia: Proteidae). Brimleyana 1:15–22.

Ashton, R. E., Jr., and R. Franz. 1979. *Bufo quercicus*. Catalogue of American Amphibians and Reptiles 222.1–222.2.

Asquith, A., and R. Altig. 1987. Phototaxis and activity patterns of *Siren intermedia*. Southwestern Naturalist 32:146–148.

Auburn, J. S., and D. H. Taylor. 1979. Polarized light perception and orientation in larval Bullfrogs *Rana catesbeiana*. Animal Behavior 27:658–668.

Audo, M. C., T. M. Mann, T. L. Polk, C. Loudenslager, W. J. Diehl, and R. Altig. 1995. Food deprivation during different periods of tadpole *Hyla chrysoscelis* ontogeny affects metamorphic performance differently. Oecologia 103:518–522.

Austin, R. M., Jr., and C. D. Camp. 1992. Larval development of Black-bellied Salamanders, *Desmognathus quadramaculatus*, in northeastern Georgia. Herpetologica 48:313–317.

Awan, A. R., and G. R. Smith. 2007. The effect of group size on the activity of Leopard Frog (*Rana pipiens*) tadpoles. Journal of Freshwater Ecology 22:355–357.

Babbitt, K. J. 1995. *Bufo terrestris* (Southern Toad). Oophagy. Herpetological Review 26:30.

——. 2001. Behaviour and growth of Southern Leopard Frog (*Rana sphenocephala*) tadpoles: Effects of food and predation risk. Canadian Journal of Zoology 79:809–814.

Babbitt, K. J., and F. Jordan. 1996. Predation on *Bufo terrestris* tadpoles: Effects of cover and predator density. Copeia 1996:485–488.

Babbitt, K. J., and W. E. Meshaka Jr. 2000. Benefits of eating conspecifics: Effects of background diet on survival and metamorphosis in the Cuban Treefrog (*Osteopilus septentrionalis*). Copeia 2000:469–474.

Babbitt, K. J., and G. W. Tanner. 1997. Effects of cover and predator identity on predation of *Hyla squirella* tadpoles. Journal of Herpetology 31:128–130.

——. 1998. Effects of cover and predator size on survival and development of *Rana utricularia* tadpoles. Oecologia 114:258–262.

Bachmann, M. D., R. G. Carlton, J. M. Burkholder, and R. G. Wetzel. 1986. Symbiosis between salamander eggs and green algae: Microelectrode measurements inside eggs demonstrate effect of photosynthesis on oxygen concentration. Canadian Journal of Zoology 64:1586–1588.

Bailey, J. R. 1937. Notes on plethodont salamanders of the southeastern United States. Occasional Papers of the Museum of Zoology, University of Michigan (364): 1–10.

Baker, C. L. 1945. The natural history and morphology of amphiumae. Report of the Reelfoot Lake Biological Station 9:55–91.

Baker, J. K. 1957. *Eurycea troglodytes*, a new blind cave salamander from Texas. Texas Journal of Science 9:326–328.

——. 1961. Distribution of and a key to the neotenic *Eurycea* of Texas. Southwestern Naturalist 6:27–32.

——. 1966. *Eurycea troglodytes*. Catalogue of American Amphibians and Reptiles 23.1–23.2.

Baker, L. C. 1937a. Mating habits and life history of *Amphiuma tridactylum* Cuvier and effects of pituitary injections. Journal of the Tennessee Academy of Science 12:9–21.

——. 1937b. Mating and life history of *Amphiuma tridactylum*. Journal of the Tennessee Academy of Science 12:206–218.

Baldwin, K. F., and A. J. K. Calhoun. 2002. *Ambystoma laterale* (Blue-spotted Salamander) and *Ambystoma maculatum* (Spotted Salamander). Predation. Herpetological Review 33:44–45.

Baldwin, K. S., and R. A. Stanford. 1987. *Ambystoma californiense* (California Tiger Salamander). Predation. Herpetological Review 18:33.

Balfour, P. S., and E. W. Stitt. 2003. *Ambystoma californiense* (California Tiger Salamander). Predation. Herpetological Review 34:44.

Ballinger, R. E., and J. D. Lynch. 1983. How to know the amphibians and reptiles. Dubuque, IA: Wm. C. Brown.

Bancroft, G. T., J. S. Godley, D. T. Gross, N. N. Rojas, D. A. Sutphen, and R. W. McDiarmid. 1983. Large-scale operations management test of use of the White Amur for control of problem aquatic plants. The Herpetofauna of Lake Conway: Species Accounts. Miscellaneous Paper A-83–5. U.S. Army Engineer Waterways Experiment Station, Vicksburg, MS. 354 pp.

Banta, A. M. 1912. The influence of cave conditions upon pigment development in larvae of *Ambystoma tigrinum*. American Naturalist 46:244–248.

Banta, A. M., and R. A. Gortner. 1914. A milky white amphibian egg jell. Biological Bulletin 27:259–261.

Banta, A. M., and W. L. McAtee. 1906. The life history of the Cave Salamander, *Spelerpes maculicaudus* (Cope). Proceedings of the United States National Museum 30:67–73.

Barbour, R. W., and R. M. Hays. 1957. The genus *Desmognathus* in Kentucky. *Desmognathus fuscus welteri* Barbour. American Midland Naturalist 58:352–357.

Barden, R. B., and J. Kezer. 1944. The eggs of certain plethodontid salamanders obtained by pituitary gland implantation. Copeia 1944:115–118.

Bardwell, J. H., C. M. Ritzi, and J. A. Parkhurst. 2007. Dietary selection among different size classes of larval *Ambystoma jeffersonianum* (Jefferson Salamander). Northeastern Naturalist 14:293–299.

Barinaga, M. 1990. Where have all the froggies gone? Science 247:1033–1034.

Barr, G. E., and K. J. Babbitt. 2002. Effects of biotic and abiotic factors on the distribution and abundance of larval Two-lined Salamanders (*Eurycea bislineata*) across spatial scales. Oecologia 133:176–185.

Bart, H. L., Jr., M. A. Bailey, R. E. Ashton Jr., and P. E. Moler. 1997. Taxonomic and nomenclatural status of the upper Black Warrior River Waterdog. Journal of Herpetology 31:192–201.

Bart, H. L., and R. W. Holzenthal. 1985. Feeding ecology of *Necturus beyeri* in Louisiana. Journal of Herpetology 19:402–410.

Baumann, W. L., and M. Huels. 1982. Nests of the Two-lined Salamander, *Eurycea bislineata*. Journal of Herpetology 16:81–83.

Beachy, C. K. 1993a. Differences in variation in egg size for several species of salamanders (Amphibia: Caudata) that use different larval environments. Brimleyana 18:71–82.

———. 1993b. Guild structure in streamside salamander communities: A test for interactions among larval plethodontid salamanders. Journal of Herpetology 27:465–468.

———. 1995a. Age at maturation, body size, and life-history evolution in the salamander family Plethodontidae. Herpetological Review 26:179–181.

———. 1995b. Effects of larval growth history on metamorphosis in a stream-dwelling salamander (*Desmognathus ochrophaeus*). Journal of Herpetology 29:375–382.

———. 1997. Effects of predatory larval salamanders (*Desmognathus quadramaculatus*) on metamorphic parameters of prey salamanders (*Eurycea wilderae*) in a stream community. Copeia 1997:131–137.

Beachy, C. K., and R. C. Bruce. 2003. Life history of a small form of the plethodontid salamander *Desmognathus quadramaculatus*. Amphibia-Reptilia 24:13–26.

Beachy, C. K., T. H. Surges, and M. Reyes. 1999. Effects of developmental and growth history on metamorphosis in the Gray Treefrog, *Hyla versicolor* (Amphibia, Anura). Journal of Experimental Zoology 283:522–530.

Bechler, D. L. 1988. Courtship behavior and spermatophore deposition by the subterranean salamander, *Typhlomolge rathbuni* (Caudata, Plethodontidae). Southwestern Naturalist 33:124–126.

Beck, C. W. 1997. Effects of change in resource level on age and size at metamorphosis in *Hyla squirella*. Oecologia 112:187–192.

Beck, C. W., and J. D. Condgon. 2003. Energetics of metamorphic climax in the Southern Toad (*Bufo terrestris*). Oecologia 137:344–351.

Beiswenger, R. E. 1975. Structure and function in aggregations of tadpoles of the American Toad, *Bufo americanus*. Herpetologica 31:222–233.

———. 1977. Diel patterns of aggregative behavior in tadpoles of *Bufo americanus*, in relation to light and temperature. Ecology 58:98–108.

Beiswenger, R. E., and F. H. Test. 1966. Effects of environmental temperature on movements of tadpoles of the American Toad, *Bufo terrestris americanus*. Papers of the Michigan Academy of Science, Arts, and Letters 51:127–141.

Berger-Bishop, L. E., and R. N. Harris. 1996. A study of caudal allometry in the salamander *Hemidactylium scutatum* (Caudata: Plethodontidae). Herpetologica 52:515–525.

Berkhouse, C. S., and J. N. Fries. 1995. Critical thermal maxima of juvenile and adult San Marcos Salamanders (*Eurycea nana*). Southwestern Naturalist 40:430–434.

Bernardo, J. 2000. Early life histories of Dusky Salamanders, *Desmognathus imitator* and *D. wrighti*, in a headwater seepage in Great Smoky Mountains National Park, USA. Amphibia-Reptilia 21:403–407.

Bernardo, J., and S. J. Agosta. 2003. Determinants of clinal variation in life history of Dusky Salamanders (*Desmognathus ocoee*): Prey abundance and ecological limits on foraging time restrict opportunities for larval growth. Journal of Zoology 259:411–421.

Berven, K. A. 1987. The heritable basis of variation in larval developmental patterns within populations of the Wood Frog (*Rana sylvatica*). Evolution 41:1088–1097.

———. 1990. Factors affecting population fluctuations in larval and adult stages of the Wood Frog (*Rana sylvatica*). Ecology 71:1599–1608.

Berven, K. A., and R. S. Boltz. 2001. Interactive effects of leech (*Desserobdella picta*) infection on Wood Frog (*Rana sylvatica*) tadpole fitness traits. Copeia 2001:907–915.

Berven, K. A., and B. G. Chadra. 1988. The relationship among egg size, density, and food level on larval development in the Wood Frog (*Rana sylvatica*). Oecologia 75:67–72.

Besharse, J. C., and R. A. Brandon. 1976. Effects of continuous light and darkness on the eyes of the troglobitic salamander *Typhlotriton spelaeus*. Journal of Morphology 149:527–546.

Besharse, J. C., and J. R. Holsinger. 1977. *Gyrinophilus subterraneus*, a new troglobitic salamander from southern West Virginia. Copeia 1977:624–634.

Bevelhimer, M. S., D. J. Stevenson, N. R. Giffen, and K. Ravenscroft. 2008. Annual surveys of larval *Ambystoma cingulatum* reveal large differences in dates of pond residency. Southeastern Naturalist 7:311–322.

Birchfield, G. L., and R. C. Bruce. 2000. Morphometric variation among larvae of four species of lungless salamanders (Caudata: Plethodontidae). Herpetologica 56:332–342.

Bishop, D. C., J. G. Palis, K. M. Enge, D. J. Printiss, and D. J. Stevenson. 2006. Capture rate, body size, and survey recommendations for larval *Ambystoma cingulatum* (Flatwoods Salamander). Southeastern Naturalist 5:9–16.

Bishop, S. C. 1920. Notes on the habits and development of the Four-toed Salamander, *Hemidactylium scutatum* (Schlegel). Bulletin of the New York State Museum (219–220): 251–282.

——. 1924. Notes on salamanders. Bulletin of the New York State Museum (253): 87–102.

——. 1925. The life of the Red Salamander. Natural History 25:385–389.

——. 1926. Notes on the habits and development of the Mudpuppy *Necturus maculosus* (Rafinesque). Bulletin of the New York State Museum (268): 5–60.

——. 1928. A new subspecies of the Red Salamander from Louisiana. Occasional Papers of the Boston Society of Natural History 5:247–249.

——. 1932. The spermatophores of *Necturus maculosus* Rafinesque. Copeia 1932:1–3.

——. 1941a. The salamanders of New York. Bulletin of the New York State Museum (324): 1–365.

——. 1941b. Notes on some salamanders with descriptions of several new forms. Occasional Papers of the Museum of Zoology, University of Michigan (451): 1–21.

——. 1943. Handbook of salamanders. The salamanders of the United States, of Canada, and of Lower California. Ithaca, NY: Comstock Publishing.

——. 1944. A new neotenic plethodont salamander, with notes on related species. Copeia 1944:1–5.

——. 1994. Handbook of salamanders. The salamanders of the United States, of Canada, and of Lower California. Ithaca, NY: Cornell University Press.

Bishop, S. C., and H. P. Chrisp. 1933. The nests and young of the Allegheny Salamander, *Desmognathus fuscus ochrophaeus* (Cope). Copeia 1933:194–198.

Bishop, S. C., and M. R. Wright. 1937. A new neotenic salamander from Texas. Proceedings of the Biological Society of Washington 50:141–143.

Bizer, J. R. 1978. Growth rates and size at metamorphosis of high elevation populations of *Ambystoma tigrinum*. Oecologia 34:175–184.

Black, J. H. 1969. Ethoecology of tadpoles of *Bufo americanus charlesmithi* in a temporary pond in Oklahoma. Journal of Herpetology 3:198.

——. 1970. A possible stimulus for the formation of some aggregations in tadpoles of *Scaphiopus bombifrons*. Proceedings of the Oklahoma Academy of Science 49:13–14.

——. 1971. The formation of tadpole holes. Herpetological Review 3:7.

——. 1974a. Tadpoles nests in Oklahoma. Oklahoma Geological Notes (3): 105.

——. 1974b. Larval spadefoot survival. Journal of Herpetology 8:371–373.

——. 1975. The formation of "tadpole nests" by anuran larvae. Herpetologica 31:76–79.

Blanchard, F. N. 1922. Discovery of the eggs of the Four-toed Salamander in Michigan. Occasional Papers of the Museum of Zoology, University of Michigan (126): 1–3.

——. 1923. The life history of the Four-toed Salamander. American Naturalist 57:262–268.

——. 1936. The number of eggs produced and laid by the Four-toed Salamander, *Hemidactylium scutatum* (Schlegel), in southern Michigan. Papers of the Michigan Academy of Science, Arts and Letters 21:567–573.

Blaustein, A. R., and R. K. O'Hara. 1987. Aggregation behaviour in *Rana cascadae* tadpoles: Association preferences among wild aggregations and responses to non-kin. Animal Behavior 35:1549–1555.

Blaustein, A. R., and D. B. Wake. 1990. Declining amphibian populations: A global phenomenon? Trends in Ecology and Evolution 5:203–204.

Blaustein, A. R., T. Yoshikawa, K. Asoh, and S. C. Walls. 1993. Ontogenetic shifts in tadpole kin recognition—loss of signal and perception. Animal Behavior 46:525–538.

Bleakney, S. 1957. The egg-laying habits of the salamander, *Ambystoma jeffersonianum*. Copeia 1957:141–142.

——. 1958. Cannibalism in *Rana sylvatica* tadpoles—a well known phenomenon. Herpetologica 14:34.

Bles, E. I. 1906. The life-history of *Xenopus laevis* Daud. Transactions of the Royal Society of Edinburgh 41:789–821.

Bolek, M. G. 1998. *Ambystoma laterale* (Blue-spotted Salamander). Courtship and egg laying behavior. Herpetological Review 29:162.

Bolek, M. G., and J. Janovy Jr. 2004. *Rana catesbeiana* (Bullfrog). Gigantic tadpole. Herpetological Review 35:376–377.

Bond, A. N. 1960. An analysis of the response of salamander gills to changes in the oxygen concentration of the medium. Developmental Biology 2:1–11.

Boone, M. D., D. E. Scott, and P. H. Niewiarowski. 2002. Effects of hatching time for larval ambystomatid salamanders. Copeia 2002:511–517.

Borchelt, R. 1990. Frogs, toads, and other amphibians in distress. National Research Council News Report 40:2–5.

Borland, J. R., and R. Rugh. 1943. Evidences of cannibalism in the tadpole of the frog *Rana pipiens*. American Naturalist 77:282–285.

Bowles, B. D., M. S. Sanders, and R. S. Hansen. 2006. Ecology of the Jollyville Plateau Salamander (*Eurycea tonkawae*: Plethodontidae) with an assessment of the potential effects of urbanization. Hydrobiologia 553:111–120.

Brady, M. K. 1924. Eggs of *Desmognathus phoca* (Matthes). Copeia 1924:29.

Bragg, A. N. 1936. Notes on the breeding habits, eggs, and embryos of *Bufo cognatus* with a description of the tadpole. Copeia 1936:14–20.

——. 1937a. Observations of *Bufo cognatus* with special reference to breeding habits and eggs. American Midland Naturalist 18:273–284.

——. 1937b. A note on the metamorphosis of the tadpoles of *Bufo cognatus*. Copeia 1937: 227–228.

——. 1940a. Observations on the ecology and natural history of Anura. II. Habits, habitat, and breeding of *Bufo woodhousii woodhousii* (Girard) in Oklahoma. American Midland Naturalist 24:306–321.

——. 1940b. Observations on the ecology and natural history of Anura. I. Habits, habitat and breeding of *Bufo cognatus* Say. American Naturalist 74:424–438.

——. 1941a. Tadpoles of *Scaphiopus bombifrons* and *Scaphiopus hammondi*. Wasmann Collector 4:92–94.

——. 1941b. Keys to the separation of adults and tadpoles within *Scaphiopus*. American Naturalist 78:256–257.

——. 1942. Observations on the ecology and natural history of Anura. X. The breeding habits of *Pseudacris streckeri* Wright and Wright in Oklahoma including a description of the eggs and tadpoles. Wasmann Collector 5:47–62.

——. 1943. Observations on the ecology and natural history of Anura. XVI. Life-history of *Pseudacris clarkii* (Baird) in Oklahoma. Wasmann Collector 5:129–140.

——. 1944a. Egg laying in leopard frogs. Proceedings of the Oklahoma Academy of Science 24:13–14.

——. 1944b. Breeding habits, eggs, and tadpoles of *Scaphiopus hurterii*. Copeia 1944: 231–241.

——. 1944c. The spadefoot toads in Oklahoma with summary of our knowledge of the group. I. American Naturalist 78:517–533.

——. 1945a. The spadefoot toads in Oklahoma with summary of our knowledge of the group. II. American Naturalist 79:52–72.

——. 1945b. Breeding and tadpole behavior in *Scaphiopus hurterii* near Norman, Oklahoma, spring, 1945. Wasmann Collector 6:69–78.

——. 1945c. Aggregational phenomena in *Scaphiopus hurterii* tadpoles. Proceedings of the Oklahoma Academy of Science 26:19.

——. 1946a. Aggregation with cannibalism in tadpoles of *Scaphiopus bombifrons* with some general remarks on the probable evolutionary significance of such phenomena. Herpetologica 3:89–97.

——. 1946b. Aggregations in tadpoles of spadefoot toads. Anatomical Record 94:351–352.

——. 1947a. Tadpole behavior in pools and streams. Proceedings of the Oklahoma Academy of Science 27:59–61.

——. 1947b. Comments on the mouthparts of tadpoles of *Bufo cognatus* and *Bufo compactilis*. Science 106:166.

——. 1948a. Additional instances of social aggregation in tadpoles. Wasmann Collector 7:65–79.

——. 1948b. Observations on the life history of *Pseudacris triseriata* (Wied) in Oklahoma. Wasmann Collector 7:149–168.

——. 1948c. Food resources utilized by tadpoles of *Scaphiopus hurterii* in temporary pools. Anatomical Record 101:706.

——. 1950. Observations on *Microhyla* (Salientia, Microhylidae). Wasmann Journal of Biology 8:113–118.

——. 1951a. A note on the egg-masses of *Scaphiopus hurterii* Strecker. Proceedings of the Oklahoma Academy of Science 30:18–19.

——. 1951b. Mass movement at metamorphosis in the Savannah Spadefoot, *Scaphiopus hurterii* Strecker. Proceedings of the Oklahoma Academy of Science 31:26–27.

——. 1952. A metamorphic aggregation in tadpoles of *Bufo*. Proceedings of the Oklahoma Academy of Science 33:13.

——. 1953. A study of *Rana areolata* in Oklahoma. Wasmann Journal of Biology 11:273–318.

——. 1954. Aggregational behavior and feeding reactions in tadpoles of the Savannah Spadefoot. Herpetologica 10:97–102.

——. 1955a. Taxonomic and physiological factors in the embryonic development of certain toads. Copeia 1955:62.

——. 1955b. Supplementary notes on the young of *Rana areolata*. Herpetologica 11:87–88.

——. 1955c. The tadpole of *Bufo debilis debilis*. Herpetologica 11:211–212.

——. 1956. Dimorphism and cannibalism in tadpoles of *Scaphiopus bombifrons* (Amphibia, Salientia). Southwestern Naturalist 1:105–108.

——. 1957a. Variation in colors and color patterns in tadpoles in Oklahoma. Copeia 1957: 36–39.

——. 1957b. Aggregational feeding and metamorphic aggregations in tadpoles of *Scaphiopus hurteri* observed in 1954. Wasmann Journal of Biology 15:61–68.

——. 1957c. Amphibian eggs produced singly or in masses. Herpetologica 13:212.

——. 1958a. On metamorphic and postmetamorphic aggregations in spadefoots. Herpetologica 13:273.

——. 1958b. The eggs of the Dwarf American Toad. Herpetologica 13:273–275.

——. 1959. Behavior of tadpoles of Hurter's Spadefoot during an exceptionally rainy season. Wasmann Journal of Biology 17:23–42.

——. 1960. Experimental observations on the feeding of spadefoot tadpoles. Southwestern Naturalist 5:201–207.

——. 1961. The behavior and comparative developmental rates in nature of tadpoles of a spadefoot, a toad and a frog. Herpetologica 17:80–84.

——. 1962a. *Saprolegnia* on tadpoles again in Oklahoma. Southwestern Naturalist 7:79–80.

——. 1962b. Predation on arthropods by spadefoot tadpoles. Herpetologica 18:144.

——. 1962c. Predator-prey relationships in two species of spadefoot tadpoles with notes on some other features of their behavior. Wasmann Journal of Biology 20:81–97.

——. 1964a. Further study of predation and cannibalism in spadefoot tadpoles. Herpetologica 20:17–24.

——. 1964b. A hypothesis to explain the almost exclusive use of temporary water by breeding spadefoot toads. Proceedings of the Oklahoma Academy of Science 44:24–25.

——. 1965a. Mass movements resulting in aggregations of tadpoles of the Plains Spadefoot, some of them in response to light and temperature (Amphibia: Salientia). Wasmann Journal of Biology 22:299–305.

——. 1965b. Gnomes of the night. Philadelphia: University of Pennsylvania Press.

——. 1966. Longevity of the tadpole stages in the Plains Spadefoot (Amphibia: Salientia). Wasmann Journal of Biology 24:71–73.

——. 1967. Recent studies on the spadefoot toads. Bios 38:75–84.

——. 1968. The formation of feeding schools in tadpoles of spadefoots. Wasmann Journal of Biology 26:11–16.

Bragg, A. N., and W. N. Bragg. 1957. Parasitism of spadefoot tadpoles by *Saprolegnia*. Herpetologica 13:191.

——. 1958. Parasitism of spadefoot tadpoles by *Saprolegnia*. Herpetologica 14:34.

——. 1959. Variation in the mouth parts in tadpoles of *Scaphiopus* (*Spea*) *bombifrons* Cope (Amphibia: Salientia). Southwestern Naturalist 3:55–59.

Bragg, A. N., and J. Bresler. 1952. Viability of the eggs of *Bufo cognatus*. Proceedings of the Oklahoma Academy of Science 32:13–14.

Bragg, A. N., and S. Hayes. 1963. A study of labial teeth rows in tadpoles of Couch's Spadefoot. Wasmann Journal of Biology 21:149–154.

Bragg, A. N., and O. M. King. 1961. Aggregational and associated behavior in tadpoles of the Plains Spadefoot. Wasmann Journal of Biology 18:273–289.

Bragg, A. N., R. Mathews, and R. Kingsinger Jr. 1964. The mouth parts of tadpoles of Hurter's Spadefoot. Herpetologica 19:284–285.

Bragg, A. N., A. O. Weese, H. A. Dundee, H. T. Fisher, A. Richards, and C. B. Clark. 1950. Researches on the Amphibia of Oklahoma. Norman: University of Oklahoma Press. 154 pp.

Brame, A. H., Jr. 1956. The number of eggs laid by the California Newt. Herpetologica 12:325–326.

——. 1968. The number of egg masses and eggs laid by the California Newt, *Taricha torosa*. Journal of Herpetology 2:169–170.

Branch, L. C., and R. Altig. 1981. Nocturnal stratification of three species of *Ambystoma* larvae. Copeia 1981:870–873.

——. 1983. Survival and behavior of four species of *Ambystoma* larvae under hypoxic conditions. Comparative Biochemistry and Physiology 74A:395–397.

Branch, L. C., and D. H. Taylor. 1977. Physiological and behavioral responses of larval Spotted Salamanders (*Ambystoma maculatum*) to various concentrations of oxygen. Comparative Biochemistry and Physiology 58A:269–274.

Brandon, R. A. 1961. A comparison of the larvae of five northeastern species of *Ambystoma* (Amphibia, Caudata). Copeia 1961:377–383.

——. 1964. An annotated and illustrated key to multistage larvae of Ohio salamanders. Ohio Journal of Science 64:252–258.

——. 1965a. *Typhlotriton, T. nereus, T. spelaeus*. Catalogue of American Amphibians and Reptiles 20.1–20.2.

——. 1965b. A new race of the neotenic salamander *Gyrinophilus palleucus*. Copeia 1965:346–352.

——. 1967a. *Gyrinophilus palleucus*. Catalogue of American Amphibians and Reptiles 32. 1–632.2.

——. 1967b. *Gyrinophilus porphyriticus*. Catalogue of American Amphibians and Reptiles 33.1–33.3.

——. 1967c. *Haideotriton, H. wallacei*. Catalogue of American Amphibians and Reptiles 39.1–39.2.

——. 1967d. Food and an intestinal parasite of the troglobitic salamander *Gyrinophilus palleucus necturoides*. Herpetologica 23:52–53.

——. 1970. *Typhlotriton, T. spelaeus*. Catalogue of American Amphibians and Reptiles 84.1–84.2.

——. 1971a. North American troglobitic salamanders: Some aspects of modification in cave habitats, with special reference to *Gyrinophilus palleucus*. Bulletin of the National Speleological Society 33:1–21.

——. 1971b. Correlation of seasonal abundance with feeding and reproductive activity in the Grotto Salamander (*Typhlotriton spelaeus*). American Midland Naturalist 86:93–100.

Brandon, R. A., and D. J. Bremer. 1966. Neotenic newts, *Notophthalmus viridescens louisianensis*, in southern Illinois. Herpetologica 22:213–217.

——. 1967. Overwintering of larval Tiger Salamanders in southern Illinois. Herpetologica 23:67–68.

Brandt, B. B. 1936. The frogs and toads of eastern North Carolina. Copeia 1936:215–223.

Braswell, A. L., and R. E. Ashton Jr. 1985. Distribution, ecology, and feeding habits of *Necturus lewisi* (Brimley). Brimleyana 10:13–35.

Brattstrom, B. H. 1962. Thermal control of aggregation behavior in tadpoles. Herpetologica 18:38–46.

Brattstrom, B. H., and J. W. Warren. 1955. Observations on the ecology and behavior of the Pacific Treefrog, *Hyla regilla*. Copeia 1955:181–191.

Breden, F. 1988. The natural history and ecology of Fowler's Toad, *Bufo woodhousei fowleri* (Amphibia: Bufonidae) in the Indiana Dunes National Lakeshore. Fieldiana Zoology 49:1–16.

Breden, F., and C. H. Kelly. 1982. The effect of conspecific interactions on metamorphosis in *Bufo americanus*. Ecology 63:1682–1689.

Breder, C. M. 1946. Amphibians and reptiles of the Rio Chucunaque drainage, Darien, Panama, with notes on their life histories and habits. Bulletin of the American Museum of Natural History 86:375–436.

Brenes, R., and N. B. Ford. 2006. Seasonality and movements of the Gulf Coast Waterdog (*Necturus beyeri*) in eastern Texas. Southwestern Naturalist 51:152–156.

Bresler, J. 1954. The development of labial teeth of salientian larvae in relation to temperature. Copeia 1954:207–211.

Bresler, J., and A. N. Bragg. 1954. Variations in the rows of labial teeth in tadpoles. Copeia 1954:255–257.

Briggler, J. T., and W. E. Moser. 2008. *Necturus maculosus* (Red River Mudpuppy). Host. Herpetological Review 39:205.

Briggs, J. L. 1987. Breeding biology of the Cascade Frog, *Rana cascadae*, with comparisons to *R. aurora* and *R. pretiosa*. Copeia 1987:241–245.

Brimley, C. S. 1920. Reproduction of the Marbled Salamander. Copeia 1920:25.

——. 1923. The Dwarf Salamander at Raleigh, N.C. Copeia 1923:81–83.

——. 1924. The water dogs of North Carolina. Journal of the Elisha Mitchell Scientific Society 40:166–168.

Britson, C. A., and R. E. Kissell Jr. 1996. Effects of food type on developmental characteristics of an ephemeral pond-breeding anuran, *Pseudacris triseriata feriarum*. Herpetologica 52:374–382.

Brockelman, W. Y. 1969. An analysis of density effects and predation in *Bufo americanus* tadpoles. Ecology 50:632–644.

Brode, W. E. 1961. Observations on the development of *Desmognathus* eggs under relatively dry conditions. Herpetologica 17:202–203.

Brodie, E. D., Jr., and D. R. Formanowicz Jr. 1987. Antipredator mechanisms of larval anurans: Protection of palatable individuals. Herpetologica 43:369–373.

Brodie, E. D., Jr., D. R. Formanowicz Jr., and E. D. Brodie III. 1978. The development of noxiousness of *Bufo americanus* tadpoles to aquatic insect predators. Herpetologica 34: 302–306.

Brodman, R. 1995. Annual variation in the breeding success of two syntopic species of *Ambystoma* salamanders. Journal of Herpetology 29:111–113.

———. 1996. Effects of intraguild interactions in fitness and microhabitat use of larval *Ambystoma* salamanders. Copeia 1996:372–378.

———. 1999. Food and space dependent effects during the interactions of two species of larval salamanders. Journal of Freshwater Ecology 14:431–437.

———. 2004. Intraguild predation on congeners affects size, aggression, and survival among *Ambystoma* salamander larvae. Journal of Herpetology 38:21–26.

———. 2008. Ecology and natural history observations of the salamander, *Siren intermedia nettingi* (Western Lesser Siren), in northern Indiana. Herpetological Review 39:414–419.

Brodman, R., and J. Jaskula. 2002. Activity and microhabitat use during interactions among five species of pond-breeding salamander larvae. Herpetologica 58:346–354.

Brophy, T. E. 1980. Food habits of sympatric larval *Ambystoma tigrinum* and *Notophthalmus viridescens*. Journal of Herpetology 14:1–6.

Brophy, T. R., and T. K. Pauley. 2001. *Eurycea cirrigera* (Southern Two-lined Salamander). Larval habitat. Herpetological Review 32:98–99.

———. 2006. *Eurycea cirrigera* (Southern Two-lined Salamander). Gill morphology. Herpetological Review 37:197.

Brown, B. C. 1942. Notes on *Eurycea neotenes*. Copeia 1942:176.

———. 1967a. *Eurycea latitans*. Catalogue of American Amphibians and Reptiles 34.1–34.2.

———. 1967b. *Eurycea nana*. Catalogue of American Amphibians and Reptiles 35.1–35.2.

———. 1967c. *Eurycea neotenes*. Catalogue of American Amphibians and Reptiles 36.1–36.2.

Brown, H. A. 1967. High temperature tolerance of the eggs of a desert anuran, *Scaphiopus hammondii*. Copeia 1967:365–370.

———. 1969. The heat resistance of some anuran tadpoles (Hylidae and Pelobatidae). Copeia 1969:138–147.

———. 1975a. Reproduction and development of the Red-legged Frog, *Rana aurora*, in northwestern Washington. Northwestern Science 49:241–252

———. 1975b. Temperature and development of the Tailed Frog, *Ascaphus truei*. Comparative and Biochemical Physiology 50A:397–405.

———. 1975c. Embryonic temperature adaptations of the Pacific Treefrog, *Hyla regilla*. Comparative and Biochemical Physiology 51A:863–873.

———. 1976. The time-temperature relation of embryonic development in the Northwestern Salamander, *Ambystoma gracile*. Canadian Journal of Zoology 54:552–558.

———. 1977. Oxygen consumption of a large, cold-adapted frog egg (*Ascaphus truei*) (Amphibia: Ascaphidae). Canadian Journal of Zoology 55:343–348.

———. 1989a. Developmental anatomy of the Tailed Frog (*Ascaphus truei*): A primitive frog with large eggs and slow development. Journal of Zoology 217:525–537.

———. 1989b. Tadpole development and growth of the Great Plains Spadefoot Toad, *Scaphiopus intermontanus*, from central Washington. Canadian Field Naturalist 103:531–534.

———. 1990. Morphological variation and age-class determination in overwintering tadpoles of the Tailed Frog, *Ascaphus truei*. Journal of Zoology 220:171–184.

Brown, L. E. 1973. *Bufo houstonensis*. Catalogue of American Amphibians and Reptiles 133.1–133.2.

——. 1992. *Rana blairi*. Catalogue of American Amphibians and Reptiles 536.1–536.6.

Brown, L. E., W. L. McClure, F. E. Potter Jr., N. J. Scott Jr., and R. A. Thomas. 1984. Recovery plan for the Houston Toad (*Bufo houstonensis*). Albuquerque: United States Fish and Wildlife Service.

Brown, M. G. 1942. An adaptation in *Ambystoma opacum* embryos to development on land. American Naturalist 76:222–223.

Bruce, R. C. 1970. The larval life of the Three-lined Salamander, *Eurycea longicauda guttolineata*. Copeia 1970:776–779.

——. 1971. Life cycle and population structure of the salamander *Stereochilus marginatus* in North Carolina. Copeia 1971:234–246.

——. 1972a. The larval life of the Red Salamander, *Pseudotriton ruber*. Journal of Herpetology 6:43–51.

——. 1972b. Variation in the life cycle of the salamander *Gyrinophilus porphyriticus*. Herpetologica 28:230–245.

——. 1974. Larval development of the salamanders *Pseudotriton montanus* and *P. ruber*. American Midland Naturalist 92:173–190.

——. 1975. Reproductive biology of the Mud Salamander, *Pseudotriton montanus*, in western South Carolina. Copeia 1975:129–137.

——. 1976. Population structure, life history and evolution of paedogenesis in the salamander *Eurycea neotenes*. Copeia 1976:242–249.

——. 1978a. Life-history patterns of the salamanders *Gyrinophilus porphyriticus* in the Cowee Mountains, North Carolina. Herpetologica 34:53–64.

——. 1978b. A comparison of the larval periods of Blue Ridge and Piedmont Mud Salamanders (*Pseudotriton montanus*). Herpetologica 34:325–332.

——. 1978c. Reproductive biology of the salamander *Pseudotriton ruber* in the southern Blue Ridge Mountains. Copeia 1978:417–423.

——. 1979. Evolution of paedogenesis in salamanders of the genus *Gyrinophilus*. Evolution 33:998–1000.

——. 1980. A model of the larval period of the Spring Salamander, *Gyrinophilus porphyriticus*, based on size-frequency distributions. Herpetologica 36:78–86.

——. 1982a. Larval periods and metamorphosis in two species of salamanders of the genus *Eurycea*. Copeia 1982:117–127.

——. 1982b. Egg-laying, larval periods and metamorphosis of *Eurycea bislineata* and *E. junaluska* at Santeetlah Creek, North Carolina. Copeia 1982:755–762.

——. 1985a. Larval period and metamorphosis in the salamander *Eurycea bislineata*. Herpetologica 41:19–28.

——. 1985b. Larval periods, population structure and the effects of stream drift in larvae of the salamanders *Desmognathus quadramaculatus* and *Leurognathus marmoratus* in a southern Appalachian stream. Copeia 1985:847–854.

——. 1988a. An ecological life table for the salamander *Eurycea wilderae*. Copeia 1988:15–26.

——. 1988b. Life history variation in the salamander *Desmognathus quadramaculatus*. Herpetologica 44:218–227.

——. 1989. Life history of the salamander *Desmognathus monticola*, with a comparison of the larval periods of *D. monticola* and *D. ochrophaeus*. Herpetologica 45:144–155.

——. 1996. Life-history perspective of adaptive radiation in desmognathine salamanders. Copeia 1996:783–790.

——. 2003. Ecological distribution of the salamanders *Gyrinophilus* and *Pseudotriton* in a southern Appalachian watershed. Herpetologica 59:301–310.

Bruce, R. C., and C. K. Beachy. 2003. Life history of a small form of the plethodontid salamander *Desmognathus quadramaculatus*. Amphibia-Reptilia 24:13–26.

Bruce, R. C., C. K. Beachy, P. G. Lenzo, S. P. Pronych, and R. J. Wassersug. 1994. Effects of lung reduction on rheotactic performance in amphibian larvae. Journal of Experimental Zoology 268:377–380.

Bruce, R. C., and N. G. Hairston Jr. 1990. Life-history correlates of body-size differences between two populations of the salamander, *Desmognathus monticola*. Journal of Herpetology 24:126–134.

Bruneau, M., and E. Magnin. 1980. Larval development in *Rana catesbeiana* Shaw (Amphibia, Anura) of the Laurentian region of Quebec. Canadian Journal of Zoology 58:169–174.

Brunkow, P. E., and J. P. Collins. 1998. Group size structure affects patterns of aggression in larval salamanders. Behavioral Ecology 8:508–514.

Buchholz, D. R., and T. B. Hayes. 2000. Larval period comparison for the Spadefoot Toads *Scaphiopus couchii* and *Spea multiplicata* (Pelobatidae: Anura). Herpetologica 56:455–468.

Burger, W. L. 1950. Novel aspects of the life history of two ambystomas. Journal of the Tennessee Academy of Science 25:252–257.

Burger, W. L., H. M. Smith, and F. E. Potter Jr. 1950. Another neotenic *Eurycea* from the Edwards Plateau. Proceedings of the Biological Society of Washington 63:51–58.

Burton, T. M. 1976. An analysis of the feeding ecology of the salamanders (Amphibia, Urodela) of the Hubbard Brook Experimental Forest, New Hampshire. Journal of Herpetology 10:187–204.

Burton, T. M., and G. E. Likens. 1975. Salamander populations and biomass in the Hubbard Brook Experimental Forest, New Hampshire. Copeia 1975:541–546.

Bury, R. B., and M. J. Adams. 1999. Variation in age at metamorphosis across a latitudinal gradient for the Tailed Frog, *Ascaphus truei*. Herpetologica 55:283–290.

Bury, R. B., P. Loafman, D. Rofkar, and K. I. Mike. 2001. Clutch size and nests of Tailed Frogs from the Olympic Peninsula, Washington. Northwestern Science 75:419–422.

Butterfield, B. P., W. E. Meshaka, and S. E. Trauth. 1989. Fecundity and egg mass size of the Illinois Chorus Frog, *Pseudacris streckeri illinoensis* (Hylidae), from northeastern Arkansas. Southwestern Naturalist 34:556–557.

Cagle, F. R. 1948. Observations on a population of the salamander *Amphiuma tridactylum* Cuvier. Ecology 29:479–491.

———. 1954. Observations on the life history of the salamander *Necturus louisianensis*. Copeia 1954:257–260.

Cahn, A. R., and W. Shumway. 1926. Color variations in larvae of *Necturus maculosus*. Copeia 1926:106–107.

Caldwell, J. P. 1982. Disruptive selection: A tail color polymorphism in *Acris* tadpoles in response to differential predation. Canadian Journal of Zoology 60:2818–2827.

———. 1987. Demography and life history of two species of chorus frogs (Anura: Hylidae) in South Carolina. Copeia 1987:114–127.

Caldwell, J. P., J. H. Thorpe, and T. O. Jervey. 1980. Predator-prey relationships among larval dragonflies, salamanders and frogs. Oecologia 46:285–289.

Caldwell, R. S., and W. C. Houtcooper. 1973. Food habits of larval *Eurycea bislineata*. Journal of Herpetology 7:386–388.

Calef, G. W. 1973. Natural mortality of tadpoles in a population of *Rana aurora*. Ecology 54:741–758.

Camara, J., and B. W. Buttrey. 1961. Intestinal protozoa from tadpoles and adults of the Mink Frog, *Rana septentrionalis* Baird. Proceedings of the South Dakota Academy of Science 41:59–60.

Camp, C. D. 2000. *Desmognathus ocoee* (Ocoee Salamander). Reproductive failure. Herpetological Review 31:166.

Camp, C. D., C. E. Condee, and G. Lovell. 1990. Oviposition, larval development, and metamorphosis in the Wood Frog, *Rana sylvatica* (Anura: Ranidae) in Georgia. Brimleyana 16:17–21.

Camp, C. D., and J. L. Marshall. 2006. Reproductive life history of *Desmognathus folkertsi* (Dwarf Black-bellied Salamander). Southeastern Naturalist 5:669–684.

Camp, C. C., T. J. Krieger, and J. A. Mitchem. 2010a. *Desmognathus quadramaculatus* (Black-bellied Salamander). Record clutch and egg size. Herpetological Review 41:330.

Camp, C. C., J. A. Mitchem, T. J. Krieger, and C. E. Skelton. 2010b. *Desmognathus folkertsi* (Dwarf Black-bellied Salamander). Reproduction. Herpetological Review 41:330.

Camp, C. D., W. E. Peterman, J. R. Milanovich, T. Lamb, and D. B. Wake. 2009. A new genus and species of lungless salamander (family Plethodontidae) from the Appalachian highlands of the south-eastern United States. Journal of Zoology 2009:1–9.

Cargo, D. G. 1960. Predation of eggs of the Spotted Salamander, *Ambystoma maculatum*, by the leech, *Macrobdella decora*. Chesapeake Science 1:119–120.

Carlyle, J. C., D. E. Sanders, and M. V. Plummer. 1998. *Eurycea lucifuga* (Cave Salamander). Reproduction. Herpetological Review 29:37–38.

Carpenter, C. C. 1953. Aggregation behavior of tadpoles of *Rana p. pretiosa*. Herpetologica 9:77–78.

Carr, A. F. 1939. *Haideotriton wallacei*, a new subterranean salamander from Georgia. Occasional Papers of the Boston Society of Natural History 8:333–336.

——. 1952. Handbook of turtles: The turtles of the United States, Canada, and Baja California. Ithaca, NY: Cornell University Press.

Carroll, S. L., E. L. Ervin, and R. N. Fisher. 2005. *Taricha torosa torosa* (Coast Range Newt): Overwintering larvae. Herpetological Review 36:297.

Casterlin, M. E., and W. W. Reynolds. 1979. Behavioural thermoregulation in *Rana pipiens* tadpoles. Journal of Thermal Biology 3:143–146.

Cecala, K. K., S. J. Price, and M. E. Dorcas. 2007. Diet of larval Red Salamanders (*Pseudotriton ruber*) examined using a nonlethal technique. Journal of Herpetology 41:741–745.

Cecil, S. G., and J. J. Just. 1979. Survival rate, population density and development of a naturally occurring anuran larva (*Rana catesbeiana*). Copeia 1979:447–453.

Chadwick, C. S. 1944. Observations on the life cycle of the Common Newt of western North Carolina. American Midland Naturalist 32:491–494.

——. 1950. Observations on the behavior of the larvae of the common American Newt during metamorphosis. American Midland Naturalist 43:392–398.

Chalmers, R. J., and C. S. Loftin. 2006. *Hemidactylium scutatum* (Four-toed Salamander). Morphology/phenology. Herpetological Review 37:69–71.

Chan, L. M. 2001. *Bufo canorus* (Yosemite Toad). Larval predation. Herpetological Review 32:101.

Chaney, A. H. 1951. The food habits of the salamander *Amphiuma tridactylum*. Copeia 1951:45–49.

Chapel, W. L. 1939. Field notes on *Hyla wrightorum* Taylor. Copeia 1939:225–227.

Chazal, A. C., J. D. Krenz, and D. E. Scott. 1996. Relationship of larval density and heterozygosity to growth and survival of juvenile Marbled Salamanders (*Ambystoma opacum*). Canadian Journal of Zoology 74:1122–1129.

Chippindale, P. T., A. H. Price, and D. M. Hillis. 1993. A new species of perennibranchiate salamander (*Eurycea*: Plethodontidae) from Austin, Texas. Herpetologica 49:248–259.

Chippindale, P. T., A. H. Price, J. J. Wiens, and D. M. Hillis. 2000. Phylogenetic relationships and systematic revision of central Texas hemidactyliine plethodontid salamanders. Herpetological Monographs (14): 1–80.

Chivers, D. P., E. L. Wildy, and A. R. Blaustein. 1997. Eastern Long-toed Salamander (*Ambystoma macrodactylum columbianum*) larvae recognize cannibalistic conspecifics. Ethology 103:187–197.

Christman, S. P., and L. R. Franz. 1973. Feeding habits of the Striped Newt, *Notophthalmus perstriatus*. Journal of Herpetology 7:133–135.

Clarke, S. F. 1880. Development of *Ambystoma punctatum*. Studies from the Biology Laboratory of Johns Hopkins University 1:105–125.

Claussen, D. L. 1973. The thermal relations of the Tailed Frog, *Ascaphus truei*, and the Pacific Treefrog, *Hyla regilla*. Comparative Biochemistry and Physiology 44A:137–153.

Cliburn, J. W. 1972. Observations on the egg mass and young of *Ambystoma maculatum* in Mississippi. Journal of the Mississippi Academy of Science 18:48–50.

Cliburn, J. W., and S. D. Carey. 1975. Foods of neotenic *Ambystoma talpoideum* and associated adults. Journal of the Mississippi Academy of Science 20:49–52.

Cliburn, J. W., and B. Q. Ward. 1963. Occurrence of *Oophila amblystomatis* (a symbiotic alga) in *Ambystoma maculatum* of the lower Gulf Coastal Plain. American Midland Naturalist 69:508–509.

Cline, G. R., and R. Tumlison. 2001. Distribution and relative abundance of the Oklahoma Salamander (*Eurycea tynerensis*). Proceedings of the Oklahoma Academy of Science 81:1–10.

Cline, G. R., R. Tumlison, and P. Zwank. 1989. Biology and ecology of the Oklahoma Salamander, *Eurycea tynerensis*: Literature review and comments on research needs. Bulletin of the Chicago Herpetological Society 24:164–168.

Coatney, C. E., Jr. 1982. Home range and nocturnal activity of the Ozark Hellbender, *Cryptobranchus alleganiensis alleganiensis*, in a Pennsylvania stream. Journal of Herpetology 5:121–126.

Cochran, P. A. 1995. *Necturus maculosus* (Mudpuppy). Reproduction. Herpetological Review 26:198.

Cochran, P. A., and J. D. Lyons. 1985. *Necturus maculosus*. Juvenile ecology. Herpetological Review 16:53.

Collette, B. B., and F. R. Gehlbach. 1961. The salamander *Siren intermedia intermedia* LeConte in North Carolina. Herpetologica 17:203–204.

Collins, J. P. 1979. Intrapopulational variation in the body size at metamorphosis and timing of metamorphosis in the Bullfrog, *Rana catesbeiana*. Ecology 60:738–749.

——. 1981. Distribution, habitats, and life history variation in the Tiger Salamander, *Ambystoma tigrinum*, in east-central and southeast Arizona. Copeia 1981:666–675.

Collins, J. P., and J. E. Cheek. 1983. Effect of food density on development of typical and cannibalistic salamander larvae in *Ambystoma tigrinum nebulosum*. American Zoologist 23:77–84.

Collins, J. P., and J. R. Holomuzski. 1984. Intraspecific variation in diet within and between trophic morphs in larval Tiger Salamanders (*Ambystoma tigrinum nebulosum*). Canadian Journal of Zoology 62:168–174.

Collins, J. P., and M. A. Lewis. 1979. Overwintering tadpoles and breeding season variation in the *Rana pipiens* complex in Arizona. Southwestern Naturalist 24:371–373.

Collins, J. P., and H. M. Wilbur. 1979. Breeding habits and habitats of the amphibians of the Edwin S. George Reserve, Michigan, with notes on the local distribution of fishes. Occasional Papers of the Museum of Zoology, University of Michigan (686): 1–34.

Collins, J. P., K. E. Zerba, and M. J. Sredl. 1993. Shaping intraspecific variation: Development, ecology and the evolution of morphology and life history variation in Tiger Salamanders. Genetica 89:167–183.

Conant, R., and J. T. Collins. 1998. A field guide to reptiles and amphibians: Eastern and central North America. Boston: Houghton Mifflin.

Conrads, L. M. 1969. Demography and ecology of the Fern Bank Salamanders, *Eurycea pterophila*. Thesis, Southwest Texas State University.

Cook, D., and M. R. Jennings. 2001. *Rana aurora draytonii* (California Red-legged Frog). Predation. Herpetological Review 32:182–183.

Cook, M. L., and B. C. Brown. 1974. Variation in the genus *Desmognathus* (Amphibia: Plethodontidae) in the western limits of its range. Journal of Herpetology 8:93–105.

Cope, E. D. 1885. The retrograde metamorphosis of *Siren*. American Naturalist 19:1226–1227.

Corkran, C. C., and C. Thoms. 1996. Amphibians of Oregon, Washington and British Columbia. Edmonton: Lone Pine Press.

Corn, P. S. 1981. Field evidence for a relationship between color and developmental rate in the Northern Leopard Frog (*Rana pipiens*). Herpetologica 37:155–160.

Corn, P. S., and L. J. Livo. 1989. Leopard Frog and Wood Frog reproduction in Colorado and Wyoming. Northwestern Naturalist 70:1–9.

Cory, L., and J. J. Manion. 1953. Predation on eggs of the Wood Frog, *Rana sylvatica*, by leeches. Copeia 1953:66.

Crawshaw, L. I., R. N. Rausch, L. P. Wollmuth, and E. J. Bauer. 1992. Seasonal rhythms of development and temperature selection in larval Bullfrogs, *Rana catesbeiana* Shaw. Physiological Zoology 65:346–359.

Cronin, J. T., and J. Travis. 1986. Size-limited predation on larval *Rana areolata* (Anura: Ranidae) by two species of backswimmer (Insecta: Hemiptera: Notonectidae). Herpetologica 42:171–174.

Cross, J. F., and R. D. Hoyt. 1972. The effect of combined temperature and photoperiod on early development rate of *Rana pipiens* eggs. Transactions of the Kentucky Academy of Science 33:27–32.

Crossland, M. R., M. N. Hearden, L. Pizzatto, R. A. Alford, and R. Shine. 2011. Why be a cannibal? The benefits of cane toad, *Rhinella marina* (= *Bufo marinus*), tadpoles of consuming conspecific eggs. Animal Behaviour 82P:775–782.

Crother, B. I., ed. 2012. Scientific and standard English names of amphibians and reptiles of North America north of Mexico, with comments regarding confidence in our understanding. SSAR Herpetological Circular 39:1–92.

Crump, M. L. 1984. Intraclutch egg size variability in *Hyla crucifer* (Anura: Hylidae). Copeia 1984:302–308.

Cuellar, O. 1994. Ecological observations on *Rana pretiosa* in western Utah. Alytes 12:109–121.

Culver, D. C. 1973. Feeding behavior of the salamander *Gyrinophilus porphyriticus* in caves. International Journal of Speleology 5:369–377.

Cunningham, J. D. 1964. Observations on the ecology of the Canyon Tree Frog, *Hyla californiae*. Herpetologica 20:55–61.

Cupp, P. V., Jr. 1980. Thermal tolerance of five salientian amphibians during development and metamorphosis. Herpetologica 36:234–244.

Dahl, A., M. P. Donovan, and T. D. Schwaner. 2000. Egg mass deposition by Arizona Toads, *Bufo microscaphus*, along a narrow canyon stream. Western North American Naturalist 60:456–458.

Dalrymple, G. H. 1970. Caddis fly larvae feeding upon eggs of *Ambystoma tigrinum*. Herpetologica 26:128–129.

Daniel, J. F. 1937a. The living egg-cell of *Triturus torosus*. Cytologia 1937:641–643.

———. 1937b. The early embryology of *Triturus torosus*. University of California Publications in Zoology 43:321–343.

Daugherty, C. H., and A. L. Sheldon. 1982. Age-determination, growth, and life history of a Montana population of the Tailed Frog (*Ascaphus truei*). Herpetologica 38:461–468.

Davic, R. D. 2005. Using limb morphology to distinguish Two-lined Salamander larvae (*Eurycea*) from Northern Dusky Salamander larvae (*Desmognathus*). Herpetological Review 36:9–12.

Davis, M. S., and G. W. Folkerts. 1986. Life history of the Wood Frog, *Rana sylvatica* LeConte (Amphibia: Ranidae), in Alabama. Brimleyana 12:29–50.

Davis, W. B., and E. T. Knapp. 1953. Notes on the salamander *Siren intermedia*. Copeia 1953:119–121.

Dayton, G. H., and R. E. Jung. 1999. *Scaphiopus couchii* (Couch's Spadefoot). Predation. Herpetological Review 30:164.

Dayton, G. H., D. Saenz, K. A. Baum, R. B. Langerhans, and T. J. DeWitt. 2005. Body shape, burst speed and escape behavior of larval anurans. Oikos 111:582–591.

Dayton, G. H., and S. D. Wapo. 2002. Cannibalistic behavior in *Scaphiopus couchii*: More evidence for larval anuran oophagy. Journal of Herpetology 36:531–532.

DeBenedictis, P. A. 1974. Interspecific competition between tadpoles of *Rana pipiens* and *Rana sylvatica*: An experimental field study. Ecological Monographs 44:129–151.

Delis, P. R., and A. P. Summers. 1996. *Hyla gratiosa* (Barking Treefrog). Reproduction. Herpetological Review 27:18–19.

Dempster, W. T. 1930. The growth of the larvae of *Ambystoma maculatum* under natural conditions. Biological Bulletin 58:182–192.

——. 1933. Growth in *Ambystoma punctatum* during embryonic and early larval period. Journal of Experimental Zoology 64:495–511.

De Neff, S. J., and D. M. Sever. 1977. Ontogenetic changes in phototactic behavior of *Ambystoma tigrinum tigrinum* (Amphibia: Urodela). Proceedings of the Indiana Academy of Science 86:478–481.

Dennis, D. M. 1962. Notes on the nesting habits of *Desmognathus fuscus fuscus* (Raf.) in Licking County, Ohio. Journal of the Ohio Herpetological Society 3:28–35.

Denoël, M., H. H. Whiteman, and S. A. Wissinger. 2006. Temporal shift of diet in alternative cannibalistic morphs of the Tiger Salamander. Biological Journal of the Linnean Society 89:373–382.

Dent, J. N., and J. S. Kirby-Smith. 1963. Metamorphic physiology and morphology of the Cave Salamander *Gyrinophilus palleucus*. Copeia 1963:119–130.

Desroches, J.-F., and D. Pouliot. 2005. *Hemidactylium scutatum* (Four-toed Salamander). Nests. Herpetological Review 2005:51–52.

Desroches, J.-F., and D. Rodrigue. 2004. Amphibiens et reptiles du Québec et des Maritimes. Waterloo, Quebec: Éditions Michel Quintin.

Dethlefsen, E. S. 1948. A subterranean nest of the Pacific Giant Salamander, *Dicamptodon ensatus* (Eschscholtz). Wasmann Collector 7:81–84.

DeVlaming, V. L., and R. B. Bury. 1970. Thermal selection in tadpoles of the Tailed Frog, *Ascaphus truei*. Journal of Herpetology 4:179–189.

Dickman, M. 1968. The effect of grazing by tadpoles on the structure of a periphyton community. Ecology 49:1188–1190.

Dodd, C. K., Jr. 1980. Notes on the feeding behavior of the Oklahoma Salamander, *Eurycea tynerensis* (Plethodontidae). Southwestern Naturalist 25:111–113.

——. 1993. Cost of living in an unpredictable environment: The ecology of Striped Newts *Notophthalmus perstriatus* during a prolonged drought. Copeia 1993:605–614.

——. 1995. The ecology of a sandhills population of the Eastern Narrow-mouthed Toad, *Gastrophryne carolinensis*, during a drought. Bulletin of the Florida Museum of Natural History, Biological Series 38:11–41.

——. 1998. *Desmognathus auriculatus* at Devil's Millhopper State Geological Site, Alachua County, Florida. Florida Scientist 61:38–45.

——. 2004. The amphibians of Great Smoky Mountains National Park. Knoxville: University of Tennessee Press.

Dodson, S. I., and V. E. Dodson. 1971. The diet of *Ambystoma tigrinum* larvae from western Colorado. Copeia 1971:614–624.

Doody, J. S. 1996. Larval growth rate of known age *Ambystoma opacum* in Louisiana under natural conditions. Journal of Herpetology 30:294–297.

Drake, D. L., A. Drayer, and S. E. Trauth. 2007. *Bufo americanus* (American Toad). Algal symbiosis. Herpetological Review 38:435–436.

Dreyer, T. F. 1915. The morphology of the tadpole of *Xenopus laevis*. Transactions of the Royal Society of South Africa 4:241–258.

Dudley, R., V. A. King, and R. J. Wassersug. 1991. The implications of shape and metamorphosis for drag forces on a generalized pond tadpole (*Rana catesbeiana*). Copeia 1991: 252–257.

Duellman, W. E. 1951. Notes on the reptiles and amphibians of Greene County, Ohio. Ohio Journal of Science 51:335–341.

——. 1968. *Smilisca baudinii*. Catalogue of American Amphibians and Reptiles 59.1–59.2.

——. 1970. The hylid frogs of Middle America. Monographs of the Museum of Natural History, University of Kansas (1): 1–753, 72 plates.

Duellman, W. E., and R. I. Crombie. 1970. *Hyla septentrionalis*. Catalogue of American Amphibians and Reptiles 92.1–92.4.

Duellman, W. E., and A. Schwartz. 1958. Amphibians and reptiles of southern Florida. Bulletin of the Florida State Museum 3:181–324.

Duellman, W. E., and L. Trueb. 1966. Neotropical hylid frogs, genus *Smilisca*. University of Kansas Publications of the Museum of Natural History 17:281–375.

——. 1986. Biology of amphibians. New York: McGraw-Hill.

Duellman, W. E., and J. T. Wood. 1954. Size and growth of the Two-lined Salamander, *Eurycea bislineata rivicola*. Copeia 1954:92–96.

Duncan, R. B. 1999. *Ambystoma tigrinum stebbinsi* (Sonoran Tiger Salamander). Predation. Herpetological Review 30:159.

Dundee, H. A. 1947. Notes on salamanders collected in Oklahoma. Copeia 1947:117–120.

——. 1965a. *Eurycea multiplicata*. Catalogue of American Amphibians and Reptiles 21.1–21.2.

——. 1965b. *Eurycea tynerensis*. Catalogue of American Amphibians and Reptiles 22.1–22.2.

——. 1971. Cryptobranchidae, *C. alleganiensis*. Catalogue of American Amphibians and Reptiles 101.1–101.4.

——. 1998. *Necturus punctatus*. Catalogue of American Amphibians and Reptiles 663.1–663.5.

Dundee, H. A., and D. A. Rossman. 1989. The amphibians and reptiles of Louisiana. Baton Rouge: Louisiana State University Press.

Dunn, E. R. 1915. The transformation of *Spelerpes ruber* (Daudin). Copeia 1915:28–30.

——. 1917a. The breeding habits of *Ambystoma opacum* (Gravenhorst). Copeia 1917:40–43.

——. 1917b. The salamanders of the genus *Desmognathus* and *Leurognathus*. Proceedings of the United States Museum of Natural History 37:593–634.

——. 1924. *Siren*, a herbivorous salamander? Science 59:145.

——. 1941. Notes on *Dendrobates auratus*. Copeia 1941:88–93.

Dupré, R. K., and J. W. Petranka. 1985. Ontogeny of temperature selection in larval amphibians. Copeia 1985:462–467.

Durflinger Moreno, M. C., C. Guyer, and M. A. Bailey. 2006. Distribution and population biology of the Black Warrior Waterdog, *Necturus alabamensis*. Southeastern Naturalist 5:69–84.

DuShane, B. P., and C. Hutchinson. 1944. Differences in size and developmental rate between eastern and midwestern embryos of *Ambystoma maculatum*. Ecology 25:414–422.

Eagleson, G. W. 1976. A comparison of life histories and growth patterns of the salamander *Ambystoma gracile* (Baird) from permanent low-altitude and montane lakes. Canadian Journal of Zoology 54:2098–2111.

Eakin, R. M. 1947. Stages in the normal development of *Hyla regilla*. University of California Publications in Zoology 51:245–257.

Easteal, S. 1986. *Bufo marinus*. Catalogue of American Amphibians and Reptiles 395.1–395.4.

Eaton, B. R., and C. A. Pazakowski. 1999. *Rana sylvatica* (Wood Frog). Predation. Herpetological Review 30:164.

Eaton, T. H., Jr. 1941. Notes on the life history of *Dendrobates auratus*. Copeia 1941:93–95.

——. 1953. Salamanders of Pitt County, North Carolina. Journal of the Elisha Mitchell Scientific Society 69:49–53.

——. 1954. *Desmognathus perlapsus* Neill in North Carolina. Herpetologica 10:41–43.

——. 1956. Larvae of some Appalachian plethodontid salamanders. Herpetologica 12:303–311.

Eaton, T. H., Jr., and R. M. Imagawa. 1948. Early development of *Pseudacris clarkii*. Copeia 1948:263–266.

Eichle, V. B., and L. S. Gray Jr. 1976. The influence of environmental lighting on the growth and prometamorphic development of larval *Rana pipiens*. Development, Growth, and Differentiation 118:177–182.

Eidietis, L. 2005. Size-related performance variation in the Wood Frog (*Rana sylvatica*) tadpole: Tactile-stimulated startle response. Canadian Journal of Zoology 83:1117–1127.

Eklöv, P., and E. Werner. 2000. Multiple predator effects on size-dependent behavior and mortality of two species of anuran larvae. Oikos 88:250–258.

Elliott, M. J., S. N. Smith, and J. B. Jensen. 2002. *Hemidactylium scutatum* (Four-toed Salamander). Reproduction. Herpetological Review 33:45.

Emerson, E. T. 1905. General anatomy of *Typhlomolge rathbuni*. Proceedings of the Boston Society of Natural History 32:43–76.

Enge, K. M. 1998. *Amphiuma means* (Two-toed Amphiuma). Diet. Herpetological Review 29:162.

Epp, K. J., and C. R. Gabor. 2008. Innate and learned predator recognition mediated by chemical signals in *Eurycea nana*. Ethology 114:607–615.

Ervin, E. L., C. D. Smith, and S. V. Christopher. 2005. *Spea hammondii* (Western Spadefoot). Reproduction. Herpetological Review 36:309–310.

Evans, T., M. Evans, and P. Verrell. 2005. *Dicamptodon aterrimus* (Idaho Giant Salamander). Reproduction. Herpetological Review 36:295.

Eycleshymer, A. C. 1893. The early development of *Amblystoma*, with observations on some other vertebrates. Journal of Morphology 10:343–418.

——. 1906. The habits of *Necturus maculosus*. American Naturalist 40:123–136.

Eycleshymer, A. C., and J. M. Wilson. 1910. Normal plates of the development of *Necturus maculosus*, Normentafeln zue Entwicklunsgeschichte der Wirbeltiere. Vienna: Gustav Fischer.

Fanning, S. A. 1966. A synopsis and key to the tadpoles of Florida. Thesis, Florida State University, Tallahassee.

Faragher, S. G., and R. G. Jaeger. 1998. Tadpole bullies: Examining mechanisms of competition in a community of larval anurans. Canadian Journal of Zoology 76:144–153.

Farner, D. S., and J. Kezer. 1953. Notes on the amphibians and reptiles of Crater Lake National Park. American Midland Naturalist 50:448–462.

Fauth, J. E. 1999. Interactions between branchiate Mole Salamanders (*Ambystoma talpoideum*) and Lesser Sirens (*Siren intermedia*): Asymmetrical competition and intraguild predation. Amphibia-Reptilia 20:119–132.

Fauth, J. E., B. W. Buchanan, S. E. Wise, S. M. Welter, and M. J. Komoroski. 1996. *Cryptobranchus alleganiensis alleganiensis* (Hellbender). Coloration. Herpetological Review 27:135.

Fauth, J. E., and W. J. Resetarits. 1991. Interactions between the salamanders *Siren intermedia* and the keystone predator *Notophthalmus viridescens*. Ecology 72:827–838.

——. 1999. Biting in the salamander *Siren intermedia intermedia*: Courtship component or agonistic behavior. Journal of Herpetology 33:493–496.

Fedak, M. A. 1971. A comparative study of the life histories of *Necturus lewisi* Brimley and *Necturus punctatus* Gibbes (Caudata: Proteidae) in North Carolina. Thesis, Duke University, Durham, NC.

Fedewa, L. A. 2006. Fluctuating gram-negative microflora in developing anurans. Journal of Herpetology 40:131–135.

Feinberg, J. A., and K. Hoffman. 2004. *Scaphiopus holbrookii* (Eastern Spadefoot). Coloration and behavior. Herpetological Review 35:377.

Fellers, G. M., A. E. Launer, G. Rathbun, S. Bobzien, J. Alvarez, D. Sterner, R. B. Seymour, and M. Westphal. 2001. Overwintering tadpoles in the California Red-legged Frog (*Rana aurora draytonii*). Herpetological Review 32:156–157.

Feminella, J. W., and C. P. Hawkins. 1994. Tailed Frog tadpoles differentially alter their feeding behavior in response to non-visual cues from four predators. Journal of the North American Benthological Society 13:310–320.

Feral, D., M. A. Camann, and H. H. Welsh Jr. 2005. *Dicamptodon tenebrosus* larvae within hyporheic zones of intermittent streams in California. Herpetological Review 36:26–27.

Ferguson, D. E. 1963. *Ambystoma macrodactylum*. Catalogue of Amphibians and Reptiles 4.1–4.2.

Ferguson, H. M. 2000. Larval colonisation and recruitment in the Pacific Giant Salamander (*Dicamptodon tenebrosus*) in British Columbia. Canadian Journal of Zoology 78:1238–1242.

Fernandez, P. J., Jr., and J. P. Collins. 1988. Effect of environmental and ontogeny on color pattern variation in Arizona Tiger Salamanders (*Ambystoma tigrinum nebulosum* Hallowell). Copeia 1988:928–938.

Field, K. J., and D. Groebner. 2005. *Rana chiricahuensis* (Chiricahua Leopard Frog). Reproduction. Herpetological Review 36:306–307.

Figiel, C. R., Jr., and R. D. Semlitsch. 1990. Population variation in survival and metamorphosis of larval salamanders (*Ambystoma maculatum*) in the presence and absence of fish predation. Copeia 1990:818–826.

———. 1991. Effects of nonlethal injury and habitat complexity on predation in tadpole populations. Canadian Journal of Zoology 69:830–834.

Fitch, H. S. 1936. Amphibians and reptiles of the Rogue River Basin, Oregon. American Midland Naturalist 17:634–652.

———. 1956. A field study of the Kansas Ant-Eating Frog, *Gastrophryne olivacea*. University of Kansas Publications of the Museum of Natural History 8:275–306.

Fitch, K. L. 1959. Observations on the nesting habits of the Mudpuppy, *Necturus maculosus* Rafinesque. Copeia 1959:339–340.

Fitzpatrick, L. C. 1973. Energy allocation in the Allegheny Mountain Salamander, *Desmognathus ochrophaeus*. Ecological Monographs 43:43–58.

Floyd, R. B. 1985. Effects of photoperiod and starvation on the temperature tolerance of larvae of the Giant Toad, *Bufo marinus*. Copeia 1985:625–631.

Foard, T., and D. L. Auth. 1990. Food habits and gut parasites of the salamander *Stereochilus marginatus*. Journal of Herpetology 24:428–431.

Fontenot, C. L., Jr. 1999. Reproductive biology of the aquatic salamander *Amphiuma tridactylum* in Louisiana. Journal of Herpetology 33:100–105.

Ford, P. L., and N. J. Scott Jr. 1996. Descriptions of *Bufo* tadpoles from the southwestern coast of Jalisco, Mexico. Journal of Herpetology 30:253–257.

Forester, D. C., and D. V. Lykens. 1988. The ability of Wood Frog eggs to withstand prolonged terrestrial stranding: An empirical study. Canadian Journal of Zoology 66:1733–1735.

Formanowicz, D. R., Jr., and E. D. Brodie Jr. 1982. Relative palatabilities of members of a larval amphibian community. Copeia 1982:91–97.

Fortman, J., and R. Altig. 1973. Characters of F1 hybrid tadpoles between six species of *Hyla*. Copeia 1973:411–415.

Foster, M. S., and R. W. McDiarmid. 1982. Study of aggregative behavior of *Rhinophrynus dorsalis* tadpoles: Design and analysis. Herpetologica 38:395–404.

Fouquette, M. J., Jr. 1969. Rhinophrynidae, *R. dorsalis*. Catalogue of American Amphibians and Reptiles 78.1–78.2.

———. 1970. *Bufo alvarius*. Catalogue of American Amphibians and Reptiles 93.1–93.4.

Fouquette, M. J., Jr., and M. J. Littlejohn. 1960. Patterns of oviposition in two species of hylid frogs. Southwestern Naturalist 5:92–96.

Fowler, J. A. 1946a. The eggs of *Pseudotriton montanus montanus*. Copeia 1946:105.

———. 1946b. Partial neoteny in a Common Newt. Proceedings of the Biological Society of Washington 59:166.

Fox, S. 2008. Oophagy and larval cannibalism without polyphenism in tadpoles of the Great Basin Spadefoot (*Spea intermontana*). Herpetological Review 390:151–154.

Franklin, C. J. 2000. *Eurycea quadridigitata* (Dwarf Salamander). Larval predation. Herpetological Review 31:98.

Franz, L. R., Jr. 1965. The eggs of the Long-tailed Salamander from a Maryland cave. Herpetologica 20:216.

———. 1967. Notes on the Long-tailed Salamander, *Eurycea longicauda longicauda* (Green), in Maryland caves. Bulletin of the Maryland Herpetological Society 3:1–6.

——. 1970a. Egg development of the Tailed Frog under natural conditions. Bulletin of the Maryland Herpetological Society 6:27–30.

——. 1970b. Food of larval Tailed Frogs. Bulletin of the Maryland Herpetological Society 6:49–51.

——. 1970c. Additional notes on the feeding of larval Giant Salamanders, *Dicamptodon ensatus*. Bulletin of the Maryland Herpetological Society 6:51–52.

Franz, R., and C. J. Chantell. 1978. *Limnaoedus, L. ocularis*. Catalogue of American Amphibians and Reptiles 209.1–209.2.

Franz, R., and H. Harris. 1965. Mass transformation and movement of larval Long-tailed Salamanders, *Eurycea longicauda longicauda* (Green). Journal of the Ohio Herpetological Society 5:32.

Freda, J. 1983. Diet of larval *Ambystoma maculatum* in New Jersey. Journal of Herpetology 17:177–179.

Freeman, J. R. 1967. Feeding behavior of the Narrow-striped Dwarf Siren *Pseudobranchus striatus axanthus*. Herpetologica 23:313–314.

Freeman, S. L., and R. C. Bruce. 2001. Larval period and metamorphosis of the Three-lined Salamander, *Eurycea guttolineata* (Amphibia: Plethodontidae), in the Chattooga River watershed. American Midland Naturalist 145:194–200.

Freiberg, R. W. 1951. An ecological study of the Narrow-mouthed Toad (*Microhyla*) in northeastern Kansas. Transactions of the Kansas Academy of Science 54:374–386.

Frese, P. W., and E. Britzke. 2001. *Siren intermedia nettingi* (Western Lesser Siren). Predation. Herpetological Review 32:99.

Frese, P. W., A. Mathis, and R. Wilkinson. 2003. Population characteristics, growth, and spatial activity of *Siren intermedia* in an intensively managed wetland. Southwestern Naturalist 48:534–542.

Frick, M. G. 1999. *Siren lacertina* (Greater Siren). Parasitism. Herpetological Review 30:162.

Fries, J. N. 2002. Upwelling flow velocity preferences of captive adult San Marcos Salamanders. North American Journal of Aquaculture 64:113–116.

Fritts, T. H. 1966. Reproductive potential of *Amphiuma means*. Copeia 1966:598–599.

Frost, D. R. 2011. Amphibian species of the world: An online reference. Version 5.5 (31 January 2011), http://research.amnh.org/vz/herpetology/amphibia/.

Gage, S. H. 1891. Life-history of the Vermilion-spotted Newt (*Diemyctylus viridescens* Raf.). American Naturalist 25:1084–1110.

Gaige, H. T. 1917. Description of a new salamander from Washington. Occasional Papers of the Museum of Zoology, University of Michigan (40): 1–3.

——. 1920. Observations upon the habits of *Ascaphus truei* Stejneger. Occasional Papers of the Museum of Zoology, University of Michigan (84): 1–11.

Garcia, T. S., R. Straus, and A. Sih. 2003. Temperature and ontogenetic effects on color change in the larval salamander species *Ambystoma barbouri* and *Ambystoma texanum*. Canadian Journal of Zoology 81:710–715.

Garcia, T. S., L. L. Thurman, J. C. Rowe, S. M. Selego, and W. Koenig. 2012. Antipredator behaviour of American Bullfrogs (*Lithobates catesbeianus*) in a novel environment. Ethology 118:867–875.

Gardner, J. D. 1995. *Pseudacris regilla* (Pacific Chorus Frog). Reproduction. Herpetological Review 26:32.

Gascon, C., and D. Planas. 1986. Spring pond water chemistry and the reproduction of the Wood Frog, *Rana sylvatica*. Canadian Journal of Zoology 64:543–550.

Gates, W. R. 1988. *Pseudacris nigrita*. Catalogue of American Amphibians and Reptiles 416.1–416.3.

Gatz, A. J., Jr. 1973. Algal entry into the eggs of *Ambystoma maculatum*. Journal of Herpetology 7:137–138.

Gaudin, A. J. 1964. The tadpole of *Hyla californiae* Gorman. Texas Journal of Science 16:80–84.

———. 1965. Larval development of the tree frogs *Hyla regilla* and *Hyla californiae*. Herpetologica 21:117–130.

Gee, J. H., and R. C. Waldick. 1995. Ontogenetic buoyancy changes and hydrostatic control in larval anurans. Copeia 1995:861–870.

Gehlbach, F. R. 1967. *Ambystoma tigrinum*. Catalogue of American Amphibians and Reptiles 52.1–52.4.

Gehlbach, F. R., R. Gordon, and J. B. Jordan. 1973. Aestivation in the salamander, *Siren intermedia*. American Midland Naturalist 89:455–463.

Gehlbach, F. R., and S. E. Kennedy. 1978. Population ecology of a highly productive aquatic salamander (*Siren intermedia*). Southwestern Naturalist 232:423–430.

Gehlbach, F. R., S. E. Kennedy, and B. Walker. 1970. Acoustic behavior of the aquatic salamander, *Siren intermedia*. Bioscience 20:1107–1108.

Gervasi, S. G., and J. Foufopoulos. 2008. Costs of plasticity: Responses to desiccation decrease post-metamorphic immune function in a pond-breeding amphibian. Functional Ecology 22:100–108.

Ghioca, D., and L. M. Smith. 2008. Population structure of *Ambystoma tigrinum mavortium* in playa wetlands: Landuse influence and variations in polymorphism. Copeia 2008:286–293.

Gibson, J. D., and D. A. Merkle. 2005. *Ambystoma maculatum* (Spotted Salamander). Reproduction. Herpetological Review 36:294.

Gibbons, J. W., and R. D. Semlitsch. 1991. Guide to the reptiles and amphibians of the Savannah River Site. Athens: University of Georgia Press.

Gilbert, M., R. LeClair Jr., and R. Fortin. 1994. Reproduction of the Northern Leopard Frog (*Rana pipiens*) in floodplain habitat in the Richelieu River, P. Quebec, Canada. Journal of Herpetology 28:465–470.

Gilbert, P. W. 1941. Eggs and nests of *Hemidactylium scutatum* in the Ithaca region. Copeia 1941:47.

———. 1942. Observations on the eggs of *Ambystoma maculatum* with special reference to the green alga found within the egg envelopes. Ecology 23:215–227.

———. 1944. The alga-egg relationship in *Ambystoma maculatum*, a case of symbiosis. Ecology 25:366–369.

Gilbertson, H., and D. J. Watermolen. 1998. Notes on Wrinkled Frog (*Rana rugosa*) tadpoles from Hawaii. Bulletin of the Chicago Herpetological Society 33:57–59.

Glass, B. P. 1951. Age at maturity of neotenic *Ambystoma tigrinum mavortium* Baird. American Midland Naturalist 46:391–394.

Godley, J. S. 1983. Observations on the courtship, nests and young of *Siren intermedia* in southern Florida. American Midland Naturalist 110:215–219.

Goff, L. J., and J. R. Stein. 1978. Ammonia: Basis for algal symbiosis in salamander egg masses. Life Science 22:1463–1468.

Goin, C. J. 1942. Description of a new race of *Siren intermedia* Le Conte. Annals of the Carnegie Museum 29:211–217.

———. 1947. Notes on the eggs and early larvae of three Florida salamanders. Natural History Miscellanea (10): 1–4.

———. 1950. A study of the salamander *Ambystoma cingulatum*, with the description of a new subspecies. Annals of the Carnegie Museum 31:299–321.

———. 1951. Notes on the eggs and early larvae of three more Florida salamanders. Annals of the Carnegie Museum 32:253–263.

———. 1957. Description of a new salamander of the genus *Siren* from the Rio Grande. Herpetologica 13:37–42.

———. 1961. The growth and size of *Siren lacertina*. Herpetologica 17:139.

Goin, C. J., and J. W. Crenshaw. 1949. Description of a new race of the salamander *Pseudobranchus striatus* (Le Conte). Annals of the Carnegie Museum 31:277–280.

Goin, C. J., and L. H. Ogren. 1956. Parasitic copepods (Argulidae) on amphibians. Journal of Parasitology 42:154.

Golden, D. R., G. R. Smith, and J. E. Rettig. 2001. Effects of age and group size on habitat selection and activity level in *Rana pipiens* tadpoles. Herpetological Journal 11:69–73.

Goldsworthy, M. O. 2007. *Ascaphus truei* (Tailed Frog). Nest site. Herpetological Review 38:68–69.

Gomez-Mestre, U., J. C. Touchon, and K. M. Warkentin. 2006. Amphibian embryos and parental defenses and a larval predator reduces egg mortality from water mold. Ecology 87:2570–2581.

Goodale, H. D. 1911. The early development of *Spelerpes bilineatus* (Green). American Journal of Anatomy 12:173–247.

Goodwin, O. K., and J. T. Wood. 1953. Notes on egg-laying of the Four-toed Salamander, *Hemidactylium scutatum* (Schlegel), in eastern Virginia. Virginia Journal of Science 1953:65–66.

Gordon, R. E. 1966. Some observations on the biology of *Pseudotriton ruber schencki*. Journal of the Ohio Herpetological Society 5:163–164.

Gosner, K. L. 1959. Systematic variations in tadpole teeth with notes on food. Herpetologica 15:203–210.

———. 1960. A simplified table for staging anuran embryos and larvae with notes on identification. Herpetologica 16:183–190.

Gosner, K. L., and I. H. Black. 1954. Larval development in *Bufo woodhousei fowleri* and *Scaphiopus holbrooki holbrooki*. Copeia 1954:251–255.

———. 1957a. Larval development in New Jersey Hylidae. Copeia 1957:31–36.

———. 1957b. The effects of acidity on the development and hatching of New Jersey frogs. Ecology 38:256–262.

———. 1958a. Notes on larval toads in the eastern United States with special reference to natural hybridization. Herpetologica 14:133–140.

———. 1958b. Notes on the life history of Brimley's Chorus Frog. Herpetologica 13:249–254.

———. 1967. *Hyla andersonii*. Catalogue of American Amphibians and Reptiles 54.1–54.2.

———. 1968. *Rana virgatipes*. Catalogue of American Amphibians and Reptiles 67.1–67.2.

Gosner, K. L., and D. A. Rossman. 1959. Observations on the reproductive cycle of the Swamp Chorus Frog, *Pseudacris nigrita*. Copeia 1959:263–266.

———. 1960. Eggs and larval development of the tree frogs *Hyla crucifer* and *Hyla ocularis*. Herpetologica 16:225–232.

Gradwell, N. 1971. *Ascaphus* tadpole: Experiments on the suction and gill irrigation mechanisms. Canadian Journal of Zoology 49:307–332.

———. 1973. On the functional morphology of suction and gill irrigation in the tadpole of *Ascaphus*, and notes on hibernation. Herpetologica 29:84–93.

Graham, S. P., E. K. Time, S. K. Hoss, M Alcorn, and J. Deitloff. 2010. Notes on reproduction in the Brownback Salamander (*Eurycea aquatica*). IRCF Reptiles & Amphibians 17:168–172.

Green, H. T. 1925. The egg laying of the Purple Salamander. Copeia 1925:32.

Green, N. B. 1938. The breeding habits of *Pseudacris brachyphona* (Cope) with a description of the eggs and tadpole. Copeia 1938:79–82.

———. 1952. A key to the eggs of West Virginia Salientia. Proceedings of the West Virginia Academy of Science 24:36–38.

———. 1968. Egg-laying situations and early larval behavior in *Eurycea lucifuga*. Journal of Herpetology 1:119–120.

Green, N. B., P. Grant Jr., and B. Dowler. 1967. *Eurycea lucifuga* in West Virginia: Its distribution, ecology and life history. Proceedings of the West Virginia Academy of Science 39:297–304.

Green, N. B., and T. K. Pauley. 1987. Amphibians and reptiles in West Virginia. Pittsburgh: University of Pittsburgh Press.

Greuter, K. L. M., and R. J. Forstner. 2003. *Bufo houstonensis* (Houston Toad). Growth. Herpetological Review 34:355–356.

Grismer, L. L. 2002. Amphibians and reptiles of Baja California including its Pacific Islands and the islands in the Sea of Cortez. Berkeley: University of California Press.

Grobman, A. B. 1943. Notes on salamanders with the description of a new species of *Cryptobranchus*. Occasional Papers of the Museum of Zoology, University of Michigan (470): 1–13.

Gunzburger, M. S. 2003. Evaluation of the hatching trigger and larval ecology of the salamander *Amphiuma means*. Herpetologica 59:459–468.

Gunzburger, M. S., and J. Travis. 2004. Evaluating predation pressure on Green Treefrog larvae across a habitat gradient. Oecologia 140:422–429.

——. 2005a. Critical literature review of the evidence for unpalatability of amphibian eggs and larvae. Journal of Herpetology 39:547–571.

——. 2005b. Effects of multiple predator species on Green Treefrog (*Hyla cinerea*) tadpoles. Canadian Journal of Zoology 83:996–1002.

——. 2007. Egg clutch characteristics of the Barking Treefrog, *Hyla gratiosa*, from North Carolina and Florida. Herpetological Review 38:22–24.

Guy, C. J., R. E. Ratajczak, and G. D. Grossman. 2004. Nest-site selection by Southern Two-lined Salamanders (*Eurycea cirrigera*) in the Georgia Piedmont. Southeastern Naturalist 3:75–88.

Hagen, C., and C. R. Wilson. 1999. *Ambystoma gracile* (Northwestern Salamander). Maximum size. Herpetological Review 30:36.

Hall, J. A. 1998. *Scaphiopus intermontanus*. Catalogue of American Amphibians and Reptiles 650.1–650.16.

Hall, J. A., J. H. Larsen Jr., and R. E. Fitzner. 1997. Postembryonic ontogeny of the Spadefoot Toad, *Scaphiopus intermontanus* (Anura: Pelobatidae): External morphology. Herpetological Monographs (11): 124–178.

——. 2002. Morphology of the prometamorphic larva of the spadefoot toad, *Scaphiopus intermontanus* (Anura: Pelobatidae), with an emphasis on the lateral line system and mouthparts. Journal of Morphology 252:114–130.

Hall, R. F. 1977. A population analysis of two species of streamside salamanders, genus *Desmognathus*. Herpetologica 33:109–113.

Hamel, J. 2009. *Anaxyrus americanus* (American Toad). Egg cannibalism. Herpetological Review 40:67–68.

Hamilton, W. J., Jr. 1940. The feeding habits of larval newts, with reference to availability and predilection of food items. Ecology 21:351–356.

——. 1948. The egg-laying process in the Tiger Salamander. Copeia 1948:212.

——. 1950. Notes on the food of the Congo Eel, *Amphiuma*. Natural History Miscellanea (62): 1–3.

——. 1955. Notes on the ecology of the Oak Toad in Florida. Herpetologica 11:205–210.

Hampton, P. S. 2006. *Notophthalmus viridescens louisianensis* (Eastern Newt). Population of gilled adults. Herpetological Review 37:71.

Hampton, S. H., and E. P. Volpe. 1963. Development and interpopulation variability of the mouthparts of *Scaphiopus holbrooki*. American Midland Naturalist 70:319–328.

Hanlin, H. G. 1978. Food habits of the Greater Siren, *Siren lacertina*, in an Alabama Coastal Plain pond. Copeia 1978:358–360.

Hanlin, H. G., and R. H. Mount. 1978. Reproduction and activity of the Greater Siren, *Siren lacertina* (Amphibia: Sirenidae) in Alabama. Journal of the Alabama Academy of Science 49:31–39.

Hardy, J. D., Jr. 1969a. A summary of recent studies of the salamander *Ambystoma mabeei*. Ref. 69-20. Solomons, MD: Chesapeake Biological Laboratory.

——. 1969b. Reproductive activity, growth, and movements of *Ambystoma mabeei* Bishop in North Carolina. Bulletin of the Maryland Herpetological Society 5:65–76.

Hardy, J. D., Jr., and J. D. Anderson. 1970. *Ambystoma mabeei*. Catalogue of American Amphibians and Reptiles 81.1–891.2.

Hardy, L. M., and M. C. Lucas. 1991. A crystalline protein is responsible for dimorphic egg jellies in the Spotted Salamander, *Ambystoma maculatum* (Shaw) (Caudata: Ambystomatidae). Comparative Biochemistry and Physiology 100A:653–660.

Hardy, L. M., and R. W. McDiarmid. 1969. The amphibians and reptiles of Sinaloa, Mexico. University of Kansas Publications, Museum of Natural History 18:39–252.

Harlan, R. A., and R. F. Wilkinson. 1981. The effects of progressive hypoxia and rocking activity on blood oxygen tension for Hellbenders, *Cryptobranchus alleganiensis*. Journal of Herpetology 15:383–387.

Harris, J. P., Jr. 1959a. The natural history of *Necturus*: I. Habitats and habits. Field and Laboratory 27:11–20.

——. 1959b. The natural history of *Necturus*: III. Food and feeding. Field and Laboratory 27:105–111.

——. 1961. The natural history of *Necturus*, IV. Reproduction. Journal of the Graduate Research Center (Field and Laboratory) 29:69–81.

Harris, R. N. 1987. Density-dependent paedomorphosis in the salamander *Notophthalmus viridescens dorsalis*. Ecology 68:705–712.

——. 1989. Ontogenetic changes in size and shape of the facultatively paedomorphic salamander *Notophthalmus viridescens dorsalis*. Copeia 1989:35–42.

Harris, R. N., R. A. Alford, and H. M. Wilbur. 1988. Density and phenology of *Notophthalmus viridescens dorsalis* in a natural pond. Herpetologica 44:234–242.

Harris, R. N., and D. E. Gill. 1980. Communal nesting, brooding behavior, and embryonic survival of the Four-toed Salamander *Hemidactylium scutatum*. Herpetologica 36:141–144.

Harris, R. N., W. W. Hames, I. T. Knight, C. A. Carrero, and T. J. Vess. 1995. An experimental analysis of joint nesting in the salamander *Hemidactylium scutatum* (Caudata: Plethodontidae): The effects of population density. Animal Behavior 50:1309–1316.

Harrison, J. R. 1973. Observations on the life history and ecology of *Eurycea quadridigitata* (Holbrook). Herpetological Information Search Service News-Journal 1:57–58.

Harrison, J. R., III, and S. I. Guttman. 2003. A new species of *Eurycea* (Caudata: Plethodontidae) from North and South Carolina. Southeastern Naturalist 2:259–178.

Harrison, R. G. 1969. Harrison stages and description of the normal development of the Spotted Salamander, *Ambystoma maculatum* (Linn.). *In* R. G. Harrison, ed., Organization and development of the embryo, pp. 44–66. New Haven: Yale University Press.

Hassinger, D. D. 1970. Notes on the thermal properties of frog eggs. Herpetologica 26:49–51.

Hassinger, D. D., J. D. Anderson, and G. H. Dalrymple. 1970. The early life history and ecology of *Ambystoma tigrinum* and *Ambystoma opacum* in New Jersey. American Midland Naturalist 84:474–495.

Hawkins, C. P., L. J., Gottschalk, and S. S. Brown. 1988. Densities and habitat of Tailed Frog tadpoles in small streams near Mt. St. Helens following the 1980 eruption. Journal of the North American Benthological Society 7:246–252.

Hay, O. P. 1888. Observations on *Amphiuma* and its young. American Naturalist 22:315–321.

——. 1889a. Observations on the habits of some amblystomas. American Naturalist 23:602–612.

——. 1889b. Notes on the life history of *Chorophilus triseriatus*. American Naturalist 23:770–774.

Healy, W. R. 1973. Life history variation and growth of juvenile *Notophthalmus viridescens* from Massachusetts. Copeia 1973:641–647.

——. 1974. Population consequences of alternative life histories in *Notophthalmus v. viridescens*. Copeia 1974:221–229.

Hecht, M. K. 1958. A synopsis of the mudpuppies of eastern North America. Proceedings of the Staten Island Institute of Arts and Letters 21:1–38.

Hecnar, S. J., and D. R. Hecnar. 1999. *Pseudacris triseriata* (Western Chorus Frog). Reproduction. Herpetological Review 30:38.

Hedeen, S. E. 1971. Growth of the tadpoles of the Mink Frog, *Rana septentrionalis*. Herpetologica 27:160–165.

———. 1975. Premetamorphic growth of *Rana catesbeiana* in southwestern Ohio. Ohio Journal of Science 75:182–183.

———. 1977. *Rana septentrionalis*. Catalogue of American Amphibians and Reptiles 202.1–202.2.

Hekkala, E. R., R. A. Saumure, J. R. Jaeger, H.-W. Herrmann, M. J. Sredl, D. F Bradford, D. Drabeck, and M. J. Blum. 2011. Resurrecting an extinct species: Archival DNA, taxonomy, and conservation of the Vegas Valley leopard frog. Conservation Genetics 12:1379–1385.

Hellman, R. E. 1953. A comparative study of eggs and tadpoles of *Hyla phaeocrypta* and *Hyla versicolor*. Publications of the Research Division of Ross Allen's Reptile Institute 1:61–74.

Henderson, B. A. 1973. The specialized feeding behavior of *Ambystoma gracile* in Marion Lake, British Columbia. Canadian Field Naturalist 87:151–154.

Hendricks, F. S. 1973. Intestinal contents of *Rana pipiens* Schreber (Ranidae) larvae. Southwestern Naturalist 18:93–114.

Hendricks, L. J., and J. Kezer. 1958. An unusual population of a blind cave salamander and its fluctuation during one year. Herpetologica 14:41–43.

Henry, W. V., and V. C. Twitty. 1940. Contributions to the life histories of *Dicamptodon ensatus* and *Ambystoma gracile*. Copeia 1940:247–250.

Herman, T. A., and K. M. Enge. 2007. *Hemidactylium scutatum* (Four-toed Salamander). Reproduction. Herpetological Review 38:176.

Herreid, C. F. II., and S. Kinney. 1966. Survival of Alaskan Woodfrog (*Rana sylvatica*) larvae. Ecology 47:1039–1040.

Hess, Z. J., and R. N. Harris. 2000. Eggs of *Hemidactylium scutatum* (Caudata: Plethodontidae) are unpalatable to insect predators. Copeia 2000:596–600.

Heyer, W. R. 1970. Studies on the genus *Leptodactylus* (Amphibia, Leptodactylidae). II. Diagnosis and distribution of the *Leptodactylus* of Costa Rica. Revista Biologia Tropical 16:171–205.

———. 1971. *Leptodactylus labialis*. Catalogue of American Amphibians and Reptiles 104.1–104.3.

Heyer, W. R., M. A. Donnelly, R. W. McDiarmid, L.-A. C. Hayek, and M. S. Foster, eds. 1994. Measuring and monitoring biological diversity: Standard methods for amphibians. Biological Diversity Handbook Series. Washington, DC: Smithsonian Institution Press.

Hews, D. K. 1988. Alarm response in larval Western Toads, *Bufo boreas*: Release of larval chemicals by a natural predator and its effect on predator capture efficiency. Animal Behavior 36:125–133.

Hews, D. K., and A. R. Blaustein. 1985. An investigation of the alarm response in *Bufo boreas* and *Rana cascadae* tadpoles. Behavioral Neural Biology 43:47–57.

Hicks, T. L., D. E. Mangan, A. P. McIntyre, and M. P. Hayes. 2008. *Rhyacotriton kezeri* (Columbia Torrent Salamander). Larval diet. Herpetological Review 39:456–457.

Hillis, D. M. 1982. Morphological differentiation and adaptation of the larvae of *Rana berlandieri* and *Rana sphenocephala* (*Rana pipiens* complex) in sympatry. Copeia 1982:168–174.

Hillis, D. M., D. A. Chamberlain, T. P. Wilcox, and P. T. Chippindale. 2001. A new species of subterranean blind salamander (Plethodontidae: Hemidactyliini: *Eurycea: Typhlomolge*) from Austin, Texas, and a systematic revision of the central Texas paedomorphic salamanders. Herpetologica 57:266–289.

Hillis, D. M., and R. Miller. 1976. An instance of overwintering of larval *Ambystoma maculatum* in Maryland. Bulletin of the Maryland Herpetological Society 12:65–66.

Hillis, R. E., and E. D. Bellis. 1971. Some aspects of the ecology of the Hellbender, *Cryptobranchus alleganiensis*, in a Pennsylvania stream. Journal of Herpetology 5:121–126.

Hilton, W. A. 1909. General features of the early development of *Desmognathus fusca*. Journal of Morphology 20:533–547.

Hinckley, M. H. 1880. Notes on the eggs and tadpoles of *Hyla versicolor*. Proceedings of the Boston Society of Natural History 21:104–107.

——. 1881. On some differences in the mouth-structure of tadpoles of the anourous Batrachians found in Milton, Mass. Proceedings of the Boston Society of Natural History 21: 307–314.

——. 1882a. Notes on the development of *Rana sylvatica* LeConte. Proceedings of the Boston Society of Natural History 22:104–107.

——. 1882b. On some differences in the mouth-structure of tadpoles of the anurous batrachians found in Milton, Massachusetts. Proceedings of the Boston Society of Natural History 21:307–314.

Hoff, K. S., M. J. Lannoo, and R. J. Wassersug. 1985. Kinematics of midwater prey capture by *Ambystoma* (Caudata: Ambystomatidae) larvae. Copeia 1985:247–251.

Hoff, K., and R. Wassersug. 1985. Do tadpoles swim like fishes? *In* H.-R. Duncker and G. Fleischer, eds., Functional morphology in vertebrates, pp. 31–34. Fortschritte der Zoologie 30. Stuttgart: Gustav Fischer-Verlag.

Hoffman, E. A., and D. W. Pfennig. 1999. Proximate causes of cannibalistic polyphenism in larval Tiger Salamanders. Ecology 80:1076–1080.

Hoffman, R. L. 1988. *Hyla femoralis*. Catalogue of American Amphibians and Reptiles 436.1–436.3.

Hoffman, R. L., and G. L. Larson. 1999. *Ambystoma gracile* (Northwestern Salamander). Predation and cannibalism. Herpetological Review 30:159.

Holbrook, C. T., and J. W. Petranka. 2004. Ecological interactions between *Rana sylvatica* and *Ambystoma maculatum*: Evidence of interspecific competition and facultative intraguild predation. Copeia 2004:932–939.

Holomuzski, J. R. 1986a. Intraspecific predation and habitat use by Tiger Salamanders (*Ambystoma tigrinum nebulosum*). Journal of Herpetology 20:439–441.

——. 1986b. Predator avoidance and diet patterns of microhabitat use by larval Tiger Salamanders. Ecology 67:737–748.

——. 1989. Predation risk and macroalgae use by the stream-dwelling salamander *Ambystoma texanum*. Copeia 1989:22–28.

——. 1991. Macrohabitat effects on egg deposition and larval growth, survival and instream dispersal in *Ambystoma barbouri*. Copeia 1991:687–694.

——. 1997. Habitat-specific life-histories and foraging by stream-dwelling American Toads. Herpetologica 53:445–453.

——. 1998. Grazing effects of Green Frog tadpoles (*Rana clamitans*) in a woodland pond. Journal of Freshwater Ecology 13:1–8.

Hom, C. L. 1987. Reproductive ecology of female Dusky Salamanders, *Desmognathus fuscus* (Plethodontidae), in the southern Appalachians. Copeia 1987:768–777.

Hossack, B. R. 2006. *Rana luteiventris* (Columbia Spotted Frog). Reproduction. Herpetological Review 37:208–209.

Hovingh, P., B. Benton, and D. Bornholdt. 1985. Aquatic parameters and life history observations of the Great Basin Spadefoot Toad in Utah. Great Basin Naturalist 45:22–30.

Howard, J. H., and R. L. Wallace. 1985. Life history characteristics of populations of the Long-toed Salamander (*Ambystoma macrodactylum*) from different altitudes. American Midland Naturalist 113:361–373.

Hoy, P. R. 1871. The development of *Amblystoma lurida* Sager. American Naturalist 5: 578–579.

Hoyt, D. L. 1960. Mating behavior and eggs of the Plains Spadefoot. Herpetologica 16: 199–201.

Hubbs, C., and N. Armstrong. 1961. Minimum developmental temperature tolerance of two anurans, *Scaphiopus couchi* and *Microhyla carolinensis*. Texas Journal of Science 13:358–362.

Hudson, R. G. 1955. Observations on the larvae of the salamander *Eurycea bislineata bislineata*. Herpetologica 11:202–204.

Huey, R. B. 1980. Sprint velocity of tadpoles (*Bufo boreas*) through metamorphosis. Copeia 1980:537–540.

Hulse, A. C. 1978. *Bufo retiformis.* Catalogue of American Amphibians and Reptiles 207.1–207.2.

Humphries, A. A., Jr. 1966. Observations on the deposition, structure and cytochemistry of the jelly envelopes of the egg of the newt, *Triturus viridescens.* Developmental Biology 13: 214–230.

Humphries, W. J. 2007. Diurnal seasonal activity of *Cryptobranchus alleganiensis* (Hellbender) in North Carolina. Southeastern Naturalist 6:135–140.

Humphries, W. J., and T. K. Pauley. 2000. Seasonal changes in nocturnal activity of the Hellbender, *Cryptobranchus alleganiensis,* in West Virginia. Journal of Herpetology 34: 604–607.

———. 2005. Life history of the Hellbender, *Cryptobranchus alleganiensis,* in a West Virginia stream. American Midland Naturalist 154:135–142.

Hunter, M. L., Jr., A. J. K. Calhoun, and M. McCollough. 1999. Maine amphibians and reptiles. Orono: University of Maine Press.

Hutcherson, J. E., C. L. Peterson, and R. F. Wilkinson. 1989. Reproductive and larval biology of *Ambystoma annulatum.* Journal of Herpetology 23:181–183.

Hutchinson, C., and D. Hewitt. 1935. A study of larval growth in *Ambystoma punctatum* and *Ambystoma tigrinum.* Journal of Experimental Zoology 71:465–480.

Hutchison, V. H. 1956. Notes on the plethodontid salamanders *Eurycea lucifuga* (Rafinesque) and *Eurycea longicauda longicauda* Green. National Speleological Society Occasional Papers (3): 1–24.

———. 1958. The distribution and ecology of the Cave Salamander, *Eurycea lucifuga.* Ecological Monographs 28:1–20.

———. 1966. *Eurycea lucifuga.* Catalogue of American Amphibians and Reptiles 24.1–24.2.

———. 1971. On the *Ambystoma* egg-alga relationship. Herpetological Review 3:82.

Hutchison, V. H., and C. S. Hammen. 1958. Oxygen utilization in the symbiosis of embryos of the salamander *Ambystoma maculatum* and the alga *Oophila amblystomatis.* Biological Bulletin 115:483–489.

Hutchison, V. H., and L. G. Hill. 1978. Thermal selection of Bullfrog tadpoles (*Rana catesbeiana*) at different stages of development and acclimation temperatures. Journal of Thermal Biology 3:57–60.

Ireland, D. H., A. J. Wirsing, and D. L. Murray. 2007. Phenotypically plastic responses of Green Frog embryos to conflicting predation risk. Oecologia 152:162–168.

Ireland, P. H. 1973. Overwintering of larval Spotted Salamander, *Ambystoma maculatum* (Caudata) in Arkansas. Southwestern Naturalist 17:435–437.

———. 1974. Reproduction and larval development of the Dark-sided Salamander, *Eurycea longicauda melanopleura* (Green). Herpetologica 30:338–343.

———. 1976. Reproduction and larval development in the Gray-bellied Salamander *Eurycea multiplicata griseogaster.* Herpetologica 32:233–238.

———. 1981. A key to the aquatic salamander larvae and larviform adults of Virginia. Catesbeiana 1:3–7.

———. 1989. Larval survivorship in two populations of *Ambystoma maculatum.* Journal of Herpetology 23:209–215.

Ireland, P. H., and R. Altig. 1983. Key to the gilled salamander larvae and larviform adults of Arkansas, Kansas, Missouri and Oklahoma. Southwestern Naturalist 28:271–274.

Jackson, M. E., D. E. Scott, and R. A. Estes. 1989. Determinants of nest success in the Marbled Salamander (*Ambystoma opacum*). Canadian Journal of Zoology 67:2277–2281.

Jackson, M. E., and R. D. Semlitsch. 1993. Paedomorphosis in the salamander *Ambystoma talpoideum*: Effects of a fish predator. Ecology 74:342–350.

Jackson, N. L. 1989. Breeding dynamics of the Houston Toad. Southwestern Naturalist 34: 374–380.

Jakubanis, J., M. J. Dreslik, and C. A. Phillips. 2008. Nest ecology of the Southern Two-lined Salamander (*Eurycea cirrigera*) in eastern Illinois. Northeastern Naturalist 15:131–140.

Jennings, M. R. 1988. *Rana onca*. Catalogue of American Amphibians and Reptiles 417.1–417.2.

Jennings, M. R., and D. Cook. 1998. *Taricha torosa torosa* (Coast Range Newt). Predation. Herpetological Review 29:230.

Jennings, R. D., and N. J. Scott Jr. 1993. Ecologically correlated morphological variation in tadpoles of the Leopard Frog, *Rana chiricahuensis*. Journal of Herpetology 27:285–293.

Jensen, J. B. 1996. *Bufo terrestris* (Southern Toad). Egg toxicity. Herpetological Review 27:138–139.

Jensen, J. B., C. D. Camp, R. M. Austin Jr., R. A. Pyrson, L Giovanetto, S. Graham, M. Meadows, and D. Demarest. 2004. *Cryptobranchus alleganiensis alleganiensis* (Eastern Hellbender). Breeding season. Herpetological Review 35:156.

Jenssen, T. A. 1967. Food habits of the Green Frog, *Rana clamitans*, before and during metamorphosis. Copeia 1967:214–218.

Johnson, C. R., and C. B. Schreck. 1969. Food and feeding of larval *Dicamptodon ensatus* from California. American Midland Naturalist 81:280–281.

Johnson, J. E., S. F. Belmont, and R. S. Wagner. 2007. DNA barcoding as a means to identify organisms associated with amphibian eggs. Herpetological Conservation and Biology 3:116–127.

Johnson, J. E., and A. E. Golberg. 1975. Movements of larval Two-lined Salamanders (*Eurycea bislineata*) in Mill River, Massachusetts. Copeia 1975:588–589.

Johnson, K. E., and L. Eidietis. 2005. Tadpole body zones differ with regard to strike frequencies and kill rates by dragonfly naiads. Copeia 2005:909–913.

Johnson, L. M. 1991. Growth and development of larval Northern Cricket Frogs (*Acris crepitans*) in relation to phytoplankton abundance. Freshwater Biology 25:51–59.

Johnson, S. A. 1996. *Hyla femoralis* (Pine Woods Treefrog). Predation. Herpetological Review 27:140.

———. 2002. Life history of the Striped Newt at a north-central Florida breeding pond. Southeastern Naturalist 1:381–402.

Johnson, S. A., and R. Franz. 1999. *Notophthalmus perstriatus* (Striped Newt). Coloration. Herpetological Review 30:89.

Johnson, V. O. 1939. A supplementary note on the larvae of *Bufo woodhousii woodhousii*. Herpetologica 1:160–164.

Johnston, G. F., and R. Altig. 1986. Identification characteristics of anuran tadpoles. Herpetological Review 17:36–37.

Jones, L. L.C., R. B. Bury, and P. S. Corn. 1990. Field observation of the development of a clutch of Pacific Giant Salamander (*Dicamptodon tenebrosus*) eggs. Northwestern Naturalist 71:93–94.

Jones, L. L. C., and M. G. Raphael. 1998. *Ascaphus truei* (Tailed Frog). Predation. Herpetological Review 29:39.

Jones, M. S., J. P. Goettl, and L. J. Livo. 1999. *Bufo boreas* (Boreal Toad). Predation. Herpetological Review 30:91.

Jones, R. L. 1986. Reproductive biology of *Desmognathus fuscus* and *Desmognathus santeetlah* in the Unicoi Mountains. Herpetologica 42:323–334.

Jordan, E. O. 1893. The habits and development of newt *Diemyctylus viridescens*. Journal of Morphology 8:269–366.

Justis, C. S., and D. H. Taylor. 1976. Extraocular photoreception and compass orientation in larval Bullfrogs, *Rana catesbeiana*. Copeia 1976:98–105.

Juterbock, J. E. 1986. The nesting behavior of the Dusky Salamander, *Desmognathus fuscus*. I. Nesting phenology. Herpetologica 42:457–471.

———. 1990. Variation in larval growth and metamorphosis in the salamander *Desmognathus fuscus*. Herpetologica 46:291–303.

Kaplan, R. H. 1979. Ontogenetic variation in "ovum" size in two species of *Ambystoma*. Copeia 1979:348–350.

――. 1980a. Ontogenetic energetics in *Ambystoma*. Physiological Zoology 53:43–56.

――. 1980b. The implications of ovum size variability for offspring fitness and clutch size within several populations of salamanders *Ambystoma*. Evolution 34:51–64.

Karlin, A. A., S. I. Guttman, and D. B. Means. 1993. Population structure in the Ouachita Mountain Dusky Salamander, *Desmognathus brimleyorum* (Caudata: Plethodontidae). Southwestern Naturalist 38:36–42.

Karlstrom, E. L. 1962. The toad genus *Bufo* in the Sierra Nevada of California. University of California Publications in Zoology 62:1–104.

――. 1973. *Bufo canorus*. Catalogue of American Amphibians and Reptiles 132.1–132.2.

Karlstrom, E. L., and R. L. Livezey. 1955. The eggs and larvae of the Yosemite Toad *Bufo canorus* Camp. Herpetologica 11:221–227.

Karraker, N. E. 1999. *Rhyacotriton variegatus* (Southern Torrent Salamander). Nest site. Herpetological Review 30:160–161.

Karraker, N. E., and G. S. Beyersdorf. 1997. A Tailed Frog (*Ascaphus truei*) nest site in northwestern California. Northwestern Naturalist 78:110–111.

Karraker, N. E., L. M. Ollivier, and C. R. Hodgson. 2005. Oviposition sites of the Southern Torrent Salamander (*Rhyacotriton variegatus*) in northwestern California. Northwestern Naturalist 86:34–36.

Karraker, N. E., D. S. Pilliod, M. J. Adams, E. L. Bull, P. S. Corn, L. V. Diller, L. A. Dupuis, M. P. Hayes, B. R. Hossack, G. R. Hodgson, E. J. Hyde, K. Lohman, B. R. Norman, L. M. Ollivier, C. A. Pearl, and C. R. Peterson. 2006. Taxonomic variation in oviposition by Tailed Frogs. Northwestern Naturalist 87:87–97.

Kats, L. B., and A. Sih. 1992. Oviposition site selection and avoidance of fish by Streamside Salamanders (*Ambystoma barbouri*). Copeia 1992:468–473.

Kehr, A. I. 1997. Stage-frequency and habitat selection of a cohort of *Pseudacris ocularis* tadpoles (Hylidae: Anura) in a Florida temporary pond. Herpetological Journal 7:103–109.

Keen, W. H. 1975. Breeding and larval development of three species of *Ambystoma* in central Kentucky (Amphibia: Urodela). Herpetologica 31:18–21.

Keen, W. H., J. Travis, and J. Juilianna. 1984. Larval growth in three sympatric *Ambystoma* salamander species: Species differences and the effects of temperature. Canadian Journal of Zoology 62:1043–1047.

Kennedy, J. P. 1961. Spawning season and experimental hybridization of the Houston Toad, *Bufo houstonensis*. Herpetologica 17:239–245.

Kenny, J. S. 1969. The Amphibia of Trinidad. Studies on the Fauna of Curaçao and the Caribbean Islands 29:1–78.

Kessel, E. L., and B. B. Kessel. 1943a. The rate of growth of the young larvae of the Pacific Giant Salamander, *Dicamptodon ensatus*. Wasmann Collector 5:108–111.

――. 1943b. Rate of growth of the older larvae of the Pacific Giant Salamander, *Dicamptodon ensatus*. Wasmann Collector 5:141–142.

――. 1944. Metamorphosis of the Pacific Giant Salamander, *Dicamptodon ensatus* (Eschscholtz). Wasmann Collector 6:38–48.

Kezer, J., D. and S. Farner. 1955. Life history patterns of the salamander *Ambystoma macrodactylum* in the high Cascade Mountains of southern Oregon. Copeia 1955:127–131.

Kiesecker, J. M., and A. R. Blaustein. 1997. Influences of egg laying behavior on pathogenic infection of amphibian eggs. Conservation Biology 11:214–220.

Kiffney, P. M., and J. S. Richardson. 2001. Interactions among nutrients, periphyton, and invertebrate and vertebrate (*Ascaphus truei*) grazers in experimental channels. Copeia 2001:422–429.

King, W. 1935. Ecological observations of *Ambystoma opacum*. Ohio Journal of Science 35:4–17.

Klassen, M. A. 1998. Observations on the breeding and development of the Plains Spadefoot, *Spea bombifrons*, in southern Alberta. Canadian Field-Naturalist 112:387–392.

Klemens, M. W. 1993. Amphibians and reptiles of Connecticut and adjacent regions. Bulletin of the Connecticut State Geological and Natural History Survey (112): 1–319.

Knepton, J. C., Jr. 1954. A note on the burrowing habits of the salamander *Amphiuma means means*. Copeia 1954:68.

Knudsen, J. W. 1960. The courtship and egg masses of *Ambystoma gracile* and *Ambystoma macrodactylum*. Copeia 1960:44–46.

Koenings, C. A., C. K. Smith, E. A. Domingue, and J. W. Petranka. 2000. *Desmognathus imitator* (Imitator Salamander). Reproduction. Herpetological Review 31:38–39.

Komoroski, M. J., and J. D. Congdon. 2001. Scaling of nonpolar lipids with ovum size in the Mole Salamander, *Ambystoma talpoideum*. Journal of Herpetology 35:517–521.

Komoroski, M. J., R. Nagle, and J. D. Congdon. 1998. Relationships of lipids to ovum size in amphibians. Physiological Zoology 71:633–641.

Korky, J. K. 1978. Differentiation of the larvae of members of the *Rana pipiens* complex in Nebraska. Copeia 1978:455–459.

———. 1999. *Bufo punctatus*. Catalogue of American Amphibians and Reptiles 689.1–689.5.

Korky J. K., and J. A. Smallwood. 2011. Geographic variation in Northern Green Frog larvae, *Lithobates clamitans melanotus*, in northwestern New Jersey. Bulletin of the Maryland Herpetological Society 47:1–10.

Korky, J. K., and R. G. Webb. 1991. Geographic variation in larvae of Mountain Treefrogs of the *Hyla eximia* group (Anura: Hylidae). Bulletin of the New Jersey Academy of Science 36:7–12.

Kraus, F. 1996. *Ambystoma barbouri*. Catalogue of American Amphibians and Reptiles 621.1–621.4.

Kraus, F., and J. W. Petranka. 1989. A new sibling species of *Ambystoma* from the Ohio River drainage. Copeia 1989:94–110.

Krupa, J. J. 1986. Multiple clutch production in the Great Plains Toad. Prairie Naturalist 18:151–152.

———. 1988. Fertilization efficiency in the Great Plains Toad, *Bufo cognatus*. Copeia 1988: 800–803.

———. 1990. *Bufo cognatus*. Catalogue of American Amphibians and Reptiles 457.1–457.8.

———. 1994. Breeding biology of Great Plains Toad in Oklahoma. Journal of Herpetology 28:217–224.

———. 1995. *Bufo woodhousii* (Woodhouse's Toad). Fecundity. Herpetological Review 26:142.

Kruse, K. C., and M. G. Francis. 1977. A predation deterrent in larvae of the Bullfrog, *Rana catesbeiana*. Transactions of the American Fisheries Society 106:248–252.

Kupferberg, S. J. 1995. The role of larval diet in anuran metamorphosis. American Zoologist 37:146–159.

———. 1997a. Facilitation of periphyton production by tadpole grazing: Functional differences between species. Freshwater Biology 37:427–439.

———. 1997b. Bullfrog (*Rana catesbeiana*) invasion of a California River: The role of larval competition. Ecology 78:1736–1751.

Kupferberg, S. J., J. C. Marks, and M. E. Power. 1994. Effects of variation in natural algal and detrital diets on larval anuran (*Hyla regilla*) life-history traits. Copeia 1994:446–457.

Lamberti, G. A., S. V. Gregory, C. P. Hawkins, R. C. Wildman, L. R. Ashkenas, and D. M. Denicola. 1992. Plant-herbivore interactions in streams near Mt. St. Helens. Freshwater Biology 27:237–247.

Landberg, T., and E. Azizi. 2010. Ontogeny of escape swimming performance in the Spotted Salamander. Functional Ecology 24:576–587.

Lannoo, M. J., and M. D. Bachmann. 1984a. On flotation and air breathing in *Ambystoma tigrinum* larvae: Stimuli for and the relationship between these behaviors. Canadian Journal of Zoology 62:15–18.

——. 1984b. Aspects of cannibalistic morphs in a population of *Ambystoma t. tigrinum* larvae. American Midland Naturalist 11:103–109.

Lannoo, M. J., D. S. Townsend, and R. J. Wassersug. 1987. Larval life in the leaves: Arboreal tadpoles types, with special attention to the morphology, ecology, and behavior of the oophagous *Osteopilus brunneus* (Hylidae) larva. Fieldiana Zoology (NS) (38): 1–31.

Lantz, L. A. 1930. Notes on the breeding habits and larval development of *Ambystoma opacum* Grav. Annals and Magazine of Natural History 5:322–325.

Larson, K. L., W. Duffy, and M. J. Lannoo. 1999. "Paedocannibal" morph barred Tiger Salamanders (*Ambystoma tigrinum mavortium*) from eastern South Dakota. American Midland Naturalist 141:124–139.

Larson, W. 1968. The occurrence of neotenic salamanders, *Ambystoma tigrinum diaboli* Dunn, in Devils Lake, North Dakota. Copeia 1968:620–621.

Lawler, K. L., and J. M. Hero. 1997. Palatability of *Bufo marinus* tadpoles to a predatory fish decreases with development. Wildlife Research 24:327–334.

Lawler, S. P. 1989. Behavioural responses of predators and predation risk in four species of larval anurans. Animal Behavior 38:1039–1047.

Lawler, S. P., and P. J. Morin. 1993. Temporal overlap, competition, and priority effects in larval anurans. Ecology 74:174–182.

Lazell, J. D., and R. A. Brandon. 1962. A new stygian salamander from the southern Cumberland Plateau. Copeia 1962:300–306.

LeClair, R., Jr., and G. Laurin. 1996. Growth and body size in populations of Mink Frogs *Rana septentrionalis* from two latitudes. Ecography 19:296–304.

Lee, D. S. 1969. A food study of the salamander *Haideotriton wallacei* Carr. Herpetologica 25:175–177.

Lee, D. S., and R. Franz. 1974. Comments on the feeding behavior of larval Tiger Salamanders, *Ambystoma tigrinum*. Bulletin of the Maryland Herpetological Society 10:105–107.

Lefcort, H. 1996. Adaptive, chemically mediated flight response in tadpoles of the Southern Leopard Frog, *Rana utricularia*. Copeia 1996:455–459.

Leff, L. G., and M. D. Bachmann. 1986. Ontogenetic changes in predatory behavior of larval Tiger Salamanders (*Ambystoma tigrinum*). Canadian Journal of Zoology 64:1337–1344.

——. 1988. Basis of selective predation by the aquatic larvae of the salamander, *Ambystoma tigrinum*. Freshwater Biology 19:87–94.

Leips, J., M. G. McManus, and J. Travis. 2000. Response of treefrog larvae to drying ponds: Comparing temporary and permanent pond breeders. Ecology 81:2997–3008.

Leips, J., and J. Travis. 1994. Metamorphic responses to changing food levels in two species of hylid frogs. Ecology 75:1345–1356.

Leonard, W. P., and D. M. Darda. 1995. *Ambystoma tigrinum* (Tiger Salamander). Reproduction. Herpetological Review 26:29–30.

Leonard, W. P., L. Hallock, and K. R. McAllister. 1997. Behavior and reproduction. *Rana pretiosa* (Oregon Spotted Frog). Herpetological Review 28:86.

Licht, L. E. 1967. Growth inhibition in crowded tadpoles: Intraspecific and interspecific effects. Ecology 48:736–745.

——. 1968. Unpalatability and toxicity of toad eggs. Herpetologica 24:93-98.

——. 1969a. Palatability of *Rana* and *Hyla* eggs. American Midland Naturalist 82:296–298.

——. 1969b. Comparative breeding behavior of the Red-legged frog (*Rana aurora aurora*) and the Western Spotted Frog (*Rana pretiosa pretiosa*) in southwestern British Columbia. Canadian Journal of Zoology 47:1287–1299.

——. 1971. Breeding habits and embryonic thermal requirements of the frogs, *Rana aurora aurora* and *Rana pretiosa pretiosa*, in the Pacific Northwest. Ecology 52:116–124.

——. 1974. Survival of embryos, tadpoles, and adults of the frogs *Rana aurora aurora* and *Rana pretiosa pretiosa* sympatric in southwestern British Columbia. Canadian Journal of Zoology 52:613–627.

——. 1975a. Growth and food of larval *Ambystoma gracile* from a lowland population in southwestern British Columbia. Canadian Journal of Zoology 53:1716–1722.

——. 1975b. Comparative life history features of the Western Spotted Frog, *Rana pretiosa*, from low- and high-elevation populations. Canadian Journal of Zoology 53:1254–1257.

——. 1992. The effect of food level on growth rate and frequency of metamorphosis and paedomorphosis in *Ambystoma gracile*. Canadian Journal of Zoology 70:87–93.

Licht, L. E., and R. G. Brown. 1967. Behavioral thermoregulation and its role in the ecology of the Red-bellied Newt, *Taricha rivularis*. Ecology 48:598–611.

Licht, L. E., and D. M. Sever. 1993. Structure and development of the parotoid gland in metamorphosed and neotenic *Ambystoma gracile*. Copeia 1993:116–123.

Limbaugh, B. A., and E. P. Volpe. 1957. Early development of the Gulf Coast Toad, *Bufo valliceps* Wiegmann. American Museum Novitates (1842): 1–32.

Lindberg, A. J. 1995. *Ambystoma tigrinum tigrinum* (Eastern Tiger Salamander). Reproduction and twinning. Herpetological Review 26:142.

Liu, C.-C. 1950. Amphibians of western China. Fieldiana: Zoology Memoirs 2:1–499 + 10 plates.

Livezey, R. L. 1950. The eggs of *Acris gryllus crepitans* Baird. Herpetologica 6:139–140.

——. 1952. Some observations of *Pseudacris nigrita triseriata* (Wied) in Texas. American Midland Naturalist 47:372–381.

——. 1960. Description of eggs of *Bufo boreas exsul*. Herpetologica 16:48.

Livezey, R. L., and A. H. Wright. 1945. Descriptions of four salientian eggs. American Midland Naturalist 34:701–706.

——. 1947. A synoptic key to the salientian eggs of the United States. American Midland Naturalist 37:179–222.

Livo, L. J. 2000. *Bufo punctatus* (Red-spotted Toad). Larval coloration. Herpetological Review 31:99.

Livo, L. J., and B. C. Kondratieff. 2000. *Bufo punctatus* (Red-spotted Toad). Predation. Herpetological Review 31:168–169.

Livo, L. J., and B. A. Lambert. 2001. *Bufo boreas* (Boreal Toad). Phoretic host. Herpetological Review 32:179–180.

Loafman, P., and L. Jones. 1996. *Dicamptodon copei* (Cope's Giant Salamander). Metamorphosis and predation. Herpetological Review 27:136.

Lockhart, O. M. 2000. *Necturus maculosus* (Mudpuppy). Egg predation. Herpetological Review 31:98.

Loeb, M. L. G., J. P. Collins, and T. J. Maret. 1994. The role of prey in controlling expression of a trophic polymorphism in *Ambystoma tigrinum*. Functional Ecology 8:151–158.

Loomis, R. B., and O. Webb. 1951. *Eurycea multiplicata* collected at the restricted type locality. Herpetologica 7:141–142.

Loraine, R. K. 1984. *Hyla crucifer crucifer* (Northern Spring Peeper). Reproduction. Herpetological Review 15:16–17.

Loredo, I., and D. Van Vuren. 1996. Reproductive ecology of a population of the California Tiger Salamander. Copeia 1996:895–901.

Lowe, W. H. 2005. Factors affecting stage-specific distribution in the stream salamander *Gyrinophilus porphyriticus*. Herpetologica 61:135–144.

Lynn, W. G., and A. Edelman. 1936. Crowding and metamorphosis in the tadpole. Ecology 17:104–109.

Mabry, C. M., and J. L. Christiansen. 1991. The activity and breeding cycle of *Scaphiopus bombifrons* in Iowa. Journal of Herpetology 25:116–119.

MacCracken, J. G. 2004. *Rhyacotriton cascadae*. (Cascade Torrent Salamander). Nest. Herpetological Review 35:367.

——. 2007. Northwestern Salamander (*Ambystoma gracile*) oviposition sites and hatching success in the Cascade Mountains of southern Washington. Herpetological Conservation and Biology 2:127–134.

———. 2008. Biphasic oviposition in Northwestern Salamanders (*Ambystoma gracile*) in the Cascade Mountains of southern Washington, USA. Herpetological Review 39:29–33.

Maeda, M., and M. Matsui. 1989. Frogs and toads of Japan. Tokyo: Bun-Ichi Shuppan.

Mallory, M. A., and J. S. Richardson. 2005. Complex interactions of light, nutrients and consumer density in a stream periphyton-grazer (Tailed Frog tadpoles) system. Journal of Animal Ecology 74:1020–1028.

Mansell, B. 1971. Notes on *Haideotriton wallacei* (Carr). Bulletin of the Philadelphia Herpetological Society 19:38–39.

Marangio, M. S. 1978. The occurrence of neotenic Rough-skinned Newts (*Taricha granulosa*) in montane lakes of southern Oregon. Northwest Science 52:343–350.

Marco, A. 2001. Effects of prolonged terrestrial stranding on aquatic *Ambystoma gracile* egg masses on embryonic development. Journal of Herpetology 35:510–513.

Marco, A., and A. R. Blaustein. 1998. Egg gelatinous matrix protects *Ambystoma gracile* embryos from prolonged exposure to air. Herpetological Journal 8:207–211.

———. 2000. Symbiosis with green algae affects survival and growth of Northwestern Salamander embryos. Journal of Herpetology 34:617–621.

Mares, M. A. 1972. Notes on *Bufo marinus* tadpole aggregations. Texas Journal of Science 223:433–435.

Marshall, C. H., L. S. Doyle, and R. H. Kaplan. 1990. Intraspecific sex-specific oophagy in a salamander and a frog: Reproductive convergence of *Taricha torosa* and *Bombina orientalis*. Herpetologica 46:395–399.

Marshall, J. L. 1996. *Eurycea cirrigera* (Southern Two-lined Salamander). Nest site. Herpetological Review 27:75–76.

———. 1999. The life-history traits of *Eurycea guttolineata* (Caudata, Plethodontidae), with implications for life-history evolution. Alytes 16:97–110.

Martof, B. S. 1952. Early transformation of the Greenfrog, *Rana clamitans* Latrielle. Copeia 1952:115–116.

———. 1956. Growth and development of the Green Frog, *Rana clamitans*, under natural conditions. American Midland Naturalist 55:101–117.

———. 1962. Some aspects of the life history and ecology of the salamander *Leurognathus*. American Midland Naturalist 67:1–35.

———. 1963. *Leurognathus marmoratus*. Catalogue of American Amphibians and Reptiles 3.1–3.2.

———. 1968. *Ambystoma cingulatum*. Catalogue of American Amphibians and Reptiles 57.1–57.2.

———. 1969. Prolonged inanition in *Siren lacertina*. Copeia 1969:285–289.

———. 1970. *Rana sylvatica*. Catalogue of American Amphibians and Reptiles 86.1–86.4.

———. 1972. *Pseudobranchus, P. striatus*. Catalogue of American Amphibians and Reptiles 118.1–118.4.

———. 1973a. *Siren intermedia*. Catalogue of American Amphibians and Reptiles 127.1–127.3.

———. 1973b. *Siren lacertina*. Catalogue of American Amphibians and Reptiles 128.1–128.2.

———. 1974a. *Sirenidae*. Catalogue of American Amphibians and Reptiles 151.1–151.2.

———. 1974b. *Siren*. Catalogue of American Amphibians and Reptiles 152.1–152.2.

———. 1975a. *Pseudotriton montanus*. Catalogue of American Amphibians and Reptiles 166.1–166.2.

———. 1975b. *Pseudotriton ruber*. Catalogue of American Amphibians and Reptiles 167.1–167.3.

———. 1975c. *Hyla squirella*. Catalogue of American Amphibians and Reptiles 168.1–168.2.

Martof, B. S., and F. L. Rose. 1963. Geographic variation in southern populations of *Desmognathus ochrophaeus*. American Midland Naturalist 69:376–425.

Martof, B. S., and E. F. Thompson. 1958. Reproductive behavior of the Chorus Frog, *Pseudacris nigrita*. Behavior 13:243–258.

Maslin, T. P. 1963a. Notes on a collection of herpetozoa from the Yucatán Peninsula of Mexico. University of Colorado Studies in Biology 9:1–20.

——. 1963b. Notes on some anuran tadpoles from Yucatan, Mexico. Herpetologica 19: 122–128.

Mathis, A., M. C. O. Ferrari, N. Windel, F. Messler, and D. P. Chivers. 2008. Learning in embryos and the ghost of predation future. Proceedings of the Royal Society 275B:2603–2607.

Mathis, A., and S. Unger. 2012. Learning to avoid dangerous habitat types by aquatic salamanders, *Eurycea tynerensis*. Ethology 118:57–62.

Maurer, E. F. 1996. Environmental variation and behaviour: Resource availability during ontogeny and feeding performance in salamander larvae (*Ambystoma texanum*). Freshwater Biology 35:35–44.

Maurer, E. F., and A. Sih. 1996. Ephemeral habitats and variation in behavior and life history: Comparisons of sibling salamander species. Oikos 76:337–349.

Maxell, B. A., J. K. Werner, P. Hendricks, and D. L. Flath. 2003. Herpetology in Montana. A history, status summary, checklists, dichotomous keys, accounts for native, potentially native, and exotic species, and indexed bibliography. Northwest Fauna Series 5. Olympia, WA: Society for Northwestern Vertebrate Biology.

Mays, C. E., and M. A. Nickerson. 1971. A population study of the Ozark Hellbender salamander, *Cryptobranchus alleganiensis bishopi*. Proceedings of the Indiana Academy of Science 81:339–340.

McAlister, W. H. 1962. Variation in *Rana pipiens* Schreber in Texas. American Midland Naturalist 67:334–363.

McAllister, C. T., and L. C. Fitzpatrick. 1985. Thermal acclimation and oxygen consumption rates in neotenic Grey-belly Salamanders, *Eurycea multiplicata griseogaster* (Plethodontidae), from an Arkansas cave. Journal of Thermal Biology 10:1–4.

——. 1989. The effect of acclimation on oxygen consumption rates in the salamander, *Eurycea neotenes*. Journal of Herpetology 23:439–442.

McAllister, C. T., and S. E. Trauth. 1996. Food habits of paedomorphic Mole Salamanders, *Ambystoma talpoideum* (Caudata: Ambystomatidae), from northeastern Arkansas. Southwestern Naturalist 41:62–64.

McAllister, C. T., C. R. Tumlison, and L. M. Cooksey. 1981. *Necturus maculosus louisianensis* (Red River Mudpuppy). Coloration. Herpetological Review 12:78.

McAlpine, D. F., B. Cougle, and T. J. Fletcher. 2001. *Rana septentrionalis* (Mink Frog). Predation. Herpetological Review 32:183–184.

McAtee, W. L. 1906. Development of the color pattern in the larvae of *Spelerpes maculicaudus*. Proceedings of the United States National Museum 30:74–83.

McCallum, M. L., and J. L. McCallum. 2005. *Hyla chrysoscelis* (Cope's Gray Treefrog). Tadpole over-wintering. Herpetological Review 36:54.

McCallum, M. L., S. E. Trauth, M. N. Mary, C. McDowell, and B. A. Wheeler. 2004. Fall breeding in the Southern Leopard Frog (*Rana sphenocephala*) in northeastern Arkansas. Southeastern Naturalist 3:401–408.

McCallum, M. L., B. A. Wheeler, and S. E. Trauth. 2003. *Notophthalmus viridescens louisianensis* (Central Newt). Branchiate adult. Herpetological Review 34:46.

McClure, K. V., J. M. Mora, and G. R. Smith. 2009. Effects of light and group size on the activity of Wood Frog tadpoles (*Rana sylvatica*) and their response to a shadow stimulus. Acta Herpetologica 4:103–107.

McCollum, S. A., and J. D. Leimberger. 1997. Predator-induced morphological changes in an amphibian: Predation by dragonflies affects tadpole shape and color. Oecologia 109:615–621.

McCollum, S. A., and J. Van Buskirk. 1996. Costs and benefits of a predator-induced polyphenism in the Gray Treefrog *Hyla chrysoscelis*. Evolution 50:583–593.

McCoy, M. W. 2007. Conspecific density determines the magnitude and character of predator-induced phenotype. Oecologia 153:871–878.

McCoy, M. W., and A. H. Savitsky. 2004. Feeding ecology of larval *Ambystoma mabeei* (Urodela: Ambystomatidae). Southeastern Naturalist 3:409–416.

McCrady, E. 1954. A new species of *Gyrinophilus* (Plethodontidae) from Tennessee caves. Copeia 1954:200–206.

McDiarmid, R. W., and R. Altig. 1999. Research: Materials and techniques. *In* R. W. McDiarmid and R. Altig, eds., Tadpoles: The biology of anuran larvae, pp. 7–23. Chicago: University of Chicago Press.

McDowell, W. T. 1989. Larval period variability of *Eurycea longicauda longicauda* in southern Illinois. Bulletin of the Chicago Herpetological Society 24:75–78.

——. 1992. An observation on the hatching season of *Eurycea longicauda*. Bulletin of the Chicago Herpetological Society 27:150.

——. 1995. Larval period and reproductive biology of *Eurycea cirrigera* from Pope County, Illinois. Bulletin of the Chicago Herpetological Society 30:248–250.

——. 1997. Life history notes on *Siren intermedia* in southern Illinois. Bulletin of the Chicago Herpetological Society 32:226–228.

——. 2008. Larval period of *Eurycea lucifuga* at a cave population in southern Illinois. Bulletin of the Maryland Herpetological Society 44:1–7.

McDowell, W. T., and B. A. Shepherd. 2003. Reproductive biology and population structure of *Eurycea longicauda longicauda*. Bulletin of the Maryland Herpetological Society 39: 52–61.

McGregor, J. H. 1897. An embryo of *Cryptobranchus*. Biological Bulletin 13:39.

McKnight, M. L., and N. A. Nelson. 2007. Life history and color variants in a matriline of Oklahoma salamander (*Eurycea tynerensis*). Southeastern Naturalist 6:727–736.

McLaren, I. A. 1965. Temperature and frog eggs. Journal of General Physiology 48:1071–1079.

McLaughlin, E. W., and A. A. Humphries Jr. 1978. The jelly envelopes and fertilization of eggs of the newt, *Notophthalmus viridescens*. Journal of Morphology 158:73–90.

McWilliams, S. R., and M. Bachmann. 1989a. Predatory behavior of larval Small-mouthed Salamanders (*Ambystoma texanum*). Herpetologica 45:459–466.

——. 1989b. Foraging ecology and prey preference of pond-form larval Small-mouthed Salamanders, *Ambystoma texanum*. Copeia 1989:948–961.

Means, D. B. 1974. The status of *Desmognathus brimleyorum* Stejneger and an analysis of the genus *Desmognathus* (Amphibia: Urodela) in Florida. Bulletin of the Florida State Museum, Biological Sciences 18:1–100.

——. 1993. *Desmognathus apalachicolae*. Catalogue of American Amphibians and Reptiles 556.1–556.2.

——. 1996. *Amphiuma pholeter*. Catalogue of American Amphibians and Reptiles 622.1–622.2.

——. 1999. *Desmognathus brimleyorum*. Catalogue of American Amphibians and Reptiles 682.1–682.4.

Means, D. B., and A. A. Karlin. 1989. A new species of *Desmognathus* from the eastern Gulf Coastal Plain. Herpetologica 45:37–46.

Means, D. B., and C. J. Longden. 1976. Aspects of the biology and zoogeography of the Pine Barrens Treefrog (*Hyla andersonii*) in northern Florida. Herpetologica 32:117–130.

Mecham, J. S. 1967a. *Notophthalmus perstriatus*. Catalogue of American Amphibians and Reptiles 38.1–38.2.

——. 1967b. *Notophthalmus viridescens*. Catalogue of American Amphibians and Reptiles 53.1–53.4.

——. 1968. *Notophthalmus meridionalis*. Catalogue of American Amphibians and Reptiles 74.1–74.2.

Mecham, J. S., and R. E. Hellman. 1952. Notes on the larvae of two Florida salamanders. Quarterly Journal of the Florida Academy of Science 15:127–133.

Meeks, D. E., and J. W. Nagel. 1973. Reproduction and development of the Wood Frog, *Rana sylvatica*, in eastern Tennessee. Herpetologica 29:188–191.

Meffe, G. K., and A. L. Sheldon. 1987. Habitat use by Dwarf Waterdogs (*Necturus punctatus*) in South Carolina streams, with life history notes. Herpetologica 43:490–496.

Meisler, J. A. 2005. *Pseudacris triseriata* (Western Chorus Frog). Reproduction. Herpetological Review 36:55.

Menke, M. E., and D. L. Claussen. 1982. Thermal acclimation and hardening in tadpoles of the Bullfrog, *Rana catesbeiana*. Journal of Thermal Biology 7:215–219.

Merrell, D. J. 1977. Life history of the Leopard Frog, *Rana pipiens*, in Minnesota. Occasional Papers of the Bell Museum of Natural History, University of Minnesota (15): 1–23.

Metcalf, M. M. 1928. The Bell-toads and their opalinid parasites. American Naturalist 62: 5–21.

Metter, D. E. 1963. Stomach contents of Idaho larval *Dicamptodon*. Copeia 1963:435–436.

——. 1964. A morphological and ecological comparison of two populations of the Tailed Frog, *Ascaphus truei* Stejneger. Copeia 1964:181–195.

——. 1966. Some temperature and salinity tolerance of *Ascaphus truei* Stejneger. Journal of the Idaho Academy of Science 4:44–47.

——. 1967. Variation in the Ribbed Frog *Ascaphus truei* Stejneger. Copeia 1967:634–649.

——. 1968. *Ascaphus truei*. Catalogue of American Amphibians and Reptiles 69.1–69.2.

Metts, B. 2001. *Ambystoma maculatum* (Spotted Salamander). Reproduction. Herpetological Review 32:98.

Micken, L. 1968. Some summer observations of the Tiger Salamander, *Ambystoma tigrinum*, in Blue Lake, Madison County, Montana. Proceedings of the Montana Academy of Science 28:77–80.

——. 1971. Additional notes on neotenic *Ambystoma tigrinum melanostictum* in Blue Lake, Madison County, Montana. Proceedings of the Montana Academy of Science 31:62–64.

Middleton, H. 1971. The lateral-line morphology and histology of *Siren intermedia nettingi* Goin (Amphibia: Sirenidae). Texas Journal of Science 22:291.

Miller, B. T., and M. L. Niemiller. 2005. *Pseudotriton ruber* (Red Salamander). Reproduction. Herpetological Review 36:429.

——. 2007. Distribution and relative abundance of Tennessee Cave Salamanders (*Gyrinophilus palleucus* and *Gyrinophilus gulolineatus*) with an emphasis on Tennessee populations. Herpetology Conservation and Biology 3:1–20.

Miller, B. T., M. L. Niemiller, and R. G. Reynolds. 2008. Observations on egg-laying behavior and interactions among attending female Red Salamanders (*Pseudotriton ruber*) with comments on the use of caves by this species. Herpetological Conservation and Biology 3:203–210.

Miller, M. R., and M. E. Robbins. 1954. The reproductive cycle in *Triturus torosus*. Journal of Experimental Zoology 125:415–445.

Mills, N. E., and C. Barnhart. 1999. Effects of hypoxia on embryonic development in two *Ambystoma* and two *Rana* species. Physiological Biochemistry and Zoology 72:179–188.

Mills, N. E., M. C. Barnhart, and R. D. Semlitsch. 2001. Effects of hypoxia on egg capsule conductance in *Ambystoma* (Class Amphibia, Order Caudata). Journal of Experimental Biology 204:3747–3753.

Minton, S. A., Jr. 1972. Amphibians and reptiles of Indiana. Indiana Academy of Science Monographs 3:1–346.

Mitchell, J. C. 1986. Life history patterns in a central Virginia frog community. Virginia Journal of Science 37:262–271.

Mitchell, J. C., and W. Brown. 2005a. *Eurycea bislineata* (Northern Two-lined salamander). Larval size. Herpetological Review 36:158.

——. 2005b. *Eurycea bislineata* (Northern Two-lined salamander). Larval microhabitat. Herpetological Review 36:158.

Mitchell, J. C., K. A. Buhlmann, and R. L. Hoffman. 1996. Predation of Marbled Salamander (*Ambystoma opacum* [Gravenhorst]) eggs by the milliped *Uroblaniulus jerseyi* (Causey). Banisteria 8:55–56.

Mitchell, J. C., and W. Gibbons. 2012. Salamanders of the Southeast. Athens: University of Georgia Press. 324 p.

Mitchell, R. W., and J. R. Reddell. 1965. *Eurycea tridentifera*, a new species of troglobitic salamander from Texas and a reclassification of *Typhlomolge rathbuni*. Texas Journal of Science 17:12–27.

Mittleman, M. B. 1966. *Eurycea bislineata*. Catalogue of American Amphibians and Reptiles 45.1–45.4.

——. 1967. *Manculus, M. quadridigitatus*. Catalogue of American Amphibians and Reptiles 44.1–44.2.

Mittleman, M. B., and G. S. Myers. 1949. Geographic variation in the Ribbed Frog, *Ascaphus truei*. Proceedings of the Biological Society of Washington 62:57–68.

Mohr, C. E. 1931. Observations on the early breeding habits of *Ambystoma jeffersonianum* in central Pennsylvania. Copeia 1931:102–104.

——. 1943. The eggs of the Long-tailed Salamander, *Eurycea longicauda longicauda* (Green). Proceedings of the Pennsylvania Academy of Science 17:86.

Moler, P. E. 1985. A new species of frog (Ranidae: *Rana*) from northwestern Florida. Copeia 1985:379–383.

——, ed. 1992. Rare and endangered biota of Florida. Volume III. Amphibians and reptiles. Gainesville: University Press of Florida.

——. 1993. *Rana okaloosae*. Catalogue of American Amphibians and Reptiles 561.1–561.3.

Moler, P. E., and J. Kezer. 1993. Karyology and systematics of the salamander genus *Pseudobranchus* (Sirenidae). Copeia 1993:39–47.

Montague, J. R. 1977. Note on the embryonic development of the Dusky Salamander, *Desmognathus fuscus* (Caudata: Plethodontidae). Copeia 1977:375.

——. 1979. Note on the larval feeding behavior in *Desmognathus fuscus fuscus*, the Northern Dusky Salamander. Copeia 1979:354.

——. 1987. Yolk absorption and early larval growth in desmognathine salamanders. Journal of Herpetology 21:226–228.

Montague, J. R., and J. W. Poinski. 1978. Note on brooding behavior in *Desmognathus fuscus fuscus* (Raf.) (Amphibia, Urodela, Plethodontidae) in Columbiana County, Ohio. Journal of Herpetology 12:104.

Moore, G. A., and R. C. Hughes 1939. A new plethodontid from eastern Oklahoma. American Midland Naturalist 22:696–699.

——. 1941. A new plethodont salamander from Oklahoma. Copeia 1941:139–142.

Moore, J. A. 1939. Temperature tolerance and rates of development in eggs of Amphibia. Ecology 20:459–478.

——. 1940. Adaptive differences in the egg membranes of frogs. American Naturalist 74:89–93.

——. 1942. Embryonic temperature tolerance and the rate of development in *Rana catesbeiana*. Biological Bulletin 83:375–388.

——. 1949. Geographic variation of adaptive characteristics in *Rana pipiens* Schreber. Evolution 3:1–24.

Moore, M. K., and V. R. Townsend Jr. 1998. The interaction of temperature, dissolved oxygen and predation pressure in an aquatic predator-prey system. Oikos 81:329–336.

Moore, N. J., and A. G. Matson. 1997. *Ambystoma texanum* (Smallmouth Salamander). Reproduction. Herpetological Review 28:199.

Moore, R. D., B. Newton, and A. Sih. 1996. Delayed hatching as a response of Streamside Salamander eggs to chemical cues from predatory sunfish. Oikos 77:331–336.

Morey, S. R. 1996. Pool duration influences age and body mass at metamorphosis in the Western Spadefoot Toad: Implications for vernal pool conservation. *In* C. W. Whitham, ed., Ecology, conservation, and management of vernal pool ecosystems, pp. 86–91. Sacramento: California Native Plant Society.

Morey, S. R., and D. N. Janes. 1994. Variation in larval habitat duration influences metamorphosis in *Scaphiopus couchii*. *In* P. R. Brown and J. W. Wright, eds., Herpetology of the

North American Deserts, pp. 159–165. Special Publication 5. Van Nuys, CA: Southwest Herpetological Society.

Morey, S. R., and D. N. Reznick. 2000. A comparative analysis of plasticity in larval development in three species of spadefoot toads. Ecology 81:1736–1749.

———. 2001. Effects of larval density on postmetamorphic Spadefoot Toads (*Spea hammondii*). Ecology 82:510–522.

———. 2004. The relationship between habitat permanence and larval development in California spadefoot toads: Field and laboratory comparisons of developmental plasticity. Oikos 104:172–190.

Morgan, T. H. 1891. Some notes on the breeding habits and embryology of frogs. American Naturalist 25:753–760.

Moriya, T., K. Kito, Y. Miuashita, and K. Asami. 1996. Preference for background color of the *Xenopus laevis* tadpole. Journal of Experimental Zoology 276:335–344.

Morris, R. L., and W. W. Tanner. 1969. The ecology of the Western Spotted Frog, *Rana pretiosa pretiosa* Baird and Girard, a life history study. Great Basin Naturalist 29:45–81.

Mosimann, J. E., and T. M. Uzzell Jr. 1952. Description of the eggs of the Mole Salamander, *Ambystoma talpoideum* (Holbrook). Natural History Miscellanea (104): 1–3.

Mould, E. D., and D. M. Sever. 1984. Body size and lipid content of *Ambystoma tigrinum* during larval ontogeny. Herpetologica 40:176–181.

Muliak, S. 1937. Notes on *Leptodactylus labialis* (Cope). Copeia 1937:72–73.

Mullally, D. P. 1953. Observations on the ecology of the toad *Bufo canorus*. Copeia 1953:182–183.

Murray, T. D. 1962. A study of two breeding populations of the salamanders *Ambystoma maculatum* and *Ambystoma opacum*. Journal of the Elisha Mitchell Scientific Society 78:102.

Murphy, T. D. 1961. Predation on eggs of the salamander, *Ambystoma maculatum*, by caddisfly larvae. Copeia 1961:495–496.

Mushinsky, H. R. 1976. Ontogenetic development of microhabitat selection in salamanders: The influence of early experience. Copeia 1976:755–758.

Myers, C. W. 1958. Notes on the eggs and larvae of *Eurycea lucifuga* Rafinesque. Quarterly Journal of the Florida Academy of Science 21:125–130.

Myers, G. S. 1931. *Ascaphus truei* in Humboldt County, California, with a note on the habits of the tadpole. Copeia 1931:56–57.

———. 1942. The Black Toad of Deep Springs Valley, Inyo County, California. Occasional Papers of the Museum of Zoology, University of Michigan (460): 1–13.

———. 1943. Notes on *Rhyacotriton olympicus* and *Ascaphus truei* in Humboldt County, California. Copeia 1943:125–126.

Najvar, P. A., J. N. Fries, and J. T. Baccus. 2007. Fecundity of San Marcos Salamanders in captivity. Southwestern Naturalist 52:145–147.

Neill, W. T. 1949. Juveniles of *Siren lacertina* and *S. l. intermedia*. Herpetologica 5:19–20.

———. 1950. A new species of salamander, genus *Desmognathus*, from Georgia. Publications of the Research Division of Ross Allen's Reptile Institute 1:1–6.

———. 1951a. A new subspecies of the dusky salamander, genus *Desmognathus* from south-central Florida. Publications of the Research Division of Ross Allen's Reptile Institute 1:25–38.

———. 1951b. A new subspecies of salamander, genus *Pseudobranchus*, from the Gulf Hammock region of Florida. Publications of the Research Division of Ross Allen's Reptile Institute 1:39–46.

———. 1957. Notes on metamorphic and breeding aggregations of the Eastern Spadefoot, *Scaphiopus holbrooki* (Harlan). Herpetologica 13:185–187.

———. 1963a. *Hemidactylium scutatum*. Catalogue of American Amphibians and Reptiles 2.1–2.2.

———. 1963b. Notes on the Alabama Waterdog, *Necturus alabamensis* Vesca. Herpetologica 19:166–174.

——. 1964. A new *species* of salamander, genus *Amphiuma*, from Florida. Herpetologica 20:62–66.

Neill, W. T., and F. L. Rose. 1949. Nests and eggs of the Southern Dusky Salamander, *Desmognathus fuscus auriculatus*. Copeia 1949:234.

Neish, I. C. 1971. Comparison of size, structure, and distributional patterns of two salamander populations in Marion Lake, British Columbia. Journal of the Fisheries Research Board of Canada 28:49–58.

Nelson, C. E. 1972a. *Gastrophryne carolinensis*. Catalogue of American Amphibians and Reptiles 120.1–120.4.

——. 1972b. *Gastrophryne olivacea*. Catalogue of American Amphibians and Reptiles 122.1–122.4.

Nelson, C. E., and H. S. Cuellar. 1968. Anatomical comparison of tadpoles of the genera *Hypopachus* and *Gastrophryne*. Copeia 1968:423–424.

Netting, M. G. 1929. The food of the Hellbender *Cryptobranchus alleganiensis* (Daudin). Copeia 1929:23–24.

Netting, M. G., and C. J. Goin. 1942. Descriptions of two new salamanders from peninsular Florida. Annals of the Carnegie Museum 29:175–196.

Newman, C. E., J. A. Feinberg, L. J. Rissler, J. Burger, and H. B. Shaffer. 2012. A new species of leopard frog (Anura: Ranidae) from the urban northeastern U.S. Molecular Phylogenetics and Evolution 63:445–455.

Newman, R. A. 1987. Effects of density and predation on *Scaphiopus couchi* tadpoles in desert ponds. Oecologia 71:301–307.

——. 1988. Adaptive plasticity in development of *Scaphiopus couchi* tadpoles in desert ponds. Evolution 42:774–783.

——. 1994. Effects of changing density and food level on metamorphosis of a desert amphibian, *Scaphiopus couchii*. Ecology 75:1085–1096.

——. 1998. Ecological constraints on amphibian metamorphosis: Interaction of temperature and larval density with responses to changing food level. Oecologia 115:9–16.

Nicholls, J. C., Jr. 1949. A new salamander of the genus *Desmognathus* from east Tennessee. Journal of the Tennessee Academy of Science 24:127–129.

Nichols, R. J. 1937. Taxonomic studies on the mouth parts of larval Anura. Illinois Biological Monographs 15:1–73.

Nickerson, M. A., K. L. Krysko, and R. D. Owen. 2003. Habitat differences affecting age class distributions of the Hellbender salamander, *Cryptobranchus alleganiensis*. Southeastern Naturalist 2:619–629.

Nickerson, M. A., and C. E. Mays. 1973. The Hellbenders: North American giant salamanders. Milwaukee Museum Publications in Biology and Geology (1): 1–106.

Nie, M., J. D. Crim, and G. R. Ultsch. 1999. Dissolved oxygen, temperature, and habitat selection by Bullfrog (*Rana catesbeiana*) tadpoles. Copeia 1999:153–162.

Niemiller, M. L., B. M. Glorioso, C. Nicholas, J. Phillips, J. Rader, E. Reed, K. L. Sykes, J. Todd, G. R. Wyckoff, E. L. Young, and B. T. Miller. 2009a. Notes on the reproduction of the Streamside Salamander, *Ambystoma barbouri*, from Rutherford County, Tennessee. Southeastern Naturalist 8:37–44.

Niemiller, M. L., and B. T. Miller. 2007. Subterranean reproduction of the Southern Two-lined Salamander (*Eurycea cirrigera*) from Short Mountain, Tennessee. Herpetological Conservation and Biology 2:106–112.

Niemiller, M. L., R. G. Reynolds, J. G. Reynolds, and B. T. Miller. 2009b. *Gyrinophilus porphyriticus* (Northern Spring Salamander). Reproduction. Herpetological Review 40:67.

Nietfeldt, J. W., S. M. Jones, D. L. Droge, and R. E. Ballinger. 1980. Rate of thermal acclimation in larval *Ambystoma tigrinum*. Journal of Herpetology 14:209–211.

Nijhuis, M. J., and R. H. Kaplan. 1998. Movement patterns and life history characteristics in a population of the Cascade Torrent Salamander (*Rhyacotriton cascadae*) in the Columbia River gorge, Oregon. Journal of Herpetology 32:301–304.

Noble, G. K. 1926. The Long Island Newt: A contribution to the life-history of *Triturus viridescens*. American Museum Novitates (228): 1–11.

——. 1927. The value of life history data in the study of the evolution of the Amphibia. Annals of the New York Academy of Science 30:31–128.

——. 1929. Further observations on the life-history of the newt, *Triturus viridescens*. American Museum Novitates (348): 1–22.

——. 1930. The eggs of *Pseudobranchus*. Copeia 1930:52.

Noble, G. K., and M. K. Brady. 1930. Observations on the life history of the Marbled Salamander, *Ambystoma opacum* Gravenhorst. Zoologica 11:89–132.

Noble, G. K., and G. Evans. 1932. Observations and experiments on the life history of the salamander, *Desmognathus fuscus fuscus* (Rafinesque). American Museum Novitates (533): 1–16.

Noble, G. K., and B. C. Marshall. 1929. The breeding habits of two salamanders. American Museum Novitates (347): 1–12.

——. 1932. The validity of *Siren intermedia* Le Conte, with observations on its life history. American Museum Novitates (532): 1–17.

Noble, G. K., and R. C. Noble. 1923. The Anderson Tree Frog (*Hyla andersonii* Baird). Observations on its habits and life history. Zoologica 2:413–455.

Noble, G. K., and P. G. Putnam. 1931. Observations on the life history of *Ascaphus truei* Stejneger. Copeia 1931:97–101.

Noble, G. K., and L. B. Richards. 1930. The induction of egg-laying in the salamander, *Eurycea bislineata*, by pituitary transplants. American Museum Novitates (396): 1–3.

——. 1932. Experiments on the egg-laying of salamanders. American Museum Novitates (513): 1–25.

Noeske, T. A., and M. A. Nickerson. 1979. Diel activity and rhythms in the Hellbender, *Cryptobranchus alleganiensis* (Caudata: Cryptobranchidae). Copeia 1979:92–95.

Nokhbatolfoghahai, M., and J. R. Downie. 2005. Larval cement gland of frogs: Comparative development and morphology. Journal of Morphology 263:270–283.

——. 2007. Amphibian hatching gland cells: Pattern and distribution in anurans. Tissue and Cell 39:225–240.

——. 2008. The external gills of anuran amphibians: Comparative morphology and ultrastructure. Journal of Morphology 269:1197–1213.

Norman, W. W. 1900. Remarks on the San Marcos Salamander, *Typhlomolge rathbuni* Stejneger. American Naturalist 34:179–183.

Norris, D. O. 1989. Seasonal changes in diet of paedogenetic Tiger Salamanders (*Ambystoma tigrinum mavortium*). Journal of Herpetology 23:87–89.

Nussbaum, R. A. 1969a. Nests and eggs of the Pacific Giant Salamander, *Dicamptodon ensatus* (Eschscholtz). Herpetologica 25:257–262.

——. 1969b. A nest site of the Olympic Salamander, *Rhyacotriton olympicus* (Gaige). Herpetologica 25:277–278.

——. 1970. *Dicamptodon copei* n. sp., from the Pacific Northwest, U.S.A. (Amphibia: Caudata: Ambystomatidae). Copeia 1970:506–514.

Nussbaum, R. A., E. D. Brodie Jr., and R. M. Storm. 1983. Amphibians and reptiles of the Pacific Northwest. Moscow: University Press of Idaho.

Nussbaum, R. A., and G. W. Clothier. 1973. Population structure, growth, and size of larval *Dicamptodon ensatus* (Eschscholtz). Northwestern Science 47:218–227.

Nussbaum, R. A., and C. K. Tait. 1977. Aspects of the life history and ecology of the Olympic Salamander, *Rhyacotriton olympicus* (Gaige). American Midland Naturalist 98:176–199.

Nyman, S. 1987. *Ambystoma maculatum* (Spotted Salamander). Reproduction. Herpetological Review 18:14–15.

——. 1991. Ecological aspects of syntopic larvae of *Ambystoma maculatum* and the *A. laterale-jeffersonianum* complex in two New Jersey ponds. Journal of Herpetology 25: 505–509.

Nyman, S., R. F. Wilkinson, and J. E. Hutchison. 1993. Cannibalism and size relations in a cohort of larval Ringed Salamanders (*Ambystoma annulatum*). Journal of Herpetology 27:78–84.

O'Donnell, D. J. 1937. Natural history of the ambystomid salamanders of Illinois. American Midland Naturalist 18:1063–1071.

O'Hara, R. K., and A. R. Blaustein. 1981. An investigation of sibling recognition in *Rana cascadae* tadpoles. Animal Behavior 29:1121–1126.

——. 1988. *Hyla regilla* and *Rana pretiosa* tadpoles fail to display kin recognition behaviour. Animal Behavior 36:946–948.

Okada, Y. 1931. The tailless batrachians of the Japanese Empire. Empirical Agricultural Experiment Station. 207 p.

Oliver, J. A., and C. E. Shaw. 1953. The amphibians and reptiles of the Hawaiian Islands. Zoologica 38:65–95.

Oliver, M. G., and H. M. McCurdy. 1974. Migration, overwintering, and reproductive patterns of *Taricha granulosa* in southern Vancouver Island. Canadian Journal of Zoology 52:541–545.

Olsson, L. 1993. Pigment pattern formation in the larval salamander *Ambystoma maculatum*. Journal of Morphology 215:151–163.

Olsson, L., and J. Löfberg. 1992. Pigment pattern formation in larval ambystomatid salamanders: *Ambystoma tigrinum tigrinum*. Journal of Morphology 211:73–85.

Oplinger, C. S. 1966. Sex ratio, reproductive cycles, and time of ovulation in *Hyla crucifer crucifer* Wied. Herpetologica 22:276–283.

Organ, J. A. 1961a. The eggs and young of the Spring Salamander, *Pseudotriton porphyriticus*. Herpetologica 17:53–56.

——. 1961b. Studies on local distribution, life history, and population dynamics of the salamander genus *Desmognathus* in Virginia. Ecological Monographs 31:189–220.

Orr, L. P., and W. T. Maple. 1978. Competition avoidance mechanisms in salamander larvae of the genus *Desmognathus*. Copeia 1978:679–685.

Orton, G. L. 1939. Key to New Hampshire amphibian larvae. New Hampshire Fish and Game Department, Survey Report (4): 218–221.

——. 1942. Notes on the larvae of certain species of *Ambystoma*. Copeia 1942:170–172.

——. 1943. The tadpole of *Rhinophrynus dorsalis*. Occasional Papers of the Museum of Zoology, University of Michigan (472): 1–7.

——. 1946. Larval development of the Eastern Narrow-mouth Frog, *Microhyla carolinensis* (Holbrook) in Louisiana. Annals of the Carnegie Museum 30:241–249.

——. 1947. Note on some hylid tadpoles in Louisiana. Annals of the Carnegie Museum 30:363–383.

——. 1952. Key to the genera of tadpoles in the United States and Canada. American Midland Naturalist 47:382–395.

——. 1954. Dimorphism in larval mouth parts in spadefoots of the *Scaphiopus hammondi* group. Copeia 1954:97–100.

Pack, H. J. 1920. Eggs of the Swamp Tree Frog. Copeia 1920:7.

Palis, J. G. 1995. Larval growth, development, and metamorphosis of *Ambystoma cingulatum* on the Gulf Coastal Plain of Florida. Florida Scientist 58:352–358.

——. 1996. *Ambystoma opacum* (Marbled Salamander). Communal nesting. Herpetological Review 27:134.

——. 1997. Distribution, habitat, and status of the Flatwoods Salamander (*Ambystoma cingulatum*) in Florida, USA. Herpetological Natural History 5:53–65.

Palis, J. G., M. J. Aresco, and S. Kilpatrick. 2006. Breeding biology of a Florida population of *Ambystoma cingulatum* (Flatwoods Salamander) during a drought. Southeastern Naturalist 5:1–8.

Panek, F. M. 1978. A developmental study of *Ambystoma jeffersonianum* and *A. platineum* (Amphibia, Urodela, Ambystomatidae). Journal of Herpetology 12:265–266.

Parker, M. S. 1991. Relationship between cover availability and larval Pacific Giant Salamander density. Journal of Herpetology 25:355–357.

———. 1994. Feeding ecology of stream-dwelling Pacific Giant Salamander larvae (*Dicamptodon tenebrosus*). Copeia 1994:705–718.

Parker, M. V. 1951. Notes on the Bird-voiced Tree Frog, *Hyla phaeocrypta*. Journal of the Tennessee Academy of Science 26:208–213.

Parmalee, J. R., M. G. Knutson, and J. E. Lyon. 2002. A field guide to amphibian larvae and eggs of Minnesota, Wisconsin, and Iowa. Washington, DC: USGS, Biological Research Division, Information and Technology Report. USGS/BRD/ITR-2002–0004. iv + 39 pp.

Parris, M. J., and R. D. Semlitsch. 1998. Asymmetric competition in larval amphibian communities: Conservation implications for the Northern Crawfish Frog, *Rana areolata circulosa*. Oecologia 116:219–226.

Parris, M. J., A. Storfer, J. P. Collins, and E. W. Davidson. 2005. Life-history responses to pathogens in Tiger Salamanders (*Ambystoma tigrinum*) larvae. Journal of Herpetology 39:366–372.

Patterson, K. K. 1978. Life history aspects of paedogenic populations of the Mole Salamander, *Ambystoma talpoideum*. Copeia 1978:649–655.

Pearl, C. A. 2000. *Bufo boreas* (Western Toad). Predation. Herpetological Review 31: 233–234.

———. 2003. *Ambystoma gracile* (Northwestern Salamander). Egg predation. Herpetological Review 34:352–353.

Pearl, C. A., M. J. Adams, G. S. Schuytema, and A. V. Nebeker. 2003. Behavioral responses of anuran larvae to cues of native and introduced predators in the Pacific Northwestern United States. Journal of Herpetology 37:572–576.

Pearl, C. A., and M. P. Hayes. 2002. Predation by Oregon Spotted Frogs (*Rana pretiosa*) on Western Toads (*Bufo boreas*) in Oregon. American Midland Naturalist 147:145–152.

Pearman, P. B. 1995. Effects of pond size and consequent predator density on two species of tadpoles. Oecologia 102:1–8.

———. 2002. Interactions between *Ambystoma* salamander larvae: Evidence for competitive asymmetry. Herpetologica 58:156–165.

Peck, S. B. 1973. Feeding efficiency of the cave salamander, *Haideotriton wallacei*. International Journal of Speleology 5:15–19.

Pehek, E. L. 1995. Competition, pH, and the ecology of larval *Hyla andersonii*. Ecology 76:1786–1793.

Perrill, S. A., and R. E. Daniel. 1983. Multiple egg clutches in *Hyla regilla*, *H. cinerea*, and *H. gratiosa*. Copeia 1983:513–516.

Peterson, C. L., C. A. Ingersol, and R. F. Wilkinson. 1989a. Winter breeding of *Cryptobranchus alleganiensis bishopi* in Arkansas. Copeia 1989:1031–1035.

Peterson, C. L., J. W. Reed, and R. F. Wilkinson. 1989b. Seasonal food habits of *Cryptobranchus alleganiensis* (Caudata: Cryptobranchidae). Southwestern Naturalist 34:438–441.

Peterson, C. L., and R. F. Wilkinson. 1996. Home range size of the Hellbender (*Cryptobranchus alleganiensis*) in Missouri. Herpetological Review 27:126–127.

Peterson, C. L., R. F. Wilkinson, D. Moll, and T. Holder. 1991. Premetamorphic survival of *Ambystoma annulatum*. Herpetologica 47:96–100.

Peterson, C. L., R. F. Wilkinson, M. S. Topping, and D. E. Metter. 1989. Age and growth of the Ozark Hellbender. Copeia 1989:225–231.

Peterson, J. A., and A. R. Blaustein. 1992. Relative palatabilities of anuran larvae to natural aquatic insect predators. Copeia 1992:577–584.

Petranka, J. W. 1982. Geographic variation in the mode of reproduction and larval characteristics of the Small-mouthed Salamander (*Ambystoma texanum*) in the east-central United States. Herpetologica 38:475–485.

———. 1984a. Ontogeny of the diet and feeding behavior of *Eurycea bislineata* larvae. Journal of Herpetology 18:48–55.

——. 1984b. Incubation, larval growth, and embryonic and larval survivorship of Smallmouth Salamanders (*Ambystoma texanum*) in streams. Copeia 1984:862–868.

——. 1984c. Sources of interpopulational variation in growth responses of larval salamanders. Ecology 65:1857–1865.

——. 1985. Does age-specific mortality decrease with age in amphibian larvae? Copeia 1985:1080–1083.

——. 1989a. Responses of toad tadpoles to conflicting chemical stimuli: Predator avoidance versus "optimal" foraging. Herpetologica 45:283–292.

——. 1989b. Density-dependent growth and survival of larval *Ambystoma*: Evidence from whole-pond manipulations. Ecology 70:1752–1767.

——. 1998. Salamanders of the United States and Canada. Washington, DC: Smithsonian Institution Press.

Petranka, J., and L. Hayes. 1998. Chemically mediated avoidance of a predatory odonate (*Anax junius*) by American Toad (*Bufo americanus*) and Wood Frog (*Rana sylvatica*) tadpoles. Behavioral Ecology and Sociobiology 42:263–271.

Petranka, J. W., J. J. Just, and E. C. Crawford. 1983. Hatching of amphibian embryos; the physiological trigger. Science 217:257–259.

Petranka, J. W., and C. A. Kennedy. 1999. Pond tadpoles with generalized morphology: Is it time to reconsider their functional roles in aquatic communities. Oecologia 120:621–631.

Petranka, J. W., and J. G. Petranka. 1980. Selected aspects of the larval ecology of the Marbled Salamander *Ambystoma opacum* in the southern portion of its range. American Midland Naturalist 104:352–363.

Petranka, J. W., A. W. Rushlow, and M. E. Hopey. 1998. Predation by tadpoles of *Rana sylvatica* on embryos of *Ambystoma maculatum*: Implications of ecological role reversal by *Rana* (predator) and *Ambystoma* (prey). Herpetologica 54:1–13.

Petranka, J. W., and A. Sih. 1987. Habitat duration, length of larval period, and the evolution of a complex life cycle of a salamander *Ambystoma texanum*. Evolution 41:1347–1356.

Petranka, J. W., A. Sih, L. B. Kats, and J. R. Holomuzski. 1987. Stream drift, size-specific predation, and the evolution of ovum size in an amphibian. Oecologia 71:624–630.

Petzing, J. E., and C. A. Phillips. 1999. *Rana sphenocephala* (Southern Leopard Frog). Reproduction (Fall breeding). Herpetological Review 30:93–94.

Pfaff, C. S., and K. B. Vause. 2002. Captive reproduction and growth of the Broad-striped Dwarf Siren (*Pseudobranchus s. striatus*). Herpetological Review 33:42–44.

Pfennig, D. W. 1990. The adaptive significance of an environmentally-cued developmental switch in an anuran tadpole. Oecologia 85:101–107.

——. 1999. Cannibalistic tadpoles that pose the greatest threat to kin are most likely to discriminate kin. Proceedings of the Royal Society of London 266B:57–61.

Pfennig, D. W., and J. P. Collins. 1993. Kinship affects morphogenesis in cannibalistic salamanders. Nature 362:836–838.

Pfennig, D. W., J. P. Collins, and R. E. Ziemba. 1999. A test of alternative hypotheses for kin recognition in cannibalistic Tiger Salamanders. Behavioral Ecology 10:436–443.

Pfennig, D. W., S. G. Ho, and E. A. Hoffman. 1998. Pathogen transmission as a selective force against cannibalism. Animal Behavior 55:1255–1261.

Pfennig, D. W., A. Mabry, and D. Orange. 1991. Environmental causes of correlations between age and size at metamorphosis in *Scaphiopus multiplicatus*. Ecology 72:2240–2248.

Pfennig, D. W., and P. J. Murphy. 2002. How fluctuating competition and phenotypic plasticity mediate species divergence. Evolution 56:1217–1228.

Pfennig, D. W., P. W. Sherman, and J. P. Collins. 1994. Kin recognition and cannibalism in polyphenic salamanders. Behavioral Ecology 5:225–232.

Pfingsten, R. A. 1965. The reproduction of *Desmognathus ochrophaeus ochrophaeus* in Ohio. Journal of the Ohio Herpetological Society 5:60.

Pfingsten, R. A., and F. L. Downs. 1989. Salamanders of Ohio. Bulletin of the Ohio Biological Survey 7:1–350.

Phillips, C. A. 1992. Variation in metamorphosis in Spotted Salamanders *Ambystoma maculatum* from eastern Missouri. American Midland Naturalist 128:276–280.

Pickwell, G. 1972. Amphibians and reptiles of the Pacific states. New York: Dover.

Pierce, B. A., J. B. Mitton, L. Jacobson, and F. L. Rose. 1983. Head shape and size in cannibal and noncannibal larvae of the Tiger Salamander from west Texas. Copeia 1983:1006–1012.

Pierce, B. A., and P. H. Whitehurst. 1990. *Pseudacris clarkii*. Catalogue of American Amphibians and Reptiles 458.1–458.3.

Piersol, W. H. 1910. Spawn and larva of *Ambystoma jeffersonianum*. American Naturalist 44:732–738.

Pike, N. 1886a. Some notes on the life-history of the Common Newt. Science 20:17–25.

———. 1886b. Notes on the life history of *Amblystoma opacum*. Bulletin of the American Museum of Natural History (7): 209–212.

Pinder, A. W., and S. C. Friet. 1994. Oxygen transport in egg masses of the amphibians *Rana sylvatica* and *Ambystoma maculatum*: Convection, diffusion and oxygen production by algae. Journal of Experimental Biology 197:17–30.

Pitt, A. L., and M. A. Nickerson. 2006. *Cryptobranchus alleganiensis* (Hellbender Salamander). Larval diet. Herpetological Review 37:69.

Platz, J. E. 1988. *Rana yavapaiensis*. Catalogue of American Amphibians and Reptiles 418.1–418.2.

———. 1991. *Rana berlandieri*. Catalogue of American Amphibians and Reptiles 508.1–508.4.

Platz, J. E., and J. S. Mecham. 1984. *Rana chiricahuensis*. Catalogue of American Amphibians and Reptiles 347.1–347.2.

Pollio, C. A. 2000. *Eurycea cirrigera* (Southern Two-lined Salamander). Larval period/size-class determination. Herpetological Review 31:166–167.

Pollister, A. W., and J. A. Moore. 1937. Tables for the normal development of *Rana sylvatica*. Anatomical Record 58:489–496.

Pope, C. H. 1919a. A note on the development of *Pseudacris feriarum* (Baird). Copeia 1919: 83–84.

———. 1919b. The life-history of the Common Water-newt (*Notophthalmus viridescens*), together with observations on the sense of smell. Annals of the Carnegie Museum 15:305–368.

———. 1921. Some doubtful points in the life-history of *Notophthalmus viridescens*. Copeia 1921:14–15.

———. 1924. Notes on North Carolina salamanders, with especial reference to the egg-laying habits of *Leurognathus* and *Desmognathus*. American Museum Novitates (153): 1–15.

———. 1928. The life-history of *Triturus viridescens*—some further notes. Copeia 1928: 61–73.

———. 1964. Amphibians and reptiles of the Chicago area. Chicago Natural History Museum. 275 pp.

Pope, K. L. 1999. *Rana muscosa* (Mountain Yellow-legged Frog). Diet. Herpetological Review 30:163–164.

Porter, K. R. 1970. *Bufo valliceps*. Catalogue of American Amphibians and Reptiles 94.1–94.4.

Portnoy, J. W. 1990. Breeding biology of the Spotted Salamander *Ambystoma maculatum* (Shaw) in acidic temporary ponds at Cape Cod. Biological Conservation 53:61–75.

Potter, F. E., and S. S. Sweet. 1981. Generic boundaries in Texas cave salamanders, and a redescription of *Typhlomolge robusta* (Amphibia: Plethodontidae). Copeia 1981:64–75.

Potthoff, T. L., and J. D. Lynch. 1986. Interpopulation variability in mouthparts in *Scaphiopus bombifrons* in Nebraska (Amphibia: Pelobatidae). Prairie Naturalist 18:15–22.

Pough, F. H. 1976. Acid precipitation and embryonic mortality of Spotted Salamanders, *Ambystoma maculatum*. Science 192:68–70.

Powers, J. H. 1903. The causes of acceleration and retardation in the metamorphosis of *Ambystoma tigrinum*: A preliminary report. American Naturalist 37:385–410.

——. 1907. Morphological variation and its causes in *Amblystoma tigrinum*. University of Nebraska Studies (7): 197–274.

Preston, W. B. 1982. The amphibians and reptiles of Manitoba. Winnipeg: Manitoba Museum of Man and Nature. 128 pp.

Price, A. H., and B. K. Sullivan. 1988. *Bufo microscaphus*. Catalogue of American Amphibians and Reptiles 415.1–415.3.

Price, J. P., R. A. Browne, and M. E. Dorcas. 2012. Resistance and resilience of a stream salamander to supraseasonal drought. Herpetologica 68:312–323.

Price, S. J., and J. M. Jaskula. 2005. *Hemidactylium scutatum* (Four-toed Salamander). Nesting ecology. Herpetological Review 36:159.

Provenzano, S. F., and M. D. Boone. 2009. Effects of density on metamorphosis of Bullfrogs in a single season. Journal of Herpetology 43:49–54.

Pryor, G. S. 2003. Growth rates and digestive abilities of Bullfrog tadpoles (*Rana catesbeiana*) fed algal diets. Journal of Herpetology 37:560–566.

Pryor, G. S., and K. A. Bjorndal. 2005a. Symbiotic fermentation, digesta passage, and gastrointestinal morphology in Bullfrog tadpoles (*Rana catesbeiana*). Physiological and Biochemical Zoology 78:201–215.

——. 2005b. Effects of the nematode *Gyrinicola batrachiensis* on development, gut morphology, and fermentation in Bullfrog tadpoles (*Rana catesbeiana*): A novel mutualism. Journal of Experimental Zoology 303A:704–712.

Pryor, G. S., D. P. German, and K. A. Bjorndal. 2006. Gastrointestinal fermentation in Greater Sirens (*Siren lacertina*). Journal of Herpetology 40:112–117.

Punzo, F. 1983. Effects of environmental pH and temperature on embryonic survival capacity and metabolic rates in the Smallmouth Salamander, *Ambystoma texanum*. Bulletin of Environmental Contamination and Toxicology 31:467–473.

——. 1991. Group learning in tadpoles of *Rana heckscheri* (Anura: Ranidae). Journal of Herpetology 25:214–217.

——. 1992. Socially facilitated behavior in tadpoles of *Rana catesbeiana* and *Rana heckscheri* (Anura: Ranidae). Journal of Herpetology 26:219–222.

Pylka, J. M., and R. D. Warren. 1958. A population of *Haideotriton* in Florida. Copeia 1958:334–336.

Quinn, H., and G. Mengden. 1984. Reproduction and growth of *Bufo houstonensis* (Bufonidae). Southwestern Naturalist 29:189–195.

Rabb, G. B. 1956. Some observations on the salamander, *Stereochilus marginatus*. Copeia 1956:119.

——. 1966. *Stereochilus marginatus*. Catalogue of American Amphibians and Reptiles 25.1–25.2.

Rachowicz, R. J. 2002. Mouthpart pigmentation in *Rana muscosa* tadpoles: Seasonal changes without chytridiomycosis. Herpetological Review 33:263–265.

Rathbun, G. B. 1998. *Rana aurora draytonii* (California Red-legged Frog). Egg predation. Herpetological Review 29:165.

Raymond, L. R. 1991. Seasonal activity of *Siren intermedia* in northwestern Louisiana (Amphibia: Sirenidae). Southwestern Naturalist 36:144–147.

Raymond, L. R., and L. M. Hardy. 1990. Demography of a population of *Ambystoma talpoideum* (Caudata: Ambystomatidae) in northwestern Louisiana. Herpetologica 46:371–382.

Redmer, M. 1998. *Hyla avivoca* (Bird-voiced Treefrog). Amplexus and oviposition. Herpetological Review 29:230–231.

——. 2000. Demographic and reproductive characteristics of a southern Illinois population of the Crayfish Frog, *Rana areolata*. Journal of the Iowa Academy of Science 107:128–133.

——. 2002. Natural history of the Wood Frog (*Rana sylvatica*) in the Shawnee National Forest, southern Illinois. Bulletin of the Illinois Natural History Survey 36:163–194.

Redmer, M., L. E. Brown, and R. A. Brandon. 1999. Natural history of the Bird-voiced Treefrog (*Hyla avivoca*) and Green Treefrog (*Hyla cinerea*) in southern Illinois. Bulletin of the Illinois Natural History Survey 36:37–66.

Reese, A. M. 1904. The sexual elements of the Giant Salamander, *Cryptobranchus alleganiensis*. Biological Bulletin 6:220–223.

——. 1906. Observations on the reactions of *Cryptobranchus* and *Necturus* to light and heat. Biological Bulletin 11:93–99.

Regan, G. T. 1969. Color dimorphism in the larval Cricket Frog *Acris crepitans*. Proceedings of the Nebraska Academy of Sciences 79:10–11.

Regester, K. J., K. R. Lips, and M. R. Whiles. 2006. Energy flow and subsidies associated with the complex life cycle of ambystomatid salamanders in ponds and adjacent forest in southern Illinois. Oecologia 147:303–314.

Regester, K. J., M. R. Whiles, and K. R. Lips. 2008. Variation in the trophic basis of production and energy flow associated with emergence of larval salamander assemblages from forest ponds. Freshwater Biology 53:1754–1767.

Regester, K. J., and B. T. Miller. 2000. *Ambystoma barbouri* (Streamside Salamander). Reproduction. Herpetological Review 31:232.

Regula Meyer, L., J. Geiger-Hayes, and P. Owens. 2007. Examination of the intestinal contents of *Rana clamitans* and *Rana catesbeiana* tadpoles for symbiotic, cellulose-digesting bacteria. Herpetological Review 38:393–395.

Reilly, S. M., and G. V. Lauder. 1988. Ontogeny of aquatic feeding performance in the Eastern Newt, *Notophthalmus viridescens* (Salamandridae). Copeia 1988:87–91.

Reilly, S. M., G. V. Lauder, and J. P. Collins. 1992. Performance consequences of a trophic polyphenism: Feeding behavior in typical and cannibal phenotypes of *Ambystoma tigrinum*. Copeia 1992:672–679.

Reilly, S. M., E. O. Wiley, and D. J. Meinhardt. 1997. An integrative approach to heterochrony: The distinction between interspecific and intraspecific phenomena. Biological Journal of the Linnean Society 60:119–143.

Relyea, R. A. 2001. Morphological and behavioral plasticity of larval anurans in response to different predators. Ecology 82:523–540.

——. 2003. Predators come and predators go: The reversibility of predator-induced traits. Ecology 84:1840–1848.

Relyea, R. A., and J. R. Auld. 2004. Having the guts to compete: How intestinal plasticity explains cost of inducible defences. Ecological Letters 7:869–875.

——. 2005. Predator- and competitor-induced plasticity: How changes in foraging morphology affect phenotypic trade-offs. Ecology 86:1723–1729.

Relyea, R. A., and E. E. Werner. 1999. Quantifying the relation between predator-induced behavior and growth performance in larval anurans. Ecology 80:2117–2124.

——. 2000. Morphological plasticity in four larval anurans distributed along an environmental gradient. Copeia 2000:178–190.

Reno, H. W., F. R. Gehlbach, and R. A. Turner. 1972. Skin and aestivational cocoon of the aquatic amphibian, *Siren intermedia* Le Conte. Copeia 1972:625–631.

Resetarits, W. J., Jr. 1998. Differential vulnerability of *Hyla chrysoscelis* eggs and hatchlings to larval insect predation. Journal of Herpetology 32:440–443.

Resetarits, W. J., Jr., J. F. Rieger, and C. A. Blinckley. 2004. Threat of predation negates density effects in larval Gray Treefrogs. Oecologia 138:532–538.

Richards, C. M. 1958. The inhibition of growth in crowded *Rana pipiens* tadpoles. Physiological Zoology 31:138–151.

——. 1962. The control of tadpole growth by alga-like cells. Physiological Zoology 35:285–296.

Richardson, J. L. 2006. Novel features of an inducible defense system in larval tree frogs (*Hyla chrysoscelis*). Ecology 87:780–787.

Richardson, J. M. L. 2001. A comparative study of activity levels in larval anurans and response to the presence of different predators. Behavioral Ecology 12:51–58.

——. 2002a. A comparative study of phenotypic traits related to resource utilization in anuran communities. Evolutionary Ecology 16:101–122.

——. 2002b. Burst swim speed in tadpoles inhabiting ponds with different top predators. Evolutionary Ecology Research 4:627–642.

Richmond, N. D. 1945. Nesting of the Two-lined Salamander on the Coastal Plain. Copeia 1945:170.

——. 1947. Life history of *Scaphiopus holbrookii holbrookii* (Harlan). Part I: Larval development and behavior. Ecology 28:53–67.

Ringia, A. M., and K. R. Lips. 2007. Oviposition, early development and growth of the Cave Salamander, *Eurycea lucifuga*: Surface and subterranean influences on a troglophilic species. Herpetologica 63:258–268.

Ritke, M. E., J. G. Babb, and M. K. Ritke. 1990. Life history of the Gray Treefrog (*Hyla chrysoscelis*) in western Tennessee. Journal of Herpetology 24:135–141.

Ritter, W. E. 1897. The life-history and habits of the Pacific Coast Newt (*Diemyctylus torosus* Esch). Proceedings of the California Academy of Sciences 1:73–114.

Roberts, D. T., D. M. Schleser, and T. L. Jordan. 1995. Notes on the captive husbandry of the Texas Salamander *Eurycea neotenes* at the Dallas Aquarium. Herpetological Review 26:23–25.

Rogers, C. P. 1996. *Rana catesbeiana* (Bullfrog). Predation. Herpetological Review 27:19.

Rombough, C. J., and M. P. Hayes. 2005. *Rana boylii* (Foothill Yellow-legged frog). Predation: Eggs and hatchlings. Herpetological Review 36:163–164.

Rombaugh, C. J., and M. P. Hayes. 2008. *Rana pretiosa* (Oregon Spotted Frog). Reproduction. Herpetological Review 39:340–341.

Rondeau, S. L., and J. H. Gee. 2005. Larval anurans adjust buoyancy in response to substrate ingestion. Copeia 2005:188–195.

Rose, F. L. 1966a. Weight change during starvation in *Amphiuma means*. Herpetologica 22:312–313.

——. 1966b. Reproductive potential of *Amphiuma means*. Copeia 1966:598–599.

——. 1967. Seasonal changes in lipid levels of the salamander *Amphiuma means*. Copeia 1967:663–666.

——. 1971. *Eurycea aquatica*. Catalogue of American Amphibians and Reptiles 116.1–116.2.

Rose, F. L., and D. Armentrout. 1976. Adaptive strategies of *Ambystoma tigrinum* Green inhabiting the Llano Estacado of west Texas. Journal of Animal Ecology 45:713–729.

Rose, F. L., and F. M. Bush. 1963. A new species of *Eurycea* (Amphibia: Caudata) from the southeastern United States. Tulane Studies in Zoology 10:121–128.

Rose, S. M. 1958. Failure of survival of slowly growing members of a population. Science 128:1026.

——. 1960. A feedback mechanism of growth control in tadpoles. Ecology 41:188–199.

Rosen, P. C. 2007. *Rana yavapaiensis* (Lowland Leopard Frog). Larval cannibalism. Herpetological Review 38:195–196.

Rossman, D. A. 1958. A new race of *Desmognathus fuscus* from the south-central United States. Herpetologica 11:158–160.

Roth, A. H., and J. F. Jackson. 1987. The effect of pool size on recruitment of predatory insects and on mortality in a larval anuran. Herpetologica 43:224–232.

Rot-Nikcevic, I., R. J. Denver, and R. J. Wassersug. 2005. The influence of visual and tactile stimulation on growth and metamorphosis in anuran larvae. Functional Ecology 19:1008–1016.

Rot-Nikcevic, I., and R. J. Wassersug. 2004. Arrested development in *Xenopus laevis* tadpoles: How size constrains metamorphosis. Journal of Experimental Biology 207:2133–2145.

Rowe, C. L., and W. A. Dunson. 1995. Impacts of hydroperiod on growth and survival of larval amphibians in temporary ponds in central Pennsylvania. Oecologia 102:397–403.

Rubbo, M. J., K. Shea, and J. M. Kiesecker. 2006. The influence of multi-stage predation on population growth and the distribution of the pond-breeding salamander, *Ambystoma jeffersonianum*. Canadian Journal of Zoology 84:449–458.

Rubenstein, N. M. 1971. Ontogenetic allometry in the salamander genus *Desmognathus*. American Midland Naturalist 85:329–348.

Rudolph, D. C. 1978. Aspects of the larval ecology of five plethodontid salamanders of the western Ozarks. American Midland Naturalist 100:141–159.

Rundio, D. E., and D. H. Olson. 2001. Palatability of Southern Torrent Salamander (*Rhyacotriton variegatus*) larvae to Pacific Giant Salamander (*Dicamptodon tenebrosus*) larvae. Journal of Herpetology 35:133–136.

Russell, A. P., and A. M. Bauer. 1993. The amphibians and reptiles of Alberta. A field guide and primer to boreal herpetology. Calgary: University of Calgary Press.

Russell, K. R., A. A. Gonyaw, J. D. Strom, K. E. Diemer, and K. C. Murk. 2002. Three new nests of the Columbia Torrent Salamander, *Rhyacotriton kezeri*, in Oregon with observations of nesting behavior. Northwestern Naturalist 83:19–22.

Ruth, B. C., W. A. Dunson, C. L. Rowe, and S. B. Hedges. 1993. A molecular and functional evaluation of the egg mass color polymorphism of the Spotted Salamander, *Ambystoma maculatum*. Journal of Herpetology 27:306–314.

Ruthven, A. G. 1907. A collection of reptiles and amphibians from southern New Mexico and Arizona. Bulletin of the American Museum of Natural History 23:483–604.

Ryan, M. J. 1978. A thermal property of the *Rana catesbeiana* (Amphibia, Anura, Ranidae) egg mass. Journal of Herpetology 12:247–248.

Ryan, T. J. 1997. The larva of *Eurycea junaluska* (Amphibia: Caudata: Plethodontidae), with comments on distribution. Copeia 1997:210–215.

———. 1998. Larval life history and abundance of a rare salamander, *Eurycea junaluska*. Journal of Herpetology 32:1–17.

———. 2007. Hydroperiod and metamorphosis in Small-mouthed Salamanders (*Ambystoma texanum*). Northeastern Naturalist 14:619–628.

Ryan, T. J., and R. C. Bruce. 1998. *Eurycea junaluska* (Junaluska Salamander). Morphology. Herpetological Review 29:163.

Ryan, T. J., and G. R. Plague. 2004. Hatching asynchrony, survival, and the fitness of alternative adult morphs in *Ambystoma talpoideum*. Oecologia 140:46–51.

Ryan, T. J., and R. D. Semlitsch. 2003. Growth and the expression of alternative life cycles in the salamander *Ambystoma talpoideum* (Caudata: Ambystomatidae). Bulletin of the Journal of the Linnean Society 80:639-646.

Ryan, T. J., and G. Swenson. 2001. Does sex influence postreproductive metamorphosis in *Ambystoma talpoideum?* Journal of Herpetology 35:697–700.

Ryder, J. A. 1889. Report on a brood of larval *Amphiuma*. American Naturalist 23:927–928.

———. 1891. Notes on the development of *Engystoma*. American Naturalist 25:838–840.

Saenz, D., J. B. Johnson, C. K. Adams, and G. H. Dayton. 2003. Accelerated hatching of Southern Leopard Frog (*Rana sphenocephala* eggs in response to the presence of crayfish (*Procambarus nigrocinctus*. Copeia 2003:646–649.

Sagar, J. P., D. H. Olson, and R. A. Schmitz. 2007. Survival and growth of larval Coastal Giant Salamanders (*Dicamptodon tenebrosus*) in streams in the Oregon Coast Range. Copeia 2007:123–130.

Salthe, S. N. 1963. The egg capsules in the Amphibia. Journal of Morphology 113:161–171.

———. 1973a. *Amphiuma means*. Catalogue of American Amphibians and Reptiles 148.1–148.2.

———. 1973b. *Amphiuma tridactylum*. Catalogue of American Amphibians and Reptiles 149.1–149.3.

Sanders, A. E. 1984. *Rana heckscheri*. Catalogue of American Amphibians and Reptiles 348.1–348.2.

Sartorius, S. S., and P. C. Rosen. 2000. Breeding phenology of the Lowland Leopard Frog (*Rana yavapaiensis*): Implications for conservation and ecology. Southwestern Naturalist 45:267–273.

Saumure, R. A., and J. S. Doody. 1998. *Amphiuma tridactylum* (Three-toed Amphiuma). Ectoparasites. Herpetological Review 29:163.

Savage, J. M. 1960. Geographic variation in the tadpole of the toad *Bufo marinus*. Copeia 1960:233–235.

——. 1968. The dendrobatid frogs of Central America. Copeia 1968:745–776.

——. 1980. A synopsis of the larvae of Costa Rican frogs and toads. Bulletin of the Southern California Academy of Sciences 79:45–54.

Savage, J. M., and F. W. Schuierer. 1961. The eggs of toads of the *Bufo boreas* group, with descriptions of the eggs of *Bufo exsul* and *Bufo nelsoni*. Bulletin of the Southern California Academy of Sciences 60:93–99.

Schaaf, R. T., Jr., and P. W. Smith. 1971. *Rana palustris*. Catalogue of American Amphibians and Reptiles 117.1–117.3.

Schaub, D. L., and J. H. Larsen. 1978. The reproductive ecology of the Pacific Treefrog (*Hyla regilla*). Herpetologica 34:409–416.

Schechtman, A. M., and J. B. Olson. 1941. Unusual temperature tolerance of an amphibian egg (*Hyla regilla*). Ecology 22:409–410.

Scherff-Norris, K. L., and L. J. Vivo. 1999. *Bufo boreas* (Boreal Toad). Phoretic host. Herpetological Review 30:162.

Schiesari, L. 2006. Pond canopy cover: A resource gradient for anuran larvae. Freshwater Biology 51:412–423.

Schiesari, L., S. D. Peacor, and E. E. Werner. 2006. The growth-mortality tradeoff: Evidence from anuran larvae and consequences for species distributions. Oecologia 149:194–202.

Schiesari, L., E. E. Werner, and G. W. Kling. 2009. Carnivory and resource-based niche differentiation in anuran larvae: Implications for food web and experimental ecology. Freshwater Biology 54:572–586.

Schmidt, U. 1995. Ein schneller Brüter Pflege und Vermebrung von *Smilisca baudini*. Elaphe 2:15–19.

Schneider, C. W. 1968. Avoidance learning and the response tendencies of the larval salamander *Ambystoma punctatum* to photic stimulation. Animal Behavior 16:492–495.

Schneider, C. W., B. W. Marquette, and P. Pietsch. 1991. Measures of phototaxis and movement detection in the larval salamander. Physiological Behaviour 50:645–647.

Schoeppner, N. M., and R. A. Relyea. 2005. Damage, digestion, and defense: The roles of alarm cues and kairomones for inducing prey defenses. Ecological Letters 8:505–512.

Schoonbee, H. J., J. F. Prinsloo, and J. G. Nxiweni. 1992. Observations on the feeding habits of larvae, juvenile and adult stages of the African Clawed Frog, *Xenopus laevis*, in impoundments in Transkei. Water South Africa 18:227–236.

Schorr, M. S., R. Altig, and W. J. Diehl. 1990. Population changes of the enteric protozoans *Opalina* spp. and *Nyctotherus cordiformis* during the ontogeny of anuran tadpoles. Journal of Protozoology 37:479–481.

Schuierer, F. W. 1958. Factors affecting neoteny in the salamander, *Dicamptodon ensatus* (Eschscholtz). Bulletin of the Southern California Academy of Sciences 57:119–121.

——. 1962. Remarks upon the natural history of *Bufo exsul* Myers, the endemic toad of Deep Springs, Inyo County, California. Herpetologica 17:260–266.

Schwartz, A. 1952. A new race of *Pseudobranchus striatus* from southern Florida. Natural History Miscellanea (115): 1–9.

Schwetmen, N. H. 1967. A morphological study of the external features, viscera, integument, and skeletons of *Eurycea nana*. Thesis, Baylor University, Waco.

Scott, D. E. 1990. Effects of larval density in *Ambystoma opacum*: An experiment in large-scale field enclosures. Ecology 71:296–306.

——. 1993. Timing of reproduction of paedomorphic and metamorphic *Ambystoma talpoideum*. American Midland Naturalist 129:397–402.

——. 1994. The effect of larval density on adult demographic traits in *Ambystoma opacum*. Ecology 75:1383–1396.

Scott, D. E., E. D. Casey, M. F. Donovan, and T. K. Lynch. 2007. Amphibian lipid levels at metamorphosis correlate to post-metamorphic terrestrial survival. Oecologia 153:521–532.

Scott, D. E., and M. R. Fore. 1995. The effect food limitation on lipid levels, growth, and reproduction in the Marbled Salamander, *Ambystoma opacum*. Herpetologica 51:462–471.

Scott, N. J., Jr., and R. D. Jennings. 1985. The tadpoles of five species of New Mexican leopard frogs. Occas. Pap. Mus. Southwestern Biologist (3): 1–21.

Scroggin, J. B., and W. B. Davis. 1956. Food habits of the Texas Dwarf Siren. Herpetologica 12:231–237.

Seale, D. B. 1980. Influence of amphibian larvae on primary production, nutrient flux, and competition in a pond ecosystem. Ecology 61:1531–1550.

——. 1982. Obligate and facultative suspension feeding in anuran larvae: Feeding regulation in *Xenopus* and *Rana*. Biological Bulletin 162:214–231.

Seale, D. B., and N. Beckvar. 1980. The comparative ability of anuran larvae (genera: *Hyla*, *Bufo*, and *Rana*) to ingest suspended blue-green algae. Copeia 1980:495–503.

Seale, D. B., K. Hoff, and R. Wassersug. 1982. *Xenopus laevis* larvae (Amphibia, Anura) as model suspension feeders. Hydrobiologia 87:161–169.

Seale, D., and R. J. Wassersug. 1979. Suspension feeding dynamics of anuran larvae related to their functional morphology. Oecologia 39:259–272.

Seigel, R. A. 1983. Natural survival of eggs and tadpoles of the Wood Frog, *Rana sylvatica*. Copeia 1983:1096–1098.

Seixas, F. J. T. D., M. S. C. R. Pereira, V. R. C. D. Andrade, G. de Braudt, and S. C. D. Nascimento. 1998. Effects of particle size of the diet in the performance of *Rana catesbeiana* tadpoles. Revista Brasileira Zootecnia 27:224–230.

Sekerak, C. M., G. W. Tanner, and J. G. Palis. 1996. Ecology of Flatwoods Salamander larvae in breeding ponds in Apalachicola National Forest. Proceedings of the Annual Conference of the Southeastern Association of Fish and Wildlife Agencies 50:321–330.

Semlitsch, R. D. 1980. Growth and metamorphosis of larval Dwarf Salamanders (*Eurycea quadridigitata*). Copeia 1980:138–140.

——. 1983a. Growth and metamorphosis of larval Red Salamanders (*Pseudotriton ruber*) on the Coastal Plain of South Carolina. Herpetologica 39:48–52.

——. 1983b. Structure and dynamics of two breeding populations of the Eastern Tiger Salamander, *Ambystoma tigrinum*. Copeia 1983:608–616.

——. 1985. Reproductive strategy of a facultatively paedomorphic salamander *Ambystoma talpoideum*. Oecologia 65:305–313.

——. 1987a. Relationship of pond drying to the reproductive success of the salamander *Ambystoma talpoideum*. Copeia 1987:61–69.

——. 1987b. Interactions between fish and salamander larvae. Costs of predator avoidance or competition. Oecologia 72:481–486.

——. 1987c. Paedomorphosis in *Ambystoma talpoideum*: Effects of density, food, and pond drying. Ecology 68:994–1002.

——. 1987d. Density-dependent growth and fecundity in the paedomorphic salamander *Ambystoma talpoideum*. Ecology 68:1003–1008.

——. 1990. Effects of body size, sibship, and tail injury on the susceptibility of tadpoles to dragon fly predation. Canadian Journal of Zoology 68:1027–1030.

Semlitsch, R. D., and J. P. Caldwell. 1982. Effects of density on growth, metamorphosis, and survivorship in tadpoles of *Scaphiopus holbrooki*. Ecology 63:905–911.

Semlitsch, R. D., and J. W. Gibbons. 1985. Phenotypic variation in metamorphosis and paedomorphosis in the salamander *Ambystoma talpoideum*. Ecology 66:1123–1130.

Semlitsch, R. D., J. W. Gibbons, and T. D. Tuberville. 1995. Timing of reproduction and metamorphosis in the Carolina Gopher Frog (*Rana capito capito*) in South Carolina. Journal of Herpetology 29:612–614.

Semlitsch, R. D., R. N. Harris, and H. M. Wilbur. 1990. Paedomorphosis in *Ambystoma talpoideum*—maintenance of population variation and alternative life history pathways. Evolution 44:1604–1613.

Semlitsch, R. D., and M. A. McMillan. 1980. Breeding migrations, population size structure, and reproduction of the Dwarf Salamander, *Eurycea quadridigitata*, in South Carolina. Brimleyana 3:97–105.

Semlitsch, R. D., and S. B. Reichling. 1989. Density-dependent injury in larval salamanders. Oecologia 81:100–103.

Semlitsch, R. D, D. E. Scott, and J. H. K. Pechmann. 1988. Time and size at metamorphosis related to adult fitness in *Ambystoma talpoideum*. Ecology 69:184–192.

Semlitsch, R. D., and S. C. Walls. 1993. Competition in two species of larval salamanders: A test of geographic variation in competitive ability. Copeia 1993:587–595.

Semlitsch, R. D., and H. Wilbur. 1988. Effects of pond drying on metamorphosis and survival in the salamander *Ambystoma talpoideum*. Copeia 1988:978–983.

———. 1989. Artificial selection for paedomorphosis in the salamander *Ambystoma talpoideum*. Evolution 43:105–112.

Sever, D. M. 1983. Observations on the distribution and reproduction of the salamander *Eurycea junaluska* in Tennessee. Journal of the Tennessee Academy of Science 58:48–50.

———. 1999a. *Eurycea bislineata*. Catalogue of American Amphibians and Reptiles 683.1–683.5.

———. 1999b. *Eurycea cirrigera*. Catalogue of American Amphibians and Reptiles 684.1–684.6.

———. 1999c. *Eurycea wilderae*. Catalogue of American Amphibians and Reptiles 685.1–685.4.

Sever, D. M., and L. D. Houck. 1985. Spermatophore formation in *Desmognathus ochrophaeus* (Amphibia: Plethodontidae). Copeia 1985:394–402.

Sever, D. M., S. A. Kramer, and S. Duff. 1987. The relation between ovum variability and larval growth in *Ambystoma tigrinum* (Amphibia: Urodela). Proceedings of the Indiana Academy of Science 96:531–536.

Sever, D. M., L. C. Rania, and J. D. Krenz. 1996. Reproduction of the salamander, *Siren intermedia* Le Conte with special reference to oviducal anatomy and mode of fertilization. Journal of Morphology 227:335–348.

Severinghaus, A. E. 1930. Gill development in *Amblystoma punctatum*. Journal of Experimental Zoology 51:1–31.

Sexton, O. J., and J. R. Bizer. 1978. Life history patterns of *Ambystoma tigrinum* in montane Colorado. American Midland Naturalist 99:101–118.

Seyle, C. W., Jr., and S. E. Trauth. 1982. Life history notes: *Pseudacris ornata* (Ornate Chorus Frog). Reproduction. Herpetological Review 13:45.

Seymour, R. S. 1995. Oxygen uptake by embryos in gelatinous egg masses of *Rana sylvatica*: The roles of diffusion and convection. Copeia 1995:626–635.

Sheen, J. P., and H. H. Whiteman. 1998. Head and body size relationships in polymorphic Tiger Salamander larvae from Colorado. Copeia 1998:1089–1093.

Sherman, E., and D. Levitis. 2003. Heat hardening as a function of developmental stage in larval and juvenile *Bufo americanus* and *Xenopus laevis*. Journal of Thermal Biology 28:373–380.

Shipman, P., D. L. Cross-White, and S. F. Fox. 1999. Diet of the Quachita Dusky Salamander (*Desmognathus brimleyorum*) in southeastern Oklahoma. American Midland Naturalist 141:398–401.

Shoop, C. R. 1960. The breeding habits of the Mole Salamander, *Ambystoma talpoideum* (Holbrook), in southeastern Louisiana. Tulane Studies in Zoology 8:65–82.

——. 1964. *Ambystoma talpoideum*. Catalogue of American Amphibians and Reptiles 8.1–8.2.

——. 1965. Aspects of reproduction in Louisiana *Necturus* populations. American Midland Naturalist 74:357–367.

——. 1974. Yearly variation in larval survival of *Ambystoma maculatum*. Ecology 55:440–444.

Shoop, C. R., and G. E. Gunning. 1967. Seasonal activity and movements of *Necturus* in Louisiana. Copeia 1967:732–737.

Showalter, A. M. 1940. A green alga in salamander eggs. Virginia Journal of Science 1:210–211.

Shrode, C. J. 1972. Effect of temperature and dissolved oxygen concentration on the rate of metamorphosis of *Ambystoma tigrinum*. Journal of Herpetology 6:199–207.

Siekmann, J. M. 1949. A survey of the tadpoles of Louisiana. Thesis, Tulane University, New Orleans.

Sih, A. 1992. Integrative approaches to the study of predation: General thoughts and a case study on sunfish and salamander larvae. Annales Zoologici Fennici 29:183–198.

Sih, A., L. and B. Kats. 1991. Effects of refuge availability on the responses of salamander larvae to chemical cues from predatory Green Sunfish. Animal Behavior 42:330–332.

Sih, A., and R. Maurer. 1992. Effects of cryptic oviposition on egg survival for stream-breeding Streamside Salamanders. Journal of Herpetology 26:114–116.

Sih, A., and R. D. Moore. 1993. Delayed hatching of salamander eggs in response to enhanced larval predation risk. American Naturalist 142:947–960.

Sih, A., and J. W. Petranka. 1988. Optimal diets: Simultaneous search and handling of multiple prey loads by salamander larvae. Behavioral Ecology and Sociobiology 23:335–339.

Silverstone, P. A. 1975. A revision of the poison-arrow frogs of the genus *Dendrobates* Wagler. Natural History Museum of Los Angeles County Science Bulletin (21): 1–55.

Simmons, R., and J. D. Hardy Jr. 1959. The River-swamp Frog, *Rana heckscheri* Wright, in North Carolina. Herpetologica 15:36–37.

Sinclair, R. M. 1950. Notes on some salamanders from Tennessee. Herpetologica 6:49–51.

——. 1951. Notes on recently transformed larvae of the salamander *Eurycea longicauda gutolineata*. Herpetologica 7:68.

Skelly, D. K. 1994. Activity level and the susceptibility of anuran larvae to predation. Animal Behavior 47:465–468.

——. 1995a. A behavioral trade-off and its consequences for the distribution of treefrog larvae. Ecology 76:150–164.

——. 1995b. Competition and the distribution of Spring Peeper larvae. Oecologia 103:203–207.

——. 1996. Pond drying, predators, and the distribution of *Pseudacris* tadpoles. Copeia 1996:599–605.

——. 2004. Microgeographic countergradient variation in the Wood Frog, *Rana sylvatica*. Evolution 58:160–165.

Skelly, D. K., L. K. Freidenburg, and J. M. Kiesecker. 2002. Forest canopy and the performance of larval amphibians. Ecology 83:983–992.

Slater, J. R. 1936. Notes on *Ambystoma gracile* Baird and *Ambystoma macrodactylum* Baird. Copeia 1936:234–236.

——. 1939. Description and life history of a new *Rana* from Washington. Herpetologica 1:145–149.

Smith, B. G. 1907a. The life history and habits of *Cryptobranchus alleghaniensis*. Biological Bulletin 13:5–39.

——. 1907b. The breeding habits of *Amblystoma punctatum* Linn. American Naturalist 41:381–390.

——. 1911a. Notes on the natural history of *Amblystoma jeffersonianum, A. punctatum* and *A. tigrinum*. Bulletin of the Wisconsin Natural History Society 9:14–27.

——. 1911b. The nests and larvae of *Necturus*. Biological Bulletin 20:191–200.

——. 1912a. The embryology of *Cryptobranchus allegheniensis*, including comparisons with some other vertebrates. I. Introduction: The history of the egg before cleavage. Journal of Morphology 23:61–157.

——. 1912b. The embryology of *Cryptobranchus allegheniensis*, including comparisons with some other vertebrates. Journal of Morphology 23:455–479.

Smith, C. C. 1960. Notes on the salamanders of Arkansas. No. 1. Life history of a neotenic stream-dwelling form. Proceedings of the Arkansas Academy of Science 13:66–74.

——. 1968. A new *Typhlotriton* from Arkansas (Amphibia Caudata). Wasmann Journal of Biology 26:155–159.

Smith, C. K. 1990. Effects of variation in body size on intraspecific competition among larval salamanders. Ecology 71:1777–1788.

Smith, C. K., J. W. Petranka, and R. Barwick. 1996a. *Desmognathus quadramaculatus* (Blackbelly Salamander). Reproduction. Herpetological Review 27:136.

——. 1996b. *Desmognathus welteri* (Black Mountain Salamander). Reproduction. Herpetological Review 27:136.

Smith, G. R. 1999a. Microhabitat preferences of Bullfrog tadpoles (*Rana catesbeiana*) of different ages. Transactions of the Nebraska Academy of Science 25:73–76.

——. 1999b. Among family variation in tadpole (*Rana catesbeiana*) responses to density. Journal of Herpetology 33:167–169.

Smith, G. R., A. Boyd, C. B. Dayer, M. E. Ogle, A. J. Terlecky, and C. J. Dibble. 2010. Effects of sibship and the presence of multiple predators on the behavior of Green Frog (*Rana clamitans*) tadpoles. Ethology 116:217–225.

Smith, G. R., A. A. Burgett, K. A. Sparks, K. G. Temple, and K. W. Winter. 2007. Temporal patterns in Bullfrog (*Rana catesbeiana*) tadpole activity: A mesocosm experiment on the effects of density and Bluegill Sunfish (*Lepomis macrochirus*) presence. Herpetological Journal 17:199–203.

Smith, G. R., H. A. Dingfelder, and D. A. Vaala. 2004. Asymmetric competition between *Rana clamitans* and *Hyla versicolor* tadpoles. Oikos 105:626–632.

Smith, G. R., and B. L. Doupnik. 2005. Habitat use and activity level of large American Bullfrog tadpoles: Choices and repeatability. Amphibia-Reptilia 26:549–552.

Smith, G. R., and A. K. Jennings. 2004. Spacing of the tadpoles of *Hyla versicolor* and *Rana clamitans*. Journal of Herpetology 38:616–618.

Smith, G. R., and J. E. Rettig. 1998. Observations on egg masses of the American Toad (*Bufo americanus*). Herpetological Natural History 6:61–64.

Smith, H. M. 1934. The amphibians of Kansas. American Midland Naturalist 15:377–528.

——. 1946a. The tadpoles of *Bufo cognatus* Say. University of Kansas Museum of Natural History Publications 1:93–96.

——. 1946b. Handbook of lizards. Lizards of the United States and of Canada. Ithaca, NY: Comstock Publishing.

——. 1956. The amphibians and reptiles of Kansas. University of Kansas Museum of Natural History Publications (9): 1–356.

Smith, H. M., C. W. Nixon, and P. E. Smith. 1947. Notes on the tadpoles and natural history of *Rana areolata circulosa*. Bulletin of the Ecological Society of America 28:54.

——. 1948. A partial description of the tadpole of *Rana areolata circulosa* and notes on the natural history of the race. American Midland Naturalist 39:608–614.

Smith, H. M., and F. E. Potter Jr. 1946. A third neotenic salamander of the genus *Eurycea* from Texas. Herpetologica 3:105–109.

Smith, K. G. 2005a. Effects of nonindigenous tadpoles on native tadpoles in Florida: Evidence of competition. Biological Conservation 123:433–441.

——. 2005b. An exploratory assessment of Cuban Treefrog (*Osteopilus septentrionalis*) tadpoles as predators of native and nonindigenous tadpoles in Florida. Amphibia-Reptilia 26:571–575.

——. 2006. Keystone predators (Eastern Newts, *Notophthalmus viridescens*) reduce the impacts of an aquatic invasive species. Oecologia 148:342–349.

Smith, K. M., and D. M. Ghioca-Robrecht. 2008. Feeding ecology of polymorphic larval Barred Tiger Salamanders in playas of the southern Great Plains. Canadian Journal of Zoology 86:554–563.

Smith, L. 1920. A note on the eggs of *Ambystoma maculatum*. Copeia 1920:41.

Smith, M. J., M. M. Drew, M. Peebles, and K. Summers. 2005. Predator cues during the egg stage affect larval development in the Gray Treefrog, *Hyla versicolor* (Anura: Hylidae). Copeia 2005:169–173.

Smith, P. W. 1966a. *Pseudacris streckeri*. Catalogue of American Amphibians and Reptiles 27.1–27.2.

——. 1966b. *Hyla avivoca*. Catalogue of American Amphibians and Reptiles 28.1–28.2.

Smith, S., and G. D. Grossman. 2003. Stream microhabitat use by larval Southern Two-lined Salamanders (*Eurycea cirrigera*) in the Georgia Piedmont. Copeia 2003:531–543.

Smith-Gill, S. J., D. E. Gill, and K. A. Berven. 1979. Predicting amphibian metamorphosis. American Naturalist 113:563–585.

Snodgrass, J. W., J. W. Akerman, A. L. Bryan Jr., and J. Burger. 1999. Influence of hydroperiod, isolation, and heterospecifics on the distribution of aquatic salamanders (*Siren* and *Amphiuma*) among depression wetlands. Copeia 1999:107–113.

Snyder, R. C. 1956. Comparative features of the life histories of *Ambystoma gracile* (Baird) from populations of low and high altitudes. Copeia 1956:41–50.

——. 1960. The egg masses of neotenic *Ambystoma gracile*. Copeia 1960:267.

——. 1963. *Ambystoma gracile*. Catalogue of American Amphibians and Reptiles 6.1–6.2.

Sorensen, K. 2004. Population characteristics of *Siren lacertina* and *Amphiuma means* in north Florida. Southeastern Naturalist 3:249–258.

Sours, G. N., and J. W. Petranka. 2007. Intraguild predation and competition mediate stage-structured interactions between Wood Frog (*Rana sylvatica*) and Upland Chorus Frog (*Pseudacris feriarum*) larvae. Copeia 2007:131–139.

Spight, T. M. 1967. Population structure and biomass production by a stream salamander. American Midland Naturalist 78:437–447.

Spotila, J. R., and R. J. Beumer. 1970. The breeding habits of the Ringed Salamander, *Ambystoma annulatum* (Cope), in northwestern Arkansas. American Midland Naturalist 84:77–89.

Spotila, J. R., and P. H. Ireland. 1970. Notes on the eggs of the Gray-bellied Salamander, *Eurycea multiplicata griseogaster*. Southwestern Naturalist 14:366–368.

Sredl, M. J., and J. P. Collins. 1991. The effect of ontogeny on interspecific interactions in larval amphibians. Ecology 72:2232–2239.

——. 1992. The interaction of predation, competition, and habitat complexity in structuring an amphibian community. Copeia 1992:607–614.

Starrett, P. H. 1960. Descriptions of tadpoles of Middle America. Miscellaneous Publications of the Museum of Zoology, University of Michigan (110): 5–37.

Stauffer, H., and R. D. Semlitsch. 1993. Effects of visual, chemical and tactile cues of fish on the behavioural responses of tadpoles. Animal Behavior 46:355–364.

Stauffer, J. R., Jr., J. E. Gates, and W. L. Goodfellow. 1983. Preferred temperature of two sympatric *Ambystoma* larvae: A proximate factor in niche segregation? Copeia 1983:1001–1005.

Stebbins, R. C. 1951. Amphibians of western North America. Berkeley: University of California Press.

——. 1955. Southern occurrence of the Olympic Salamander *Rhyacotriton olympicus*. Herpetologica 11:238–239.

——. 1985. A field guide to western reptiles and amphibians. Field marks of all species in western North America, including Baja California. Boston: Houghton Mifflin.

Steele, C. A., E. D. Brodie Jr., and J. G. McCracken. 2003a. *Dicamptodon copei* (Cope's Giant Salamander). Reproduction. Herpetological Review 34:227–228.

——. 2003b. Relationships between abundance of Cascade Torrent Salamanders and forest age. Journal of Wildlife Management 67:447-453.

Steinke, J. H., and D. G. Benson Jr. 1970. The structure and polysaccharide chemistry of the jelly envelopes of the egg of the frog, *Rana pipiens*. Journal of Morphology 130:57–66.

Steinwascher, K. 1978. The effect of coprophagy on the growth of *Rana catesbeiana* tadpoles. Copeia 1978:130–134.

Steinwascher, K., and J. Travis. 1983. Influence of food quality quantity on early larval growth of two anurans. Copeia 1983:238–242.

Stejneger, L. 1892. Preliminary description of a new genus and species of blind cave salamander from North America. Proceedings of the United States National Museum 15:115–117.

——. 1896. Description of a new genus and species of blind, tailed batrachian from the subterranean water of Texas. Proceedings of the United States National Museum 18:619–621.

Stenhouse, S. L., N. G. Hairston, and A. E. Cory. 1983. Predation and competition in *Ambystoma* larvae: Field and laboratory experiments. Journal of Herpetology 17:210–220.

Stephenson, B., and P. Verrell. 2003. Courtship and mating of the Tailed frog (*Ascaphus truei*). Journal of Zoology 259:15–22.

Stevenson, H. M. 1976. Vertebrates of Florida. Identification and distribution. Gainesville: University Presses of Florida.

Stewart, M. M. 1956. The separate effects of food and temperature differences on the development of Marbled Salamander larvae. Journal of the Elisha Mitchell Scientific Society 72:47–56.

Stille, W. T. 1954. Eggs of the salamander *Ambystoma jeffersonianum* in the Chicago area. Copeia 1954:300.

Stoneburner, D. L. 1978. Salamander drift: Observations on the Two-lined Salamander (*Eurycea bislineata*]. Freshwater Biology 8:291–293.

Storer, T. I. 1925. A synopsis of the Amphibia of California. University of California Publications in Zoology 27:1–342.

Storfer, A., J. Cross, V. Rush, and J. Caruso. 1999. Adaptive coloration and gene flow as a constraint to local adaptation in the Streamside Salamander, *Ambystoma barbouri*. Evolution 53:889–898.

Storfer, A., and C. White. 2004. Phenotypically plastic responses of larval Tiger Salamanders, *Ambystoma tigrinum*, to different predators. Journal of Herpetology 38:612–615.

Storm, R. M. 1960. Notes on the breeding biology of the Red-legged Frog (*Rana aurora aurora*). Herpetologica 16:251–259.

Storz, B. L. 2004. Reassessment of the environmental mechanisms controlling developmental polyphenism in spadefoot toad tadpoles. Oecologia 141:402–410.

Storz, B. L., and J. Travis. 2007. Temporally dissociated, trait-specific modifications underlie phenotypic polyphenism in *Spea multiplicata* tadpoles, which suggest modularity. Scientific World Journal 7:715–726.

Strauss, R. E., and R. Altig. 1992. Ontogenetic body form changes in three ecological morphs of anuran tadpoles. Growth, Development and Aging 56:3–16.

Strecker, J. K., Jr. 1908a. Notes on the habits of two Arkansas salamanders and a list of batrachians and reptiles collected at Hot Springs. Proceedings of the Biological Society of Washington 21:85–90.

——. 1908b. Notes on the life history of *Scaphiopus couchii* Baird. Proceedings of the Biological Society of Washington 21:199–206.

——. 1909. Contributions to Texan herpetology. Notes on the Texan salamander (*Ambystoma texanum* Matthes). Baylor University Bulletin 12:17–20.

——. 1922. An annotated catalogue of the amphibians and reptiles of Bexar County, Texas. Bulletin No. 4, Scientific Society of San Antonio. 31 p.

——. 1926a. On the habits and variations of *Pseudacris ornata* (Holbrook). Contributions of the Baylor University Museum (7): 8–11.

——. 1926b. Chapters from the life-histories of Texas reptiles and amphibians. Part I. Contributions of the Baylor University Museum (8): 1–12.

Stuart, J. N., and C. W. Painter. 1996. Natural history notes on the Great Plains Narrowmouth Toad, *Gastrophryne olivacea*, in New Mexico. Bulletin of the Chicago Herpetological Society 31:44–47.

Stuart, L. C. 1948. The amphibians and reptiles of Alta Verapaz, Guatemala. Miscellaneous Publications of the Museum of Zoology, University of Michigan (69): 1–109.

——. 1961. Some observations on the natural history of tadpoles of *Rhinophrynus dorsalis* Dumeril and Bibron. Herpetologica 17:73–79.

Sugg, D. W., A. A. Karlin, C. R. Preston, and D. R. Heath. 1988. Morphological variation in a population of the salamander, *Siren intermedia nettingi*. Journal of Herpetology 22:243–247.

Sullivan, A. M., P. W. Frese, and A. Mathias. 2000. Does the aquatic salamander, *Siren intermedia*, respond to chemical cues from prey? Journal of Herpetology 34:607–611.

Sutherland, M. A. B., G. M. Gouchie, and R. J. Wassersug. 2009. Can visual stimulation alone induce phenotypically plastic responses in *Rana sylvatica* tadpole oral structures? Journal of Herpetology 43:165–168.

Svihla, A. 1935. Notes on the Western Spotted Frog, *Rana p. pretiosa*. Copeia 1935:119–122.

——. 1936. *Rana rugosa* (Schlegel): Notes of the life history of this interesting frog. Mid-Pacific Magazine 49:124–125.

Swart, C. C., and R. C. Taylor. 2004. Behavioral interactions between the Giant Water Bug (*Belostoma lutarium*) and tadpoles of *Bufo woodhousii*. Southeastern Naturalist 3:13–24.

Sweet, S. 1977a. *Eurycea tridentifera*. Catalogue of American Amphibians and Reptiles 199.1–199.2.

——. 1977b. Natural metamorphosis in *Eurycea neotenes*, and the generic allocation of the Texas *Eurycea* (Amphibia: Plethodontidae). Herpetologica 33:364–375.

——. 1982. A distributional analysis of epigean populations of *Eurycea neotenes* in central Texas, with comments on the origin of troglobitic populations. Herpetologica 38:430–444.

——. 1984. Secondary contact and hybridization in the Texas Cave Salamanders *Eurycea neotenes* and *E. tridentifera*. Copeia 1984:428–441.

——. 1991. Initial report on the ecology and status of the Arroyo Toad (*Bufo microscaphus californicus*) on the Los Padres National Forest of southern California, with management recommendations. Contract report to USDA, Forest Service, Los Padres National Forest, Goleta, CA.

——. 1993. Second report on the biology and status of the Arroyo Toad (*Bufo microscaphus californicus* on the Los Padres National Forest of southern California. Contract report to USDA, Forest Service, Los Padres National Forest, Goleta, CA.

Taber, C. A., R. F. Wilkinson Jr., and M. S. Topping. 1975. Age and growth of Hellbenders in the Niangua River, Missouri. Copeia 1975:633–639.

Taggart, T. W. 1997. Status of *Bufo debilis* (Anura: Bufonidae) in Kansas. Kansas Herpetological Society Newsletter 109:7–12.

Taigen, T. L., and F. H. Pough. 1981. Activity metabolism of the toad (*Bufo americanus*): Ecological consequences of ontogenetic change. Journal of Comparative Physiology 144:247–252.

Tait, C. K., and L. V. Diller. 2006. Life history of the Southern Torrent Salamander (*Rhyacotriton variegatus*) in coastal northern California. Journal of Herpetology 40:43–54.

Takahashi, M. K., and M. J. Parris. 2008. Life cycle polyphenism as a factor affecting ecological divergence within *Notophthalmus viridescens*. Oecologia 158:23–34.

Talentino, K. A., and E. Landre. 1991. Comparative development of two species of sympatric *Ambystoma* salamanders. Journal of Freshwater Ecology 6:395–401.

Tanner, V. M. 1939. A study of the genus *Scaphiopus*. Great Basin Naturalist 1:3–20.

Tanner, W. W., D. L. Fisher, and T. J. Willis. 1971. Notes on the life history of *Ambystoma tigrinum nebulosum* Hallowell in Utah. Great Basin Naturalist 31:213–222.

Tarr, T. L., and K. J. Babbitt. 2002. Effects of habitat complexity and predator identity on predation of *Rana clamitans* larvae. Amphibia-Reptilia 23:13–20.

Taylor, A. C., and J. J. Kollros. 1946. Stages in the development of *Rana pipiens* larvae. Anatomical Record 94:7–23.

Taylor, B. E., and D. E. Scott. 1997. Effects of larval density dependence on population dynamics of *Ambystoma opacum*. Herpetologica 53:132–145.

Taylor, B. E., R. A. Estes, J. H. K. Pechmann, and R. D. Semlitsch. 1988. Trophic relations in a temporary pond: Larval salamanders and their microinvertebrate prey. Canadian Journal of Zoology 66:2191–2198.

Taylor, C. L., and R. Altig. 1995. Oral disc kinematics of four rheophilous anuran tadpoles. Herpetological Natural History 3:101–106.

Taylor, C. L., R. Altig, and C. R. Boyle. 1995. Can anuran tadpoles choose among foods that vary in quality? Alytes 13:81–86.

Taylor, C. N., K. L. Oseen, and R. J. Wassersug. 2004. On the behavioural response of *Rana* and *Bufo* tadpoles to echinostomatoid cercariae: Implications to synergistic factors influencing trematode infections in anurans. Canadian Journal of Zoology 82:701–706.

Taylor, D. H. 1972. Extra-optic photoreception compass orientation in larval and adult salamanders (*Ambystoma tigrinum*). Animal Behavior 20:233–236.

Taylor, E. H. 1942. Tadpoles of Mexican Anura. University of Kansas Science Bulletin 28:37–55.

Taylor, J. T. 1983a. Orientation and flight responses of a neotenic salamander (*Ambystoma gracile*) in Oregon. American Midland Naturalist 109:40–49.

———. 1983b. Size-specific associations of larval and neotenic Northwestern Salamanders, *Ambystoma gracile*. Journal of Herpetology 17:203–209.

———. 1984. Comparative evidence for competition between the salamanders *Ambystoma gracile* and *Taricha granulosa*. Copeia 1984:672–683.

Telford, S. R., Jr. 1954. A description of the larvae of *Ambystoma cingulatum bishopi* Goin, including an extension of the range. Quarterly Journal of the Florida Academy of Sciences 17:233–238.

Test, F. H., and R. G. McCann. 1976. Foraging behavior of *Bufo americanus* tadpoles in response to high densities of micro-organisms. Copeia 1976:576–578.

Thaker, M., C. R. Gabor, and J. N. Fries. 2006. Sensory cues for conspecific associations in aquatic San Marcos salamanders. Herpetologica 62:151–155.

Thibaudeau, D. G., and R. Altig. 1988. Sequence of ontogenetic development and atrophy of the oral apparatus of six anuran tadpoles. Journal of Morphology 197:63–69.

Thiemann, G. W., and R. J. Wassersug. 2000. Patterns and consequences of behavioral responses to predators and parasites in *Rana* tadpoles. Biological Journal of the Linnean Society 71:513–528.

Thomas, L. A., and J. Allen. 1997. *Bufo houstonensis* (Houston Toad). Behavior. Herpetological Review 28:40–41.

Thompson, C. E., J. M. Walker, F. T. Waterstrat, A. P. McIntyre, and M. P. Hayes. 2011. *Rhyacotriton kezeri* (Columbia Torrent Salamander). Oviposition site. Herpetological Review 42:406–408.

Thompson, E. L., and J. E. Gates. 1982. Breeding pool segregation by the Mole Salamanders, *Ambystoma jeffersonianum* and *A. maculatum*, in a region of sympatry. Oikos 38:273–279.

Thompson, E. L., J. E. Gates, and G. J. Taylor. 1980. Distribution and breeding habitat selection of the Jefferson Salamander, *Ambystoma jeffersonianum*, in Maryland. Journal of Herpetology 14:113–120.

Thompson, H. B. 1913. Description of a new subspecies of *Rana pretiosa* from Nevada. Proceedings of the Biological Society of Washington 26:53–56.

Thurow, G. R. 1997a. Observations on *Hemidactylium scutatum* habitat and distribution. Bulletin of the Chicago Herpetological Society 32:1–6.

———. 1997b. *Rana sylvatica* (Wood Frog). Egg mass features and larval behavior. Herpetological Review 28:148.

Tilley, S. G. 1968. Size-fecundity relationships and their evolutionary implications in five des-mognathine salamanders. Evolution 22:806–816.

——. 1969. Variation in the dorsal pattern of *Desmognathus ochrophaeus* at Mt. Mitchell, North Carolina, and elsewhere in the southern Appalachian Mountains. Copeia 1969:169–175.

——. 1972. Aspects of parental care and embryonic development in *Desmognathus ochro-phaeus*. Copeia 1972:532–540.

——. 1973a. Life histories and natural selection in populations of the salamander *Desmog-nathus ochrophaeus*. Ecology 54:3–17.

——. 1973b. *Desmognathus ochrophaeus*. Catalogue of American Amphibians and Reptiles 129.1–129.4.

——. 1973c. Observations on the larval period and female reproductive ecology of *Desmog-nathus ochrophaeus* (Amphibia: Plethodontidae) in western North Carolina. American Midland Naturalist 89:394–407.

——. 1974. Structures and dynamics of populations of the salamander *Desmognathus ochro-phaeus* Cope in different habitats. Ecology 55:808–817.

——. 1980. Life histories and comparative demography of two salamander populations. Co-peia 1980:806–821.

——. 1981. A new species of *Desmognathus* (Amphibia: Caudata: Plethodontidae) from the southern Appalachian Mountains. Occasional Papers of the Museum of Zoology, University of Michigan (695): 1–23.

——. 1985. *Desmognathus imitator*. Catalogue of American Amphibians and Reptiles 359.1–359.2.

——. 2000. *Desmognathus santeetlah*. Catalogue of American Amphibians and Reptiles 703.1–703.3.

Tilley, S. G., and D. W. Tinkle. 1968. A reinterpretation of the reproductive cycle and demog-raphy of the salamander *Desmognathus ochrophaeus*. Copeia 1968:299–303.

Ting, H.-P. 1951. Duration of the tadpole stage of the Green Frog, *Rana clamitans*. Copeia 1951:82.

Tomśon, O. H., and D. E. Ferguson. 1972. Y-axis orientation in larvae and juveniles of three species of *Ambystoma*. Herpetologica 28:6–9.

Topping, M. S., and C. A. Ingersol. 1981. Fecundity in the Hellbender, *Cryptobranchus alleg-aniensis*. Copeia 1981:873–876.

Touchon, J. C., I. Gomez-Mestre, and K. M. Warkentin. 2006. Hatching plasticity in two tem-perate anurans: Responses to a pathogen and predation cues. Canadian Journal of Zoology 84:556–563.

Townsend, C. H. 1882. Habits of the *Menopoma*. American Naturalist 16:139–140.

Trapido, H., and R. T. Clausen. 1940. The larvae of *Eurycea bislineata major*. Copeia 1940:244–246.

Trapp, M. M. 1956. Range and natural history of the Ringed Salamander, *Ambystoma annula-tum* Cope (Ambystomatidae). Southwestern Naturalist 1:78–82.

Trauth, S. E. 1983. Reproductive biology and spermathecal anatomy of the Dwarf Salamander (*Eurycea quadridigitata*) in Alabama. Herpetologica 39:9–15.

——. 1988. Egg clutches of the Ouachita Dusky Salamander, *Desmognathus brimleyorum* (Caudata: Plethodontidae), collected in Arkansas during a summer drought. Southwestern Naturalist 33:234–236.

——. 1989. Female reproductive traits of the Southern Leopard Frog, *Rana sphenocephala* (Anura: Ranidae), from Arkansas. Proceedings of the Arkansas Academy of Science 43:105–108.

Trauth, S. E., M. E. Cartwright, and W. E. Meshaka. 1989a. Reproduction in the Wood Frog, *Rana sylvatica* (Anura: Ranidae), from Arkansas. Proceedings of the Arkansas Academy of Science 43:114–116.

——. 1989b. Winter breeding in the Ringed Salamander, *Ambystoma annulatum* (Caudata: Ambystomatidae), from Arkansas. Southwestern Naturalist 34:145–146.

Trauth, S. E., M. E. Cartwright, J. D. Wilhide, and D. H. Jamieson. 1995a. A review of the distribution life history of the Wood Frog, *Rana sylvatica* (Anura: Ranidae), in north-central Arkansas. Bulletin of the Chicago Herpetological Society 30:45–51.

Trauth, S. E., B. G. Cochran, D. A. Saugey, W. P. Posey, and W. A. Stone. 1993. Distribution of the Mole Salamander, *Ambystoma talpoideum* (Urodela: Ambystomatidae), in Arkansas with notes on paedomorphic populations. Proceedings of the Arkansas Academy of Science 47:154–156.

Trauth, S. E., R. L. Cox, B. P. Butterfield, D. A. Saugey, and W. E. Meshaka. 1990. Reproductive phenophases and clutch characteristics of selected Arkansas amphibians. Proceedings of the Arkansas Academy of Science 44:107–113.

Trauth, S. E., R. L. Cox Jr., J. D. Wilhide, and H. Worley. 1995b. Egg mass characteristics of terrestrial morphs of the Mole Salamander, *Ambystoma talpoideum* (Caudata: Ambystomatidae), from northeastern Arkansas and clutch comparisons with other *Ambystoma* species. Proceedings of the Arkansas Academy of Science 49:193–196.

Trauth S. E., and A. Holt. 1993. Notes on the breeding biology of Hurter's Spadefoot Toad, *Scaphiopus holbrookii hurterii*, in Arkansas. Bulletin of the Chicago Herpetological Society 28:236–239.

Trauth, S. E., W. E. Meshaka, and B. P. Butterfield. 1989c. Reproduction and larval development in the Marbled Salamander, *Ambystoma opacum* (Caudata: Ambystomatidae), from Arkansas. Proceedings of the Arkansas Academy of Science 43:109–111.

Trauth, S. E., and J. W. Robinette. 1990. Notes on distribution, mating activity, and reproduction in the Bird-voiced Treefrog, *Hyla avivoca*, in Arkansas. Bulletin of the Chicago Herpetological Society 25:218–219.

Trauth, S. E., H. W. Robinson, and M. V. Plummer. 2004. The amphibians and reptiles of Arkansas. Fayetteville: University of Arkansas Press.

Trauth, S. E., J. D. Wilde, and P. Daniel. 1992. Status of the Ozark Hellbender, *Cryptobranchus bishopi* (Urodela: Cryptobranchidae), in the Spring River, Fulton County, Arkansas. Proceedings of the Arkansas Academy of Science 46:83–86.

Travis, J. 1980a. Genetic variation for larval specific growth rate in the frog *Hyla gratiosa*. Growth 44:167–181.

——. 1980b. Phenotypic variation and the outcome of interspecific competition in hylid tadpoles. Evolution 34:40–50.

——. 1981a. A key to the tadpoles of North Carolina. Brimleyana 6:119–127.

——. 1981b. Control of larval growth variation in a population of *Pseudacris triseriata* (Anura: Hylidae). Evolution 5:423–432.

——. 1983a. Variation in development patterns of larval anurans in temporary ponds. I. Persistent variation within a *Hyla gratiosa* population. Evolution 37:496–512.

——. 1983b. Variation in growth and survival of *Hyla gratiosa* larvae in experimental enclosures. Copeia 1983:232–237.

——. 1984. Anuran size at metamorphosis: Experimental test of a model based on intraspecific competition. Ecology 65:1155–1160.

Travis, J., S. B. Emerson, and M. Blouin. 1987. A quantitative-genetic analysis of larval life-history traits in *Hyla crucifer*. Evolution 41:145–156.

——. 1988. Evaluating a hypothesis about heterochrony: Larval life history traits and hindlimb morphology in *Hyla crucifer*. Evolution 42:68–78.

Travis, J., W. H. Keen, and J. Juilianna. 1985a. The effects of multiple factors on viability selection in *Hyla gratiosa* tadpoles. Evolution 39:1087–1099.

——. 1985b. The role of relative body size in a predator-prey relationship between dragonfly naiads and larval anurans. Oikos 45:59–65.

Travis, J., and J. C. Trexler. 1986. Interactions among factors affecting growth, development and survival in experimental populations of *Bufo terrestris* (Anura: Bufonidae). Oecologia 69:110–116.

Trenham, P. C., H. B. Shaffer, W. D. Joenig, and M. R. Stromberg. 2000. Life history and demographic variation in the California Tiger Salamander (*Ambystoma californiense*). Copeia 2000:365–377.

Trowbridge, A. H., and M. S. Trowbridge. 1937. Notes on the cleavage rate of *Scaphiopus bombifrons* Cope, with additional remarks on certain aspects of the life history. American Naturalist 71:460–480.

Trowbridge, M. S. 1941. Studies on the normal development of *Scaphiopus bombifrons* Cope. I. The cleavage period. Transactions of the American Microscopical Society 60:508–526.

———. 1942. Studies on the normal development of *Scaphiopus bombifrons* Cope, II. The later embryonic and larval periods. Transactions of the American Microscopical Society 61:66–83.

Trueb, L. 1969. *Pternohyla, P. dentata, P. fodiens*. Catalogue of American Amphibians and Reptiles 77.1–77.4.

Tubbs, L. O. E., R. Stevens, M. Wells, and R. Altig. 1993. Ontogeny of the oral apparatus of the tadpole of *Bufo americanus*. Amphibia-Reptilia 14:333–340.

Tucker, J. K. 1995. Early post-transformational growth in the Illinois Chorus Frog (*Pseudacris streckerii illinoensis*). Journal of Herpetology 29:314–316.

———. 1997. Description of newly transformed froglets of the Illinois Chorus Frog (*Pseudacris streckeri illinoensis*). Transactions of the Illinois Academy of Science 90:161–166.

———. 1999. Fecundity in the Tiger Salamander (*Ambystoma tigrinum*) from west-central Illinois. Amphibia-Reptilia 20:436–438.

Tumlison, R., and G. R. Cline. 1997. Further notes on the habitat of the Oklahoma Salamander (*Eurycea tynerensis*). Proceedings of the Oklahoma Academy of Science 77:103–106.

———. 2002. Food habits of the Banded Sculpin (*Cottus carolinae*) in Oklahoma with reference to predation on the Oklahoma Salamander (*Eurycea tynerensis*). Proceedings of the Oklahoma Academy of Science 82:111–113.

———. 2003. Association between the Oklahoma Salamander (*Eurycea tynerensis*) and Ordovician-Silurian strata. Southwestern Naturalist 48:93–95.

Tumlison, R., G. R. Cline, and P. Zwank. 1990a. Surface habitat association of the Oklahoma Salamander (*Eurycea tynerensis*). Herpetologica 46:169–175.

———. 1990b. Prey selection in the Oklahoma Salamander (*Eurycea tynerensis*). Journal of Herpetology 24:222–225.

———. 1990c. Morphological discrimination between the Oklahoma Salamander (*Eurycea tynerensis*) and the Graybelly Salamander (*Eurycea multiplicata griseogaster*). Copeia 1990:242–246.

Tumlison, R., and S. E. Trauth. 2006. A novel facultative mutualistic relationship between bufonid tadpoles and flagellated green algae. Herpetology Conservation and Biology 1:51–55.

Tupa, D. D., and W. K. Davis. 1976. Population dynamics of the San Marcos Salamander, *Eurycea nana* Bishop. Texas Journal of Science 27:179–195.

Turner, F. B. 1952. The mouth parts of tadpoles of the Spadefoot Toad, *Scaphiopus hammondi*. Copeia 1952:172–175.

———. 1958. Life-history of the Western Spotted Frog in Yellowstone National Park, Wyoming. Herpetologica 14:96–100.

———. 1960. Population structure and dynamics of the Western Spotted Frog, *Rana p. pretiosa* Baird and Girard, in Yellowstone National Park, Wyoming. Ecological Monographs 30:251–278.

Turner, F. B., and P. C. Dumas. 1972. *Rana pretiosa*. Catalogue of American Amphibians and Reptiles 119.1–119.4.

Turnipseed, G., and R. Altig. 1975. Population density and age structure of three species of hylid tadpoles. Journal of Herpetology 9:287–291.

Turtle, S. L. 2000. Embryonic survivorship of the Spotted Salamander (*Ambystoma macula-* *tum*) in roadside and woodland vernal pools in southeastern New Hampshire. Journal of Herpetology 34:60–67.

Twitty, V. C. 1936. Correlated genetic and embryological experiments on *Triturus*. I. Hybridization: Development of three species of *Triturus* and their hybrid combinations. II. Transplantation: The embryological basis of species differences in pigment pattern. Journal of Experimental Zoology 74:239–302.

——. 1941. Data on the life history of *Ambystoma tigrinum californiense* Gray. Copeia 1941:1–4.

——. 1942. The species of California *Triturus*. Copeia 1942:65–76.

——. 1964. *Taricha rivularis*. Catalogue of American Amphibians and Reptiles 9.1–9.2.

——. 1966. Of scientists and salamanders. W. H. Freeman, San Francisco.

Tyler, J. D., and H. N. Buscher. 1980. Notes on a population of larval *Ambystoma tigrinum* (Ambystomatidae) from Cimarron County, Oklahoma. Southwestern Naturalist 25:391–395.

Tyler, T. J., W. J. Liss, L. M. Ganio, G. L. Larson, R. Hoffman, E. Deimling, and G. Lomnicky. 1998a. Interaction between introduced trout and larval salamanders (*Ambystoma macrodactylum*) in high-elevation lakes. Conservation Biology 12:94–105.

Tyler, T. J., W. L. Liss, R. L. Hoffman, and L. S. Ganio. 1998b. Experimental analysis of trout effects on survival, growth, and habitat use of two species of ambystomatid salamander. Journal of Herpetology 32:345–349.

Uhlenhuth, E. 1919. Observations on the distribution of the blind Texas Cave Salamander, *Typhlomolge rathbuni*. Copeia 1919:26–27.

——. 1921. Observations on the distribution and habits of the blind Texas Cave Salamander, *Typhlomolge rathbuni*. Biological Bulletin 40:73–104.

Ultsch, G. R. 1971. The relationship of dissolved carbon dioxide and oxygen to microhabitat selection in *Pseudobranchus striatus*. Copeia 1971:247–252.

——. 1973. Observations on the life history of *Siren lacertina*. Herpetologica 29:304–305.

——. 1974. Gas exchange and metabolism in the Sirenidae (Amphibia: Caudata)—I. Oxygen consumption of submerged sirenids as a function of body size and respiratory surface area. Comparative Biochemistry and Physiology 47A:485–498.

Ultsch, G. R., and S. J. Arceneaux. 1988. Gill loss in larval *Amphiuma tridactylum*. Journal of Herpetology 22:347–348.

Ultsch, G. R., E. L. Brainerd, and D. C. Jackson. 2004. Lung collapse among aquatic reptiles and amphibians during long-term diving. Comparative Biochemistry and Physiology 139A:111–115.

Ultsch, G. R., and J. T. Duke. 1990. Gas exchange and habitat selection in the aquatic salamanders *Necturus maculosus* and *Cryptobranchus alleganiensis*. Oecologia 83:250–258.

Uzzell, T. 1967a. *Ambystoma jeffersonianum*. Catalogue of American Amphibians and Reptiles 47.1–47.2.

——. 1967b. *Ambystoma laterale*. Catalogue of American Amphibians and Reptiles 48.1–48.2.

——. 1967c. *Ambystoma platineum*. Catalogue of American Amphibians and Reptiles 49.1–49.2.

——. 1967d. *Ambystoma tremblayi*. Catalogue of American Amphibians and Reptiles 50.1–50.2.

Vaglia, J. L., S. K. Babcock, and R. N. Harris. 1997. Tail development and regeneration throughout the life cycle of the Four-toed Salamander *Hemidactylium scutatum*. Journal of Morphology 233:15–30.

Valentine, B. D. 1963. The salamander genus *Desmognathus* in Mississippi. Copeia 1963: 130–139.

——. 1964a. *Desmognathus ocoee*. Catalogue of American Amphibians and Reptiles 7.1–7.2.

——. 1964b. The external morphology of the plethodontid salamander, *Haideotriton*. Journal of the Ohio Herpetological Society 4:99–102.

———. 1964c. A preliminary key to the families of salamanders and sirenids with gills or gill slits. Copeia 1964:582–583.

———. 1974. *Desmognathus quadramaculatus*. Catalogue of American Amphibians and Reptiles 153.1–153.4.

———. 1989. Larval salamanders. *In* R. A. Pfingsten and F. L. Downs, eds., Salamanders of Ohio, pp. 46–60. Bulletin of the Ohio Biological Survey 7:1–350.

Valentine, B. D., and D. M. Dennis. 1964. A comparison of the gill-arch system and fins of three genera of larval salamanders, *Rhyacotriton*, *Gyrinophilus*, and *Ambystoma*. Copeia 1964:160–201.

Valerio, C. E. 1971. Ability of some tropical tadpoles to survive without water. Copeia 1971:364–375.

Van Buskirk, J. 1988. Interactive effects of dragonfly predation in experimental pond communities. Ecology 69:857–867.

———. 2000. The costs of an inducible defense in anuran larvae. Ecology 81:2813–2821.

Van Buskirk, J., and S. A. McCollum. 2000. Influence of tail shape on tadpole swimming performance. Journal of Experimental Biology 203:2149–2155.

Van Buskirk, J., S. S. McCollum, and E. E. Werner. 1997. Natural selection for environmentally induced phenotypes in tadpoles. Evolution 51:1983–1992.

Van Buskirk, J., and R. A. Relyea. 1998a. Selection for phenotypic plasticity in *Rana sylvatica* tadpoles. Biological Journal of the Linnean Society 65:301–328.

———. 1998b. Natural selection for phenotypic plasticity: Predator-induced morphological responses in tadpoles. Biological Journal of the Linnean Society 65:301–328.

Van Buskirk, J., and D. C. Smith. 1991. Density-dependent population regulation in a salamander. Ecology 72:1747–1793.

Van Denburgh, J. 1912. Notes on *Ascaphus*, the discoglossid toad of North America. Proceedings of the California Academy of Sciences 3:259–264.

Vences, M., M. W. Penuel-Matthews, D. R. Vieites, and R. Altig. 2003. *Rana temporaria* (Common Frog) and *Bufo fowleri* (Fowler's Toad) tadpoles. Protozoan infestation. Herpetological Review 34:237–238.

Venesky, M. D., and M. J. Parris. 2009. Intraspecific variation in life history traits among two forms of *Ambystoma barbouri* larvae. American Midland Naturalist 162:202–206.

Venesky, M. D., T. E. Wilcoxen, M. A. Rensel, L. Rollins-Smith, J. L. Kerby, and M. J. Parris. 2012. Dietary protein restriction impairs growth, immunity, and disease resistance in Southern Leopard Frog tadpoles. Oecologia 169:23–31.

Vera Candioti, M. F., and R. Altig. 2010. A survey of shape variation in keratinized labial teeth of anuran larvae as related to phylogeny and ecology. Biological Journal of the Linnean Society 101:609–625.

Verma, N., and B. A. Pierce. 1994. Body mass, developmental stage, interspecific differences in acid tolerance of larval anurans. Texas Journal of Science 46:319–327.

Verrill, A. E. 1863. Notice of the eggs and young of a salamander *Desmognathus fusca* Baird, from Maine. Proceedings of the Boston Society of Natural History 9:253–255.

Viertel, B. 1999. Feeding strategies of *Xenopus laevis* larvae (Amphibia, Anura, Pipidae). *In* C. Miaud and R. Guyetant, eds., Current studies in herpetology: Proceedings of the 9th Ordinary General Meeting of the Societas Europaea Herpetologica, 25–29 August 1998, Le Bourget du Lac, France, pp. 439–444.

Viosca, P., Jr. 1924. Observations on the life history of *Ambystoma opacum*. Copeia 1924:86–88.

———. 1925. A terrestrial form of *Siren lacertina*. Copeia 1925:102–104.

———. 1937. A tentative revision of the genus *Necturus* with descriptions of three new species from the southern Gulf drainage area. Copeia 1937:120–138.

Viparina, S., and J. J. Just. 1975. The life period, growth and differentiation of *Rana catesbeiana* larvae occurring in nature. Copeia 1975:103–109.

Vogt, R. C. 1981. Natural history of amphibians and reptiles of Wisconsin. Milwaukee: Milwaukee Public Musuem.

Volpe, E. P. 1953. Embryonic temperature adaptations and relationships in toads. Physiological Zoology 26:344–354.

——. 1956. Experimental F1 hybrids between *Bufo valliceps* and *Bufo fowleri*. Tulane Studies in Zoology 4:61–75.

——. 1957a. Embryonic temperature tolerance and rate of development in *Bufo valliceps*. Physiological Zoology 30:164–176.

——. 1957b. The early development of *Rana capito sevosa*. Tulane Studies in Zoology 5:207–225.

——. 1957c. Embryonic temperature adaptations in highland *Rana pipiens*. American Naturalist 91:303–310.

——. 1959a. Experimental and natural hybridization between *Bufo terrestris* and *Bufo fowleri*. American Midland Naturalist 61:295–312.

——. 1959b. Hybridization of *Bufo valliceps* with *Bufo americanus* and *Bufo terrestris*. Texas Journal of Science 11:335–342.

Volpe, E. P., and J. L. Dobie. 1959. The larva of the Oak Toad, *Bufo quercicus* Holbrook. Tulane Studies in Zoology 7:145–152.

Volpe, E. P., and C. R. Shoop. 1963. Diagnosis of larvae of *Ambystoma talpoideum*. Copeia 1963:444–446.

Volpe, E. P., M. A. Wilkens, and J. L. Dobie. 1961. Embryonic and larval development of *Hyla avivoca*. Copeia 1961:340–349.

Vonesh, J. R., and O. de la Cruz. 2002. Complex life cycles and density dependence: Assessing the contribution of egg mortality to amphibian declines. Oecologia 133:325–333.

Voris, H. K., and J. P. Bacon Jr. 1966. Differential predation on tadpoles. Copeia 1966:594–598.

Voss, S. R. 1993a. Effect of temperature on body size, developmental stage, and timing of hatching in *Ambystoma maculatum*. Journal of Herpetology 27:329–333.

——. 1993b. Relationship between stream order and length of larval period in the salamander *Eurycea wilderae*. Copeia 1993:736–742.

Voss, W. J. 1961. Rate of larval development and metamorphosis in the Spadefoot Toad, *Scaphiopus bombifrons*. Southwestern Naturalist 6:168–174.

Vredenburg, V. 2000. *Rana muscosa* (Mountain Yellow-legged Frog). Egg predation. Herpetological Review 31:170–171.

Wagner, W. E., Jr. 1986. Tadpoles and pollen: Observations on the feeding behavior of *Hyla regilla* tadpoles. Copeia 1986:802–804.

Wahbe, R. R., and F. L. Bunnell. 2003. Relations among larval Tailed Frogs, forest harvesting, stream microhabitat, and site parameters in southwestern British Columbia. Canadian Journal of Forestry Research 33:1256–1266.

Wake, D. B., and N. Shubin. 1998. Limb development in the Pacific Giant Salamander, *Dicamptodon* (Amphibia, Caudata, Dicamptodontidae). Canadian Journal of Zoology 76:2058–2066.

Wakeman, J. M., and G. R. Ultsch. 1975. The effect of dissolved O2 and CO2 on metabolism and gas-exchange partitioning in aquatic salamanders. Physiological Zoology 48:348–359.

Waldman, B. 1981. Sibling recognition in toad tadpoles: The role of experience. Zeitshrift für Tierpsychologie 56:341–358.

——. 1982a. Adaptive significance of communal oviposition in Wood Frogs (*Rana sylvatica*). Behavior, Ecology and Sociobiology 10:169–174.

——. 1982b. Sibling association among schooling toad tadpoles: Field evidence and implications. Animal Behavior 30:700–713.

——. 1984. Kin recognition and sibling association among Wood Frog (*Rana sylvatica*) tadpoles. Behavior, Ecology and Sociobiology 14:171–180.

———. 1985a. Sibling recognition in toad tadpoles: Are kinship labels transferred among individuals? Zeitshrift für Tierpsychologie 68:41–57.

———. 1985b. Olfactory basis of kin recognition in toad tadpoles. Journal of Comparative Physiology 156A:565–577.

———. 1989. Do anuran larvae retain kin recognition abilities following metamorphosis? Animal Behavior 37:1055–1058.

Waldman, B., and K. Adler. 1979. Toad tadpoles associate preferentially with siblings. Nature 282:611–613.

Waldman, B., and M. J. Ryan. 1983. Thermal advantages of communal egg mass deposition in Wood Frogs (*Rana sylvatica*). Journal of Herpetology 17:70–72.

Walker, C. F. 1946. The amphibians of Ohio, Part I. The frogs and toads (Order Salientia). Ohio State Museum Science Bulletin 1:1–109.

Walker, D., and S. D. Busack. 2000. *Rana catesbeiana* (Bullfrog). Tadpole depth record. Herpetological Review 31:236.

Wallace, R. L., and L. V. Diller. 1998. Length of the larval cycle of *Ascaphus truei* in coastal streams of the Redwood Region, northern California. Journal of Herpetology 32:404–409.

Walls, S. C. 1991. Ontogenetic shifts in the recognition of siblings and neighbors by juvenile salamanders. Animal Behavior 42:423–434.

———. 1995. Differential vulnerability to predation and refuge use in competing larval salamanders. Oecologia 101:86–93.

———. 1996. Differences in foraging behavior explain interspecific growth inhibition in competing salamanders. Animal Behavior 52:1157–1162.

———. 1998. Density dependence in a larval salamander: The effects of interference and food limitation. Copeia 1998:926–935.

Walls, S. C., and R. Altig. 1986. Female reproductive biology and larval life history of *Ambystoma* salamanders: A comparison of egg size, hatchling size, and larval growth. Herpetologica 42:334–345.

Walls, S. C., J. J. Beatty, B. N. Tissot, D. G. Hokit, and A. R. Blaustein. 1993a. Morphological variation and cannibalism in a larval salamander (*Ambystoma macrodactylum columbianum*). Canadian Journal of Zoology 71:1543–1551.

Walls, S. C., S. S. Belanger, and A. R. Blaustein. 1993b. Morphological variation in a larval salamander: Dietary induction of plasticity in head shape. Oecologia 96:162–168.

Walls, S. C., and A. R. Blaustein. 1994. Does kinship influence density dependence in a larval salamander. Oikos 71:459–468.

———. 1995. Larval Marbled Salamanders, *Ambystoma opacum*, eat their kin. Animal Behavior 50:537–545.

Walls, S. C., C. S. Conrad, M. L. Murillo, and A. R. Blaustein. 1996. Agonistic behaviour in larvae of the Northwestern Salamander (*Ambystoma gracile*): The effects of kinship, familiarity and population source. Animal Behavior 133:965–984.

Walls, S. C., and R. G. Jaeger. 1987. Aggression and exploitation as mechanisms of competition in larval salamanders. Canadian Journal of Zoology 65:2938–2944.

———. 1989. Growth in larval salamanders is not inhibited through chemical interference competition. Copeia 1989:1049–1052.

Walls, S. C., and R. E. Roudebush. 1991. Reduced aggression toward siblings as evidence of kin recognition in cannibalistic salamanders. American Naturalist 138:1027–1038.

Walls, S. C., and R. D. Semlitsch. 1991. Visual and movement displays function as agonistic behavior in larval salamanders. Copeia 1991:936–942.

Walls, S. C., D. G. Taylor, and C. M. Wilson. 2002. Interspecific differences in susceptibility to competition and predators in a species-pair of larval amphibians. Herpetologica 58:104–118.

Walls, S. C., and M. G. Williams. 2001. The effect of community composition on persistence of prey with their predators in an assemblage of pond-breeding amphibians. Oecologia 128:134–141.

Walston, L. J., and S. J. Mullin. 2007. Population responses of Wood Frogs (*Rana sylvatica*) tadpoles to overwintered Bullfrog (*Rana catesbeiana*) tadpoles. Journal of Herpetology 41:24–31.

Walters, B. 1975. Studies of interspecific predation within an amphibian community. Journal of Herpetology 9:267–279.

Walters, P. J., and L. Greenwald. 1977. Physiological adaptations of aquatic newts (*Notophthalmus viridescens*) to a terrestrial environment. Physiological Zoology 50:88–98.

Ward, D., and O. J. Sexton. 1981. Anti-predator role of salamander egg membranes. Copeia 1981:724–726.

Warkentin, K. M. 1992a. Effects of temperature and illumination on feeding rates of Green Frog tadpoles (*Rana clamitans*). Copeia 1992:725–730.

———. 1992b. Microhabitat use and feeding rate variation in Green Frog tadpoles (*Rana clamitans*). Copeia 1992:731–740.

Warner, S. C., W. A. Dunson, and J. Travis. 1991. Interaction of pH, density, and priority effects on the survivorship and growth of two species of hylid tadpoles. Oecologia 88:331–339.

Warner, S. C., J. Travis, and W. A. Dunson. 1993. Effects of pH variation on interspecific competition between two species of hylid tadpoles. Ecology 74:183–194.

Warny, P. R., W. Meshaka Jr., and A. M. Klippel. 2012. Larval growth and transformation size of the Green Frog, *Lithobates clamitans melanota* (Rafinesque, 1820), in south-eastern New York. Herpetology Notes 5:59–62.

Wasserman, A. O. 1968. *Scaphiopus holbrookii*. Catalogue of American Amphibians and Reptiles 70.1–70.4.

———. 1970. *Scaphiopus couchii*. Catalogue of American Amphibians and Reptiles 85.1–85.4.

Wassersug, R. J. 1976a. Oral morphology of anuran larvae: Terminology and general description. Occasional Papers of the Museum of Natural History, University of Kansas (48): 1–23.

———. 1976b. Internal oral features in *Hyla regilla* (Anura: Hylidae) larvae: An ontogenetic study. Occasional Papers of the Museum of Natural History, University of Kansas (49): 1–24.

———. 1976c. The identification of leopard frog tadpoles. Copeia 1976:413–414.

———. 1989. Locomotion in amphibian larvae (or "Why aren't tadpoles built like fishes?"). American Zoologist 29:65–84.

———. 1992. The basic mechanisms of ascent and descent by anuran larvae (*Xenopus laevis*). Copeia 1992:890–894.

———. 1996. The biology of *Xenopus* tadpoles. *In* R. C. Tinsley and H. R. Kobel, eds., The biology of *Xenopus*, pp. 195–211. Oxford: Oxford University Press.

Wassersug, R. J., and M. E. Feder. 1983. The effects of aquatic oxygen concentration, body size and respiratory behaviors on the stamina of obligate aquatic (*Bufo americanus*) and facultative air-breathing (*Xenopus laevis* and *Rana berlandieri*) anuran larvae. Journal of Experimental Biology 105:173–190.

Wassersug, R. J., and C. M. Hessler. 1971. Tadpole behaviour: Aggregation in larval *Xenopus laevis*. Animal Behavior 19:386–389.

Wassersug, R. J., and K. Hoff. 1985. The kinematics of swimming in anuran larvae. Journal of Experimental Biology 119:1–30.

Wassersug, R. J., A. M. Lum, and M. J. Potel. 1981. An analysis of school structure for tadpoles (Anura: Amphibia). Behavioral Ecology and Sociobiology 9:15–22.

Wassersug, R. J., and A. M. Murphy. 1987. Aerial respiration facilitates growth in suspension-feeding anuran larvae (*Xenopus laevis*). Journal of Experimental Biology 46:141–147.

Wassersug, R. J., and E. A. Seibert. 1975. Behavioral responses of amphibian larvae to variation in dissolved oxygen. Copeia 1975:86–103.

Wassersug, R. J., and D. G. Sperry. 1977. The relationship of locomotion to differential predation on *Pseudacris triseriata* (Anura: Hylidae). Ecology 58:830–839.

Wassersug, R. J., and M. Yamashita. 2001. Plasticity and constraints on feeding kinematics in anuran larvae. Comparative Biochemistry and Physiology 131A:183–195.

Watermolen, D. J. 1995. A key to the eggs of Wisconsin's amphibians. Wisconsin Department of Natural Resources Research Report (165): 1–12.

Watermolen, D. J., and H. Gilbertson. 1996. Keys for the identification of Wisconsin's larval amphibians. Wisconsin Department of Natural Resources Endangered Research Report (109): 1–14.

Watkins, T. B. 1997. The effect of metamorphosis on the repeatability of maximal locomotor performance in the Pacific Tree Frog, *Hyla regilla*. Journal of Experimental Biology 200:2663–2668.

——. 2000. The effects of acute and developmental temperature on burst swimming speed and myofibrillar ATPase activity in tadpoles of the Pacific Tree Frog, *Hyla regilla*. Physiological and Biochemical Zoology 73:356–364.

——. 2001. A quantitative genetic test of adaptive decoupling across metamorphosis for locomotor and life-history traits in the Pacific Tree Frog, *Hyla regilla*. Evolution 55:1668–1677.

Watkins, T. B., and M. A. McPeek. 2006. Growth and predation risk in Green Frog tadpoles (*Rana clamitans*): A quantitative genetic analysis. Copeia 2006:478–488.

Watkins, T. B., and J. Vraspir. 2006. Both incubation temperature and posthatching temperature affect swimming performance and morphology of Wood Frog tadpoles (*Rana sylvatica*). Physiological and Biochemical Zoology 79:140–149.

Watney, G. M. S. 1941. Notes on the life history of *Ambystoma gracile* Baird. Copeia 1941:14–17.

Watson, S., and A. P. Russell. 2000. A posthatching developmental staging table for the Long-toed Salamander, *Ambystoma macrodactylum krausei*. Amphibia-Reptilia 21:143–154.

Webb, R. G. 1963. The larvae of the Casque-headed Frog, *Pternohyla fodiens* Boulenger. Texas Journal of Science 15:89–97.

——. 1965. Observations on breeding habits of the Squirrel Treefrog, *Hyla squirella* Bosc in Daudin. American Midland Naturalist 74:500–501.

——. 1971. Egg deposition of the Mexican Smilisca, *Smilisca baudini*. Journal of Herpetology 5:185–187.

Webb, R. G., and J. K. Korky. 1977. Variation in tadpoles of frogs of the *Rana tarahumarae* group in western Mexico (Anura: Ranidae). Herpetologica 33:73–82.

Webb, R. G., and W. L. Roueche. 1971. Life history aspects of the Tiger Salamander (*Ambystoma tigrinum mavortium*) in the Chihuahuan desert. Great Basin Naturalist 31:193–212.

Weber, J. A. 1944. Observations on the life history of *Amphiuma means*. Copeia 1944:61–62.

Weber, R. E., R. M. G. Wells, and J. E. Rossetti. 1985. Adaptations to neoteny in the salamander, *Necturus maculosus*. Blood respiratory properties and interactive effects of pH, temperature and ATP on hemoglobin oxygenation. Comparative Biochemistry and Physiology 80A:495–501.

Weigmann, D. L., and R. Altig. 1975. Anoxic tolerance of three species of salamander larvae. Comparative Biochemistry and Physiology 50A:681–684.

Weiss, P. B. 1945. The normal stages in the development of the South African Clawed Toad, *Xenopus laevis*. Anatomical Record 93:161–169.

Wells, K. D. 1977. Multiple egg clutches in the Green Frog (*Rana clamitans*). Herpetologica 32:85–87.

——. 1978. Courtship and paternal behavior in a Panamanian poison-arrow frog (*Dendrobates auratus*). Herpetologica 34:148–155.

Wells, K. S., and R. N. Harris. 2001. Activity levels and the tradeoff between growth and survival in the salamanders *Ambystoma jeffersonianum* and *Hemidactylium scutatum*. Herpetologica 57:116–127.

Welsh, H. H., Jr. 1990. Relictual amphibians and old-growth forests. Conservation Biology 4:309–319.

Welsh, H. H., Jr., and A. J. Lind. 1996. Habitat correlates of the Southern Torrent Salamander, *Rhyacotriton variegatus* (Caudata: Rhyacotritonidae), in northwestern California. Journal of Herpetology 30:385–398.

——. 2002. Multiscale habitat relationships of stream amphibians in the Klamath-Siskiyou region of California and Oregon. Journal of Wildlife Management 66:581-602.

Werner, E. E. 1991. Nonlethal effects of a predator on competitive interactions between two anuran larvae. Ecology 72:1709–1720.

——. 1992. Competitive interactions between Wood Frog and Northern Leopard Frog larvae: The influence of size and activity. Copeia 1992:26–35.

——. 1994. Ontogenetic scaling of competitive relations: Size-dependent effects and responses in two anuran larvae. Ecology 75:197–213.

Werner, E. E., and B. R. Anholt. 1996. Predator-induced behavioral indirect effects: Consequences to competitive interactions in anuran larvae. Ecology 77:157–169.

Werner, E. E., and K. S. Glennemeier. 1999. Influence of forest canopy cover on the breeding pond distributions of several amphibian species. Copeia 1999:1–12.

Werner, E. E., and M. A. McPeek. 1994. Direct and indirect effects of predators on two anuran species along an environmental gradient. Ecology 75:1368–1382.

Wernz, J. G. 1969. Spring mating of *Ascaphus*. Journal of Herpetology 3:167–169.

Wernz, J. G., and R. M. Storm. 1969. Pre-hatching stages of the Tailed Frog, *Ascaphus truei*. Herpetologica 25:86–93.

West, L. B. 1960. The nature of growth inhibitory material from crowded *Rana pipiens* tadpoles. Physiological Zoology 33:232–239.

Wheeler, B. A., E. Prosen, A. Mathis, and R. F. Wilkinson. 2003. Population declines of a long-lived salamander: A 20+-year study of Hellbenders, *Cryptobranchus alleganiensis*. Biological Conservation 109:151–156.

Whiles, M. R., J. B. Jensen, J. G. Palis, and W. G. Dyer. 2004. Diets of larval Flatwoods Salamanders, *Ambystoma cingulatum*, from Florida and South Carolina. Journal of Herpetology 38:208–214.

Whitaker, J. O., Jr. 1971. A study of the Western Chorus Frog, *Pseudacris triseriata*, in Vigo County, Indiana. Journal of Herpetology 5:127–150.

Whitaker, J. O., Jr., W. W. Cudmore, and B. A. Brown. 1982. Food of larval, subadult and adult Smallmouth Salamanders, *Ambystoma texanum*, from Vigo County, Indiana. Proceedings of the Indiana Academy of Science 90:461–464.

White, A. W., and G. H. Pyke. 2002. Captive frog egg numbers—a misleading indicator of breeding potential. Herpetofauna 32:102–109.

Whiteman, H. H., and W. S. Brown. 1996. Growth and foraging consequences of facultative paedomorphosis in the Tiger Salamander, *Ambystoma tigrinum nebulosum*. Evolutionary Ecology 10:433–446.

Whiteman, H. H., R. D. Howard, X. Spray, and J. McGrady-Steed. 1998. Facultative paedomorphosis in an Indiana population of the Eastern Tiger Salamander, *Ambystoma tigrinum tigrinum*. Herpetological Review 29:141–143.

Whiteman, H. H., R. D. Howard, and K. A. Whitten. 1995. Effects of pH on embryo tolerance and adult behavior in the Tiger Salamander, *Ambystoma tigrinum tigrinum*. Canadian Journal of Zoology 73:1529–1537.

Whiteman, H. H., J. P. Sheen, E. B. Johnson, A. VanDuesen, R. Cargille, and T. W. Saco. 2003. Heterospecific prey and trophic polyphenism in larval Tiger Salamanders. Copeia 2003:56–67.

Whiteman, H. H., S. A. Wissinger, and W. S. Brown. 1996. Growth and foraging consequences of facultative paedomorphosis in the Tiger Salamander, *Ambystoma tigrinum nebulosum*. Evolutionary Ecology 10:433–446.

Whiteman, H. H., S. A. Wissinger, M. Denoël, C. J. Mecklin, N. N. Gerlane, and J. J. Gutrich. 2012. Larval growth in polyphenic salamanders: Making the best of a bad lot. Oecologia 168:109–118.

Whitford, W. G., and A. Vinegar. 1966. Homing, survivorship, and overwintering of larvae in Spotted Salamanders, *Ambystoma maculatum*. Copeia 1966:515–519.

Whitham, J., and A. Mathis. 2000. Effects of hunger and predation risk on foraging behavior of Graybelly Salamanders, *Eurycea multiplicata*. Journal of Chemical Ecology 26:1659–1665.

Wiens, J. A. 1970. Effects of early experience on substrate pattern selection in *Rana aurora* tadpoles. Copeia 1970:543–548.

———. 1972. Anuran habitat selection: Early experience and substrate selection in *Rana cascadae* tadpoles. Animal Behavior 20:218–220.

Wilbur, H. M. 1971. The ecological relationship of the salamander *Ambystoma laterale* to its all-female, gynogenetic associate. Evolution 25:168–179.

———. 1972. Competition, predation, and the structure of the *Ambystoma-Rana sylvatica* community. Ecology 53:3–21.

———. 1976. Density-dependent aspects of metamorphosis in *Ambystoma* and *Rana sylvatica*. Ecology 57:1289–1296.

———. 1977a. Propagule size, number, and dispersion pattern in *Ambystoma* and *Asclepias*. American Naturalist 11:43–68.

———. 1977b. Density-dependent aspects of growth and metamorphosis in *Bufo americanus*. Ecology 58:196–200.

———. 1977c. Interactions of food level and population density in *Rana sylvatica*. Ecology 58:206–209.

———. 1982. Competition between tadpoles of *Hyla femoralis* and *Hyla gratiosa* in laboratory experiments. Ecology 63:278–282.

———. 1987. Regulation of structure in complex systems: Experimental temporary pond communities. Ecology 37–1452.

Wilbur, H. M., and R. A. Alford. 1985. Priority effects in experimental pond communities: Responses of *Hyla* to *Bufo* and *Rana*. Ecology 66:1106–1114.

Wilbur, H. M., and J. E. Fauth. 1990. Experimental aquatic food webs: Interactions between two predators and two prey. American Naturalist 135:176–204.

Wilbur, H. M., P. J. Morin, and R. N. Harris. 1983. Salamander predation and the structure of experimental communities: Anuran responses. Ecology 64:1423–1429.

Wilbur, H. M., and R. D. Semlitsch. 1990. Ecological consequences of tail injury in *Rana* tadpoles. Copeia 1990:18–24.

Wilder, H. H. 1891. A contribution to the anatomy of *Siren intermedia*. Zoologische Jahrbucher 4:653–696.

———. 1899. *Desmognathus fusca* (Rafinesque) and *Spelerpes bislineatus* (Green). American Naturalist 33:231–246.

———. 1904. The early development of *Desmognathus fusca*. American Naturalist 38:117–125.

Wilder, I. W. 1913. The life history of *Desmognathus fusca*. Biological Bulletin 24:251–343.

———. 1917. On the breeding habits of *Desmognathus fuscus*. Biological Bulletin 32:13–20.

———. 1924a. The relation of growth to metamorphosis in *Eurycea bislineata* (Green). Journal of Experimental Zoology 40:1–112.

———. 1924b. The developmental history of *Eurycea bislineata* in western Mass. Copeia 1924:77–80.

Wildy, E. L. 2001. The effects of food level and conspecific density on biting and cannibalism in larval Long-Toed Salamanders, *Ambystoma macrodactylum*. Oecologia 128:202–209.

Wildy, E. L., and A. R. Blaustein. 2001. Learned recognition of intraspecific predators in larval Long-toed Salamanders *Ambystoma macrodactylum*. Ethology 107:479–494.

Wildy, E. L., D. P. Chivers, and A. R. Blaustein. 1999. Shifts in life-history traits as a response to cannibalism in larval Long-toed Salamanders (*Ambystoma macrodactylum*). Journal of Chemical Ecology 25:2337–2346.

Wildy, E. L., D. P. Chivers, J. M. Kiesecker, and A. R. Blaustein. 1998. Cannibalism enhances growth in larval Long-toed Salamanders (*Ambystoma macrodactylum*). Journal of Herpetology 32:286–289.

Williams, B. K., T. A. G. Rittenhouse, and R. D. Semlitsch. 2008. Leaf litter input mediates tadpole performance across forest canopy treatments. Oecologia 155:377–384.

Wilson, D. J., and H. Lefcort. 1993. The effect of predator diet on the alarm response of Red-legged Frog, *Rana aurora*, tadpoles. Animal Behavior 46:1017–1019.

Wilson, R. S., R. S. James, and I. A. Johnston. 2000. Thermal acclimation of locomotor performance in tadpoles and adults of the aquatic frog *Xenopus laevis*. Journal of Comparative Physiology 170B:117–124.

Wiltenmuth, E. B. 1997. Agonistic behavior and use of cover by stream-dwelling larval salamanders (*Eurycea wilderae*). Copeia 1997:439–443.

Winne, C. T., and T. J. Ryan. 2001. Aspects of sex-specific differences in the expression of an alternative life cycle in the salamander *Ambystoma talpoideum*. Copeia 2001:143–149.

Wiseman, K. D., K. R. Marlow, R. E. Jackson, and J. E. Drennan. 2005. *Rana boylii* (Foothill Yellow-legged Frog). Predation. Herpetological Review 36:162–163.

Wollmuth, L. P., and L. I. Crawshaw. 1988. The effect of development and season on temperature selection in Bullfrog tadpoles. Physiological Zoology 61:461–469.

Wollmuth, L. P., L. I. Crawshaw, R. B. Forbes, and D. A. Grahn. 1987. Temperature selection during development in a montane anuran species, *Rana cascadae*. Physiological Zoology 60:472–480.

Wood, J. T. 1946. Measurements of a giant *Pseudotriton montanus montanus* larva from Great Smoky Mountains National Park. Copeia 1946:168.

——. 1948. Eggs, larvae, and attending females of *Desmognathus f. fuscus* in southwestern Ohio and southeastern Indiana. American Midland Naturalist 39:93–95.

——. 1949. *Eurycea bislineata wilderae* Dunn. Herpetologica 5:61–62.

——. 1953a. Observations on the complements of ova and nesting of the Four-toed Salamander in Virginia. American Naturalist 87:77–86.

——. 1953b. The nesting of the Two-lined Salamander, *Eurycea bislineata*, on the Virginia Coastal Plain. Natural History Miscellanea (122): 1–7.

——. 1955. The nesting of the Four-toed Salamander, *Hemidactylium scutatum* (Schlegel), in Virginia. American Midland Naturalist 53:381–389.

Wood, J. T., F. G. Carey, and R. H. Rageot. 1955. The nesting and ovarian eggs of the Dusky Salamander, *Desmognathus f. fuscus* Raf., in southeastern Virginia. Virginia Journal of Science 1955:149–153.

Wood, J. T., and W. E. Duellman. 1951. Ovarian egg complements in the salamander *Eurycea bislineata rivicola* Mittleman. Copeia 1951:181.

Wood, J. T., and M. E. Fitzmaurice. 1948. Eggs, larvae, and attending females of *Desmognathus f. fuscus* in southwestern Ohio and southeastern Indiana. American Midland Naturalist 39:93–95.

Wood, J. T., and R. H. Rageot. 1963. The nesting of the Many-lined salamander in the Dismal Swamp. Virginia Journal of Science 1963:121–125.

Wood, J. T., and R. H. Wilkinson. 1952. Observations on the egg masses of Spotted Salamanders, *Ambystoma maculatum* (Shaw), in the Williamsburg area. Virginia Journal of Science 3:68–70.

Wood, J. T., and F. E. Wood. 1955. Notes on the nests and nesting of the Carolina Mountain Dusky Salamander in Tennessee and Virginia. Journal of the Tennessee Academy of Science 30:36–39.

Woodward, B. D. 1982a. Tadpole competition in a desert anuran community. Oecologia 54:96–100.

——. 1982b. Tadpole interactions in the Chihuahuan Desert at two experimental densities. Southwestern Naturalist 27:119–121.

——. 1982c. Local intraspecific variation in clutch parameters in the Spotted Salamander (*Ambystoma maculatum*). Copeia 1982:157–160.

——. 1987a. Intra- and interspecific variation in spadefoot toad (*Scaphiopus*) clutch parameters. Southwestern Naturalist 32:127–156.

——. 1987b. Interactions between Woodhouse's Toad tadpoles (*Bufo woodhousii*) of mixed sizes. Copeia 1987:380–386.

Woodward, B. D., and P. Johnson. 1985. *Ambystoma tigrinum* (Ambystomatidae) predation on *Scaphiopus couchi* (Pelobatidae) tadpoles of different sizes. Southwestern Naturalist 30:460–461.

Woodward, B. D., and J. Travis. 1991. Paternal effects on juvenile growth and survival in Spring Peepers (*Hyla crucifer*). Evolutionary Ecology 5:40–51.

Workman, G., and K. C. Fisher. 1941. Temperature selection and the effect of temperature on movement in frog tadpoles. American Journal of Physiology 133:499–500.

Worthington, R. D. 1968. Observations on the relative sizes of three species of salamander larvae in a Maryland pond. Herpetologica 24:242–246.

——. 1969. Additional observations on sympatric species of salamander larvae in a Maryland pond. Herpetologica 25:227–229.

Worthington, R. D., and D. B. Wake. 1971. Larval morphology and ontogeny of the ambystomatid salamander, *Rhyacotriton olympicus*. American Midland Naturalist 85:349–365.

Wright, A. A., and A. H. Wright. 1933. Handbook of frogs and toads. The frogs and toads of the United States and Canada. Ithaca, NY: Comstock Publishing.

Wright, A. H. 1908. Notes on the breeding habits of *Amblystoma punctatum*. Biological Bulletin 14:284–289.

——. 1910. The Anura of Ithaca, N.Y.: A key to their eggs. Biological Bulletin 18:69–71.

——. 1914. North American Anura. Life-histories of the Anura of Ithaca, New York. Carnegie Institute of Washington, Publication 197. vii + 98 pp.

——. 1924. A new bullfrog (*Rana heckscheri*) from Georgia and Florida. Proceedings of the Biological Society of Washington 37:141–152.

——. 1929. Synopsis and description of North American tadpoles. Proceedings of the United States National Museum 74:1–70, plates 1–9.

——. 1932. Life-histories of the frogs of Okefinokee Swamp, Georgia. New York: MacMillan.

Wright, A. H., and A. A. Allen. 1908. Notes on the breeding habits of the Swamp Cricket Frog *Chorophilus triseriatus*. American Naturalist 42:39–42.

——. 1909. The early breeding habits of *Amblystoma punctatum*. American Naturalist 43:687–692.

Wright, A. H., and A. A. Wright. 1923. The tadpoles of the frogs of Okefinokee Swamp, Georgia. Anatomical Record 24:406.

——. 1924. A key to the eggs of the Salientia east of the Mississippi River. American Naturalist 58:375–381.

——. 1949. Handbook of frogs and toads. The frogs and toads of the United States and Canada. Ithaca, NY: Comstock Publishing.

——. 1957. Handbook of snakes of the United States and Canada. Vols. I and II. Ithaca, NY: Cornell University Press.

——. 1995. Handbook of frogs and toads of the United States and Canada. Ithaca, NY: Cornell University Press.

Wright, M. L., S. T. Jorey, Y. M. Myers, M. L. Fieldstad, C. M. Paquette, and M. B. Clark. 1988. Influence of photoperiod, daylength, and feeding schedule on tadpole growth and development. Development, Growth, and Differentiation 30:315–323.

Yeatman, H. C., and H. B. Miller. 1985. A naturally metamorphosed *Gyrinophilus palleucus* from the type-locality. Journal of Herpetology 19:304–306.

Youngstrom, K. A., and H. M. Smith. 1936. Description of the larvae of *Pseudacris triseriata* and *Bufo woodhousii woodhousii* (Anura). American Midland Naturalist 17:629–633.

Zerba, K. E., and J. P. Collins. 1992. Spatial heterogeneity and individual variation in diet of an aquatic top predator. Ecology 73:268–279.

Ziemba, R. E., M. T. Myers, and J. P. Collins. 2000. Foraging under the risk of cannibalism leads to divergence in body size among Tiger Salamander larvae. Oecologia 124:225–231.

Zweifel, R. G. 1955. Ecology, distribution, and systematics frogs of the *Rana boylei* group. University of California Publications in Zoology 54:207–292.

——. 1961. Larval development of the tree frogs *Hyla arenicolor* and *Hyla wrightorum*. American Museum Novitates (2056): 1–19.

——. 1968a. *Rana muscosa*. Catalogue of American Amphibians and Reptiles 65.1–65.2.

——. 1968b. *Rana tarahumarae*. Catalogue of American Amphibians and Reptiles 66.1–66.2.

——. 1968c. *Rana boylii*. Catalogue of American Amphibians and Reptiles 71.1–71.2.

——. 1968d. Reproductive biology of anurans of the arid Southwest with emphasis on adaptations of embryos to temperature. Bulletin of the American Museum of Natural History 140:1–64.

——. 1970. Description notes on larvae of toads of the *debilis* group, genus *Bufo*. American Museum Novitates (2497): 1–13.

——. 1977. Upper thermal tolerances of anuran embryos in relation to stage of development and breeding habits. American Museum Novitates (2617): 1–21.

INDEX OF COMMON NAMES

Page numbers in italics refer to figures. Page numbers preceded by "P" refer to plates.

INDEX OF SCIENTIFIC NAMES

Page numbers in italics refer to figures. Page numbers preceded by "P" refer to plates.